Tryptophan in Animal Nutrition and Human Health

Yulong Yin • Sung Woo Kim • Xiongzhuo Tang
Editors

Tryptophan in Animal Nutrition and Human Health

Springer

Editors
Yulong Yin
Tianjin Institute of Industrial Biotechnology
Chinese Academy of Sciences, National
Center of Technology Innovation for
Synthetic Biology
Tianjin, China

Institute of Subtropical Agriculture
Chinese Academy of Science
Changsha, China

Sung Woo Kim
Department of Animal Science
North Carolina State University
Raleigh, NC, USA

Xiongzhuo Tang
Animal Nutritional Genome and Germplasm
Innovation Research Center, College of
Animal Science and Technology
Hunan Agricultural University
Changsha, Hunan, China

Yuelushan Laboratory
Changsha, China

ISBN 978-981-97-4718-4 ISBN 978-981-97-4719-1 (eBook)
https://doi.org/10.1007/978-981-97-4719-1

This Springer imprint is published by the registered company Springer Nature Singapore Pte Ltd.
The registered company address is: 152 Beach Road, #21-01/04 Gateway East, Singapore 189721, Singapore

Preface

Tryptophan is an essential amino acid in mammals and its metabolic pathways play pivotal roles in the maintenance of host nutritional and physiological functions. However, it lacks a comprehensive reference book to summarize the current research progress of tryptophan-mediated nutritional intervention and its interactions with host-microbes in farmed animals, particularly for its studies in pigs.

This book integrates the most recent research progress of tryptophan and its metabolites in nutrition and health of pigs. Meanwhile, this book also summarizes the effect of tryptophan nutrition on the regulation of various physiological functions in pigs and other farmed animals like poultry and ruminants. Furthermore, the functional role of tryptophan in human diseases has also been included in this book. Finally, this book includes detailed information on the manufacturing process of industrial production of tryptophan and methodology to study tryptophan metabolism.

This book is co-edited by Professor Yulong Yin who is also Academician of the Chinese Academy of Engineering, WNR Distinguished Professor Sung Woo Kim who is also University Faculty Scholar at North Carolina State University, and by Professor Xiongzhuo Tang at Hunan Agricultural University based on their extensive and original research contributions in tryptophan metabolism and nutrition. This book aims to provide a reference book for both academic and industrial professionals in animal science, animal agriculture, biology, biotechnology, and food and feed production.

Lastly, we would like to thank all contributors who spent countless time and efforts in writing this book. We also thank Dr. Fenfen Zhou and her group members Zhijun Cheng, Yuqing Wei, Chengyuan Li, Lin Yuan, Zixian Xiao, Yanqi Lu, Yu Deng, Zhenying Hu, Xiaofei Liang, Weijian Zheng, and Xiaotao Luo from Hunan Agriculture University for their valuable support in language service. We thank the funding supports from the 2024 Hunan Key Project of Basic Research Program (2024JC0007), Yuelushan Laboratory, the Tianjin Synthetic Biotechnology Innovation Capacity Improvement Project (TSBICIP-CXRC-038-01 and E1M0660201), and the National Natural Science Foundation of China (U23A20233).

Yulong Yin
Sung Woo Kim
Xiongzhuo Tang

Contents

Chapter 1
The Origin and Metabolism of Tryptophan

Jinfeng Miao, Zhenguo Hu, Kang Xu, Zian Yuan, Yadong Zhao, Libin Wang, Yi-Zhou Gao, Junyan Xie, Yulong Yin, and Xiongzhuo Tang

---

J. Miao
College of Veterinary Medicine, Nanjing Agricultural University, Nanjing, China
e-mail: miaojinfeng@njau.edu.cn

Z. Hu · K. Xu (✉) · J. Xie
Laboratory of Animal Nutritional Physiology and Metabolic Process, Institute of Subtropical Agriculture, Chinese Academy of Science, Changsha, Hunan, China
e-mail: huzhenguo22@mails.ucas.ac.cn; xukang@isa.ac.cn; xiejunyan09@isa.ac.cn

Z. Yuan
Animal Nutritional Genome and Germplasm Innovation Research Center, College of Animal Science and Technology, Hunan Agricultural University, Changsha, Hunan, China
e-mail: 1287735078@stu.hunau.edu.cn; xiongzhuo.tang@hunau.edu.cn

X. Tang (✉)
Animal Nutritional Genome and Germplasm Innovation Research Center, College of Animal Science and Technology, Hunan Agricultural University, Changsha, Hunan, China

Yuelushan Laboratory, Changsha, China
e-mail: 1287735078@stu.hunau.edu.cn; xiongzhuo.tang@hunau.edu.cn

Y. Zhao
School of Food and Pharmacy, Zhejiang Ocean University, Zhoushan City, Zhejiang Province, China
e-mail: yadong@kth.se

L. Wang
College of Horticulture, Nanjing Agricultural University, Nanjing, Jiangsu, China
e-mail: wanglibin@njau.edu.cn

Y.-Z. Gao
The Center for Microbes, Development and Health, Shanghai Institute of Immunity and Infection, Chinese Academy of Sciences, Shanghai, China
e-mail: yinyulong@isa.ac.cn

Y. Yin (✉)
Tianjin Institute of Industrial Biotechnology, Chinese Academy of Sciences, National Center of Technology Innovation for Synthetic Biology, Tianjin, China

Institute of Subtropical Agriculture, Chinese Academy of Science, Changsha, China
e-mail: yinyulong@isa.ac.cn

Y. Yin et al. (eds.), *Tryptophan in Animal Nutrition and Human Health*,
https://doi.org/10.1007/978-981-97-4719-1_1

Abstract Tryptophan, an essential amino acid, plays various physiological roles in the host body by affecting immunity, reproduction, lactation, neuronal function, and protein synthesis. Exploring the regulatory mechanisms of Trp and its metabolites on host health promotes in-depth research and application of Trp-mediated nutrition in animals, including livestock and poultry production. This chapter describes the origin of Trp from alimentary proteins and briefly summarizes the metabolic pathways of Trp in animals.

Abbreviations

3-HAA	3-hydroxy-o-aminobenzoic acid
3-HK	3-hydroxykynurenine (3-HK)
5-HIAA	5-hydroxyindoleacetic acid
5-HT	5-hydroxytryptamine
5-HTP	5-hydroxytryptophan
AA	anthranilic acid
AADC	aromatic L-amino acid decarboxylase
ACMS	2-amino-3-carboxymuconic semialdehyde
ACMSD	2-amino-3-carboxymuconic semialdehyde decarboxylase
AHR	aryl hydrocarbon receptor
AMPs	antimicrobial peptides
ARNT	aryl hydrocarbon receptor nuclear translocator
ASL	argininosuccinate lyase
ASS	argininosuccinate synthetase
ATP	adenosine triphosphate
IAA	indole-3-acetic acid
IAAld	indole-3-acetaldehyde
IAld	indole-3-aldehyde
IDO	indoleamine-2,3-dioxygenase
IFN-γ	interferon-gamma
ILA	indole-3-lactic acid
IPA	indole-3-propionic acid
KA	kynurenic acid
KAT	kynurenine transaminase
Kyn	kynurenine
NFK	N-formylkynurenine
OS	oxidative stress
PTHrP	parathyroid hormone-related protein
PXR	pregnane X receptors
RORγt	retinoid-related orphan receptor γt
TDO	tryptophan 2,3-dioxygenase
TPH	tryptophan hydroxylase
Trp	tryptophan

WARS	tryptophanyl-tRNA synthetase
XA	xanthurenic acid
XREs	xenobiotic response elements

1.1 Introduction

A number of physiological functions depend on tryptophan (Trp) and its metabolites. 95% of Trp is metabolized by the kynurenine (KYN) pathway mediated by indoleamine-2,3-dioxygenase (IDO) to produce metabolites such as KYN, 3-hydroxy-o-aminobenzoic acid (3-HAA), and kynurenic acid (KA), while 4–6% of Trp is metabolized by intestinal microorganisms to produce indole and its derivatives. Additionally, 1–2% of Trp forms 5-hydroxytryptamine (5-HT) by tryptophan hydroxylase (TPH) in intestinal chromophores (Gao et al. 2018; Roager and Licht 2018). These metabolites play a crucial role in regulating immune cell activity and reproduction, as well as maintaining the structure and functionality of epithelial cells.

Trp plays a complex regulatory role in the immune system, primarily in three ways (Fiore and Murray 2021): (i) through local consumption of Trp, e.g., by providing amino acids for T-cell proliferation; (ii) through the production of metabolites such as KYN and indole, thus regulating body immunity; and (iii) through the regulation of immune cell metabolism, e.g., Trp plays a controlling role in activated macrophages through NAD^+ production via KP metabolism (Minhas et al. 2019). Trp can be metabolized by intestinal microorganisms to produce various aryl hydrocarbon receptor (AHR) ligands, and these metabolites can regulate intestinal homeostasis through a series of reactions upon binding to the AHR (Cervantes-Barragan et al. 2017; Zelante et al. 2013). Furthermore, pregnane X receptors (PXR) can be bound by indole propionic acid (IPA), which may regulate the function of the intestinal barrier in mice (Venkatesh et al. 2014). IA has the potential to improve the structure of the intestinal barrier and reduce inflammatory reactions by improving the differentiation of goblet cells and the formation of mucus, which may be mediated via AHR activation (Wlodarska et al. 2017). Generally speaking, these investigations demonstrate that Trp metabolites can enhance intestinal barrier function by modulating PXR and AHR signaling pathways.

The symbiotic relationship between the host and intestinal microflora affects the host's physiological function in many ways, such as regulating immune response and maintaining intestinal homeostasis. By producing a variety of metabolites, microbial communities can directly affect the physiology of the host, including the nervous and immune systems. Trp is related to shaping gut microbiota and regulating the production of antimicrobial peptides (AMPs) (Rao et al. 2021). On the one hand, supplementation of Trp in piglets can significantly improve the variety of gut bacteria (Liang et al. 2018). On the other hand, the application of Trp can decrease the number of conditional pathogens while increasing the number of Trp-metabolizing bacteria (such as *Lactobacillus*) (Liang et al. 2019). Genera carrying the tryptamine gene can metabolize Trp into indole and its derivatives, and these metabolites can act as ligands for AHR, PXR, and retinoid-related orphan receptor γt (RORγt), thus regulating the health of the body (Wyatt and Greathouse 2021). In addition, the application of Trp can improve the indole content and *Akkermansia* ratio in the colon of aging mice and reduce the level of inflammatory reaction and oxidative damage (Yin et al. 2021). Conversely, a lack of Trp in the diet will increase the body's inflammatory response, which has been connected with *Clostridium* spp. (Yusufu et al. 2021).

Trp also affects the reproductive and lactational functions of animals. Pregnancy is associated with higher basal oxygen consumption and oxidative stress (OS). Uncontrolled OS can result in undesirable pregnancy outcomes. Trp is an important antioxidant in the human placenta because it aids in inhibiting the lipid peroxidation response under oxidative stress. Controlling and minimizing OS may be possible by regulating Trp supplementation and metabolism, considering the variations in Trp metabolism at different stages of pregnancy. Tryptophan has been demonstrated in animal studies to reduce miscarriage and mortality rates in pregnant mice induced with toxicity while also increasing fetal survival rates and overall litter size. IDO, the rate-limiting enzyme of the KYN pathway, has high levels in the placenta, which can prevent fetal rejection, and IDO inhibitors can cause abortion in mice during pregnancy (Broekhuizen et al. 2021). In addition, maternal demand for Trp increases during pregnancy, mainly to maintain maternal requirements, protein deposition in maternal tissues, and fetal growth protein deposition. Trp depletion has negative effects on both the mother and fetus (Franco et al. 2014). During pregnancy, the fetal demand for Trp and the transport of Trp in the umbilical cord increase. In most cases, both total and free Trp are higher in the umbilical cord than in maternal plasma (Tricklebank et al. 1979). At birth, the levels of free and total Trp in plasma are high, but they quickly decrease within a day. 5-HT, a Trp metabolite, is particularly crucial for fetal development, maternal mental status, and uterine myometrial contraction (Doherty et al. 2011; Moiseiwitsch 2000). 5-HT also has an important physiological role in maternal lactation function. In a rat model, 5-HT affects the energy metabolism in the mammary gland. During lactation, 5-HT can affect the secretion of parathyroid hormone-related protein (PTHrP) in the mammary gland, thus promoting the formation of milk.

1.2 The Origin of Tryptophan

There are two main sources of Trp production in the host body. One is endogenously generated from the catabolism of amino acids, which makes up about two-thirds of the total amount of Trp in the body. The other is approximately one-third of exogenously absorbed amino acids derived from the dietary protein intake (Cui 2003). The main origin of Trp in farmed animals comes from diets or feed additives. The amount of Trp in naturally plant-based diets, such as cereal and their by-products, is lower than that in animal-based feed, such as fish meal and whey powder. Appropriate addition of animal-based feed can effectively increase Trp intake and reduce the incidence of deleterious symptoms caused by insufficient Trp intake. In contrast, the Trp in feed additives used for dietary supplementation in animal husbandry is mainly obtained by bacteria-based fermentation rather than natural synthesis.

Trp is an essential nutritional component of the human diet. The average daily intake of Trp by the human body is approximately 600–900 mg, which fully meets the World Health Organization's recommendation of approximately 3.5–6.0 mg/kg of body weight per day. As Trp is crucial in maintaining human physiological functions, the addition of Trp derivates or Trp-rich diet has begun to be utilized to prevent diseases or improve drug treatment. Trp can be found in a wide range of natural foods, including seeds, soybeans, milk, and vegetables. Notably, the Trp content stored in tissues is low, resulting in generally low concentrations in the body (Richard et al. 2009).

During protein synthesis, a specific aminoacyl-tRNA synthetase activates Trp and then transfers it to the corresponding tRNA by the hydrolysis of adenosine triphosphate (ATP). Human genes WARS and WARS2 encode two forms of tryptophanyl-tRNA synthetases in the cytoplasm and mitochondria, respectively. Cytosolic tryptophanyl-tRNA synthetase is induced by INF-γ and produces two active peptides, T1-TrpRS and T2-TrpRS, after proteolysis (Rubin et al. 1991). Notably, different subtypes of tryptophanyl-tRNA synthetase are produced by alternative splicing and proteolytic cleavage. These isomers seem to play a critical role in protein synthesis, innate immune regulation, angiogenesis, and IFN-γ signaling.

Phylogenetically, Trp distribution is highly conserved, representing less than 2% of the amino acids in the proteome at all taxonomic levels (McKay et al. 2018). Although the frequency of Trp residues is very low, they play vital roles in protein-protein interactions, redox processes, and transmembrane protein stability (Ma et al. 2003). For example, the indole ring thermodynamically stabilizes the supramolecular structure due to its hydrophobicity through the van der Waals forces. The aromatic nitrogen interacts with the positively charged carbon and acts as a hydrogen bond donor. The hydrophobic part of the indole ring in Trp contributes to the insertion and stabilization of transmembrane proteins at the interface between phospholipid bilayer and water (Yau et al. 1998). Moreover, the sites required for protein-protein interactions and therapeutic drug targeting are usually located in structurally conserved regions containing Trp residues (McKay et al. 2018).

Trp residues and tyrosine are present in many important oxygen-responsive enzyme clusters. In fact, these two amino acids are actively involved in the oxidation-reduction reactions of free radicals (Gray and Winkler 2016). For example, some posttranslational modifications have been found in specific Trp residues in myocardial mitochondrial proteins, and serval oxidation products such as hydroxytryptophan, N-formylkynurenine (NFK)/dihydroxytryptophan, and KYN have been identified through mass spectrometry analyses (Taylor et al. 2003). The biological functions of Trp oxidation are diverse and strictly dependent on the functions of the oxidized protein. For example, Trp oxidation is considered a natural mechanism for mitochondrial respiratory complex I to remove reactive oxygen species (Taylor et al. 2003).Additionally, oxidative stress within the body may drive Trp oxidation, with oxidized peptides and proteins being detected locally or in circulation.

One type of molecule in which Trp residues are often present in its peptide sequence is the so-called small bioactive peptides. These Trp-rich peptides represent a group of molecules associated with innate immunity, defending against viruses and fungal infections by disrupting the extracellular cell membrane function or directly acting on intracellular targets (Chan et al. 2006). In the human body, endogenous Trp-rich antimicrobial peptides against gram-positive and gram-negative bacteria are produced by the proteolysis of lysozymes. Similarly, the indolicidin-rich Trp peptides produced by bovine neutrophils have strong antibacterial activity against gram-positive and gram-negative bacteria, fungi, and HIV-1 infections. Additionally, puroindoline A and B in wheat grains exhibit antifungal activity (Comai and Gobbi 2014).

1.3 The Source of Tryptophan from Alimentary Proteins

Tryptophan is an essential amino acid for both human and animal nutrition, and it is of great importance for nutritionists to accurately determine the content of tryptophan in food (Leverton 1964). When we eat, our body's digestive system breaks down the alimentary proteins in food into individual amino acids, which are absorbed by cells and used to make other proteins and large molecules. In addition, lack of alimentary protein may impair amino acid metabolism in animals. It has been shown that alimentary protein deficiency decreases the activities of argininosuccinate synthetase (ASS) and argininosuccinate lyase (ASL) in mouse liver (Kopylchuk et al. 2020). Tryptophan is abundant in alimentary proteins, and a large number of epidemiological and experimental data suggests that some alimentary proteins and amino acids in supplements can modify various types of cholesterol in the blood (Blachier et al. 2010). Here, we briefly summarize the sources of tryptophan from different kinds of alimentary proteins.

1.3.1 Cereals

Cereals are the traditional staple food of many Asian people, including rice, wheat, millet, and soybeans. Currently, the main sources of cereal protein are rice and wheat, accounting for more than half of total cereal production, and these two grains are the main sources of nutrition for humans (Sudheesh et al. 2022). In general, the quality of grain proteins largely depends on their amino acid composition, and tryptophan, an essential amino acid that makes up grain proteins, is present in all types of cereals. Macrae et al. determined the amount of tryptophan in the proteins of nine cereal grains: wheat, rice (brown), rice (milled), maize (normal), maize (high lysine), barley, rye, millets, and sorghum. Wheat has the highest tryptophan content, while ordinary corn has the lowest, with 2.4 g and 0.5 g per 100 g of grain protein, respectively (Nichols et al. 2008; Sudheesh et al. 2022). At present, high-performance liquid chromatography is the main method for measuring tryptophan content in cereal proteins. Çevikkalp et al. successfully measured the concentration of tryptophan in six grains (in mg/100 g): durum wheat, barley, rye, kidney bean, chickpea, and red lentil (Cevikkalp et al. 2016). Consumption of tryptophan-rich grains has been shown to improve sleep in infants, including sleep disorders such as excessive nighttime wakefulness (Cubero et al. 2009). Additionally, consumption of grains with high tryptophan (60 mg/30 g) for breakfast and dinner in middle-aged and older adults has also been reported to increase sleep efficiency and duration and improve symptoms of anxiety and depression (Wenefrida et al. 2013).

1.3.2 Nuts

Nuts, including almonds, walnuts, pistachios, Brazil nuts, peanuts, pecans, cashews, hazelnuts, and chestnuts, are normally considered high-energy, high-fat foods (Arslan et al. 2020). However, nuts are actually composed of various nutrients such as proteins, carbohydrates, and fats, with a relatively high content of total protein, making nuts one of the high-quality sources of plant protein. In addition, almost all nuts, with the exception of macadamia nuts, contain the same amount of tryptophan as that in a whole egg (Brufau et al. 2006). Walnuts serve as an excellent source of dietary tryptophan, a prerequisite for the synthesis of serotonin, which improves memory (Pribis and Shukitt-Hale 2014). Peanuts are also a major source of alimentary protein, providing essential amino acids such as tryptophan, tyrosine, and methionine, deficiencies of which lead to deleterious diseases (Settaluri et al. 2012). It has been shown that mixed nuts often act as a dietary source to affect tryptophan metabolism and reduce the expression of cardiovascular risk factors as well as increase satiety (Yang et al. 2023).

1.3.3 Milk

Milk and dairy products are important nutritional supplies for the human body (Visioli and Strata 2014). Milk proteins and milk-derived peptides have been found to promote health and reduce the risk of disease, particularly in the regulation of diseases associated with metabolic syndrome (Nongonierma and Fitzgerald 2012). Among alimentary proteins, milk proteins consist of a high level of tryptophan, and tryptophan residues in milk proteins can be broken down by proteolytic or peptidolytic enzymes to release as free amino acids or peptide sequences for further absorption (Nongonierma and Fitzgerald 2015). Fermented dairy products contain various probiotics and increase their protein levels (Bornkessel et al. 2016). It has been shown that the concentration of tryptophan and 5-hydroxytryptophan increases with fermentation, whereas the concentration of KYN does not differ between milk and its fermented derivatives at the time of packaging and shelf-life (Bertazzo et al. 2016).

1.3.4 Legumes

Legumes are a well-known common protein source, and the processing of legumes through different methods can improve the digestibility of legume proteins (Drulyte and Orlien 2019; Tomé 2013). Legumes are rich in tryptophan and lysine (Mossé and Huet 1990). One study determined the amount of tryptophan and protein in nine legume seed flours: lupin, soybean, peanut, kidney bean, broad bean, lentil, wild pea, chickpea, and pea. They found that soybean and lupin exhibit the highest protein content, chickpea displays the highest free tryptophan content, while peanut manifests the lowest free tryptophan content (Comai et al. 2007). Development in legume breeding may promote the use of legumes as a major source of alimentary protein (Semba et al. 2021).

In general, tryptophan is found in a diverse range of alimentary food, such as cereals, meat, nuts, milk and dairy products, fish, seeds, legumes, etc. (Keszthelyi et al. 2009). Furthermore, diverse steps in food processing and preparation, including heating, assembling, and cleaning, may lead to oxidative degradation and the cross-linking between protein molecules, thus reducing the bioactivities of tryptophan and its derivatives (Cevikkalp et al. 2016). Therefore, it is of great importance to further develop new methods to fully utilize tryptophan nutrition from alimentary proteins in the context of food composition.

1.4 Metabolism of Tryptophan

Trp belongs to one type of aromatic amino acids and cannot be naturally synthesized in vertebrates due to the lack of chorismic acid, a key precursor for aromatic amino acid synthesis. Dietary-derived Trp acts as an intermediate molecule/precursor to participate in protein biosynthesis, and these processes mainly occur in the liver and intestines. The endogenous Trp metabolism in the body has four outcomes: (1) synthesis of the neurotransmitter serotonin by tryptophan hydroxylase-mediated oxidation and aromatic amino acid decarboxylase-mediated decarboxylation (Maffei 2020), (2) production of tryptamine upon tryptophan decarboxylase 1-catalyzed decarboxylation, (3) generation of indole and its derivatives through the gut microbiota expressing tryptophanase (Roager and Licht 2018), and (4) production of niacin via the KYN pathway (Badawy 2017) (Fig. 1.1).

There are two major Trp metabolic pathways that occur in host cells. The first and the most significant one is the KP, which accounts for about 95% of Trp degradation in the host body (Badawy 2017). KP is crucial for the biosynthesis of nicotinic acid, with about 60 mg of Trp generating 1 mg of nicotinic acid (Goldsmith

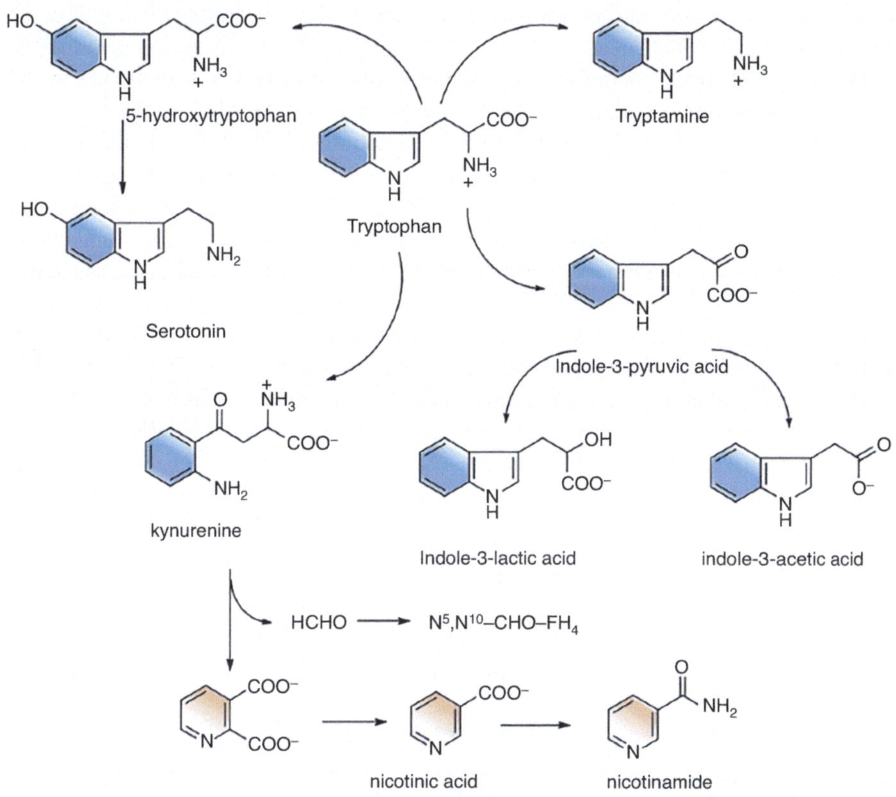

Fig. 1.1 Metabolic pathways of Trp. (Source: *Animal Biochemistry*, fourth edition, edited by Zou Sixiang, published by China Agriculture Press)

1958). Trp is first degraded into NFK through the actions of tryptophan 2,3-dioxygenase (TDO) and IDO-1/IDO-2. Then NFK is rapidly converted into KYN by formyl kynurenine formamidase. Although these enzymes catalyze similar oxidation reactions, there are differences in their biochemical structures, localizations, substrate specificity, and functions. TDO is mainly present in the liver (Bertazzo et al. 2001), but its expression also has been reported in the frontal cortex (Miller et al. 2004). Moreover, excessive TDO can utilize oxygen as a co-factor to metabolize Trp, thereby controlling plasma amino acid homeostasis. TDO activity in the liver is induced by its substrates, including Trp, glucocorticoids, and heme (Crotti et al. 2015; Ehrenshaft et al. 2015; Fedorova et al. 2010; Rexroth et al. 2012). Compared with TDO, IDO has broader substrate specificity; it acts on other indole derivatives and exerts its activity in collaboration with superoxide and oxygen (Chan et al. 2006).

IDO has two different types: IDO-1 and IDO-2 (Meininger et al. 2011). IDO-1 is most abundant in non-liver tissues, while IDO-2 is abundant in the kidneys, liver, and brain. In addition, IDO-1 acts as an important immune activator and is constitutively expressed in mature dendritic cells, secondary lymphoid organs, the female genital tract, placental and pulmonary endothelial cells, intestinal interstitial cells, and pancreatic β cells (Théate et al. 2015). However, IDO-1 expression is inducible in most tissues under pro-inflammatory conditions and plays a critical role in the formation of peripheral tolerance in autoimmune diseases, transplantation, and cancer. Furthermore, both IDO-1 and IDO-2 can catabolize L-/D-Trp and other indole-containing compounds such as 5-HTP, serotonin, melatonin, 5-HIAA, and tryptamine. In comparison, the affinity of IDO-2 to Trp is much lower than that of IDO-1 (Lee et al. 2014). Therefore, it is believed that IDO-2 may act as a negative regulator of IDO-1 through the competitive mechanism by binding heme. The activity of IDO is regulated by various factors including pro-inflammatory cytokines, interferon-gamma (Lucki 1998), lipopolysaccharide (Comai and Gobbi 2014), viruses (Jockers et al. 2016), and tumor cells (Hornyák et al. 2018). KYN is metabolized into KA and anthranilic acid (AA) by KYN transaminase (KAT) and kynureninase. It is also partially converted into 3-hydroxykynurenine (3-HK), which is further converted into xanthurenic acid (XA) by KAT (Fig. 1.2). Subsequently,

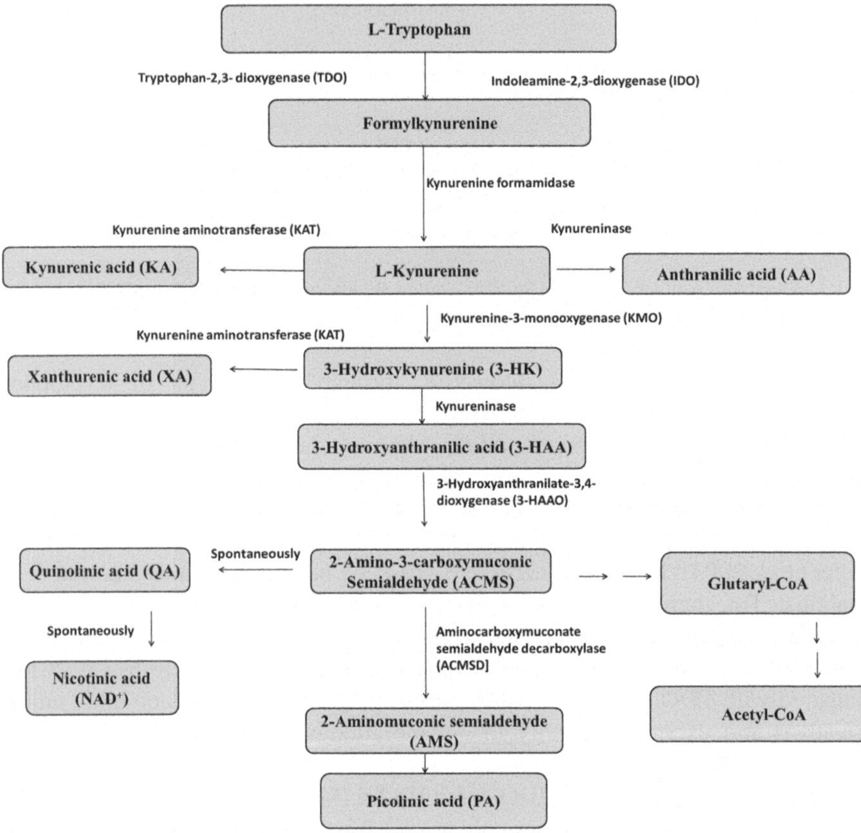

Fig. 1.2 The KYN metabolic pathways (Platten et al. 2019)

kynureninase catalyzes the hydrolysis of 3-HK to form 3-hydroxyanthranilic acid (3-HAA), which can be converted into the unstable intermediate 2-amino-3-carboxymuconic semialdehyde (ACMS) by 3-hydroxyanthranilic acid ester 3,4-dioxygenase (3-HAAO). Then, ACMS is either spontaneously converted into quinolinic acid (QA), which is further decarboxylated to form the NAD precursor nicotinic acid, or further converted into picolinic acid (PA) after enzymatic decarboxylation by 2-amino-3-carboxymuconic semialdehyde decarboxylase (ACMSD) (Posa et al. 2018) (Fig. 1.2).

The second pathway is the serotonin pathway, in which approximately 1–2% of the ingested Trp is hydrolyzed to 5-HTP by the rate-limiting enzyme TPH. There are two isoforms of TPH: TPH1 and TPH2. TPH1 is primarily found in the spleen, enterochromaffin cells, pineal gland, and thymus, while TPH2 is exclusively present in neuronal cells (Walther et al. 2003). Subsequently, 5-HTP is rapidly decarboxylated to 5-HT by aromatic L-amino acid decarboxylase (AADC). Finally, 5-HT is further metabolized to 5-HIAA by monoamine oxidase (MAO) or to melatonin under the enzymatic activity of N-acetyltransferase.

In addition to these two Trp metabolic pathways occurring in the host cells, the gut microbiota is also capable of metabolizing dietary Trp. Although most Trp is absorbed by the small intestine, the remaining Trp along with the chyme reaches the large intestine, where a variety of anaerobic microorganisms initiate the indole/AHR pathway to degrade Trp into indole and its derivatives, including indole-3-aldehyde (IAld), indole-3-acetic acid (IAA), indole-3-propionic acid (IPA), indole-3-acetaldehyde (IAAld), indole-3-lactic acid (ILA), and indoleacrylic acid (Alexeev et al. 2018; Hubbard et al. 2015). The AHR is usually distributed in the cytoplasm of host cells, binds to aromatic ligands, then enters the nucleus through nuclear pores and forms a dimer with the aryl hydrocarbon receptor nuclear translocator (ARNT). This dimer complex further binds to the xenobiotic response elements (XREs) located on the target genes to regulate their expression. Meanwhile, AHR is also capable of binding to other transcription factors in the nucleus, thereby affecting their activities and subsequently regulating their target gene expression. Some indole derivatives can bind to AHRs on the surface of intestinal cells to exert their biological functions and maintain intestinal homeostasis. In a healthy gastrointestinal tract, Trp metabolic pathways are kept in a stable and balanced state, thereby maintaining intestinal motility, neurological function, immunity, and intestinal barrier function. Impaired Trp metabolism is often associated with deleterious diseases.

Taken together, Trp is crucial in protein biosynthesis, energy metabolism, immune regulation and other physiological functions in the host body. Firstly, L-Trp acts as the intermediate molecule for various protein biosynthesis. Additionally, L-Trp can be metabolized into nicotinic acid, a component of coenzyme I and coenzyme II, affecting aerobic respiration and glycolysis for energy metabolism. Furthermore, L-Trp and its metabolites derived from either the host cell or the gut microbiota participate in a variety of biological functions in the gut. The detailed metabolic function of L-Trp in regulating intestinal development, immune function, gut microbial homeostasis, interorgan communication, diseases, and its nutritional application in livestock and poultry will be described in other chapters of this book.

References

Alexeev EE, Lanis JM, Kao DJ, Campbell EL, Kelly CJ, Battista KD et al (2018) Microbiota-derived indole metabolites promote human and murine intestinal homeostasis through regulation of interleukin-10 receptor. Am J Pathol 188(5):1183–1194

Arslan J, Gilani AU, Jamshed H, Khan SF, Kamal MA (2020) Edible nuts for memory. Curr Pharm Des 26(37):4712–4720

Badawy AAB (2017) Kynurenine pathway of tryptophan metabolism: regulatory and functional aspects. Int J Tryptophan Res 10(0):117864691769193

Bertazzo A, Ragazzi E, Biasiolo M, Costa C, Allegri G (2001) Enzyme activities involved in tryptophan metabolism along the kynurenine pathway in rabbits. Biochim Biophys Acta 1527(3):167–175

Bertazzo A, Ragazzi E, Visioli F (2016) Evolution of tryptophan and its foremost metabolites' concentrations in milk and fermented dairy products. PharmaNutrition 4(2):62–67

Blachier F, Lancha AH Jr, Boutry C, Tome D (2010) Alimentary proteins, amino acids and cholesterolemia. Amino Acids 38(1):15–22

Bornkessel S, Bröring S, Omta S (2016) Crossing industrial boundaries at the pharma-nutrition interface in probiotics: a life cycle perspective. undefined, 4, undefined

Broekhuizen M, Danser AHJ, Reiss IKM, Merkus D (2021) The function of the Kynurenine pathway in the placenta: a novel pharmacotherapeutic target? Int J Environ Res Public Health 18(21)

Brufau G, Boatella J, Rafecas M (2006) Nuts: source of energy and macronutrients. Br J Nutr 96(Suppl 2):S24–S28

Cervantes-Barragan L, Chai JN, Tianero MD, Di Luccia B, Ahern PP, Merriman J et al (2017) Lactobacillus reuteri induces gut intraepithelial CD4(+)CD8αα(+) T cells. Science 357(6353):806–810

Cevikkalp SA, Loker GB, Yaman M, Amoutzopoulos B (2016) A simplified HPLC method for determination of tryptophan in some cereals and legumes. Food Chem 193:26–29

Chan DI, Prenner EJ, Vogel HJ (2006) Tryptophan- and arginine-rich antimicrobial peptides: structures and mechanisms of action. Biochim Biophys Acta Biomembr 1758(9):1184–1202

Comai S, Gobbi G (2014) Unveiling the role of melatonin MT2 receptors in sleep, anxiety and other neuropsychiatric diseases: a novel target in psychopharmacology. J Psychiatry Neurosci 39(1):6–21

Comai S, Bertazzo A, Bailoni L, Zancato M, Costa CVL, Allegri G (2007) Protein and non-protein (free and protein-bound) tryptophan in legume seeds. Food Chem 103(2):657–661

Crotti S, Enzo MV, Bedin C, Pucciarelli S, Maretto I, Del Bianco P et al (2015) Clinical predictive circulating peptides in rectal cancer patients treated with neoadjuvant chemoradiotherapy. J Cell Physiol 230(8):1822–1828

Cubero J, Chanclón B, Sánchez S, Rivero M, Rodríguez AB, Barriga C (2009) Improving the quality of infant sleep through the inclusion at supper of cereals enriched with tryptophan, adenosine-5′-phosphate, and uridine-5′-phosphate. Nutr Neurosci 12(6):272–280

Cui (2003) Research progress of tryptophan nutrition. Feed China (15):20–22

Doherty LF, Kwon HE, Taylor HS (2011) Regulation of tryptophan 2,3-dioxygenase by HOXA10 enhances embryo viability through serotonin signaling. Am J Physiol Endocrinol Metab 300(1):E86–E93

Drulyte D, Orlien V (2019) The effect of processing on digestion of legume proteins. Food Secur 8(6)

Ehrenshaft M, Deterding LJ, Mason RP (2015) Tripping up Trp: modification of protein tryptophan residues by reactive oxygen species, modes of detection, and biological consequences. Free Radic Biol Med 89:220–228

Fedorova M, Todorovsky T, Kuleva N, Hoffmann R (2010) Quantitative evaluation of tryptophan oxidation in actin and troponin I from skeletal muscles using a rat model of acute oxidative stress. Proteomics 10(14):2692–2700

Fiore A, Murray PJ (2021) Tryptophan and indole metabolism in immune regulation. Curr Opin Immunol 70:7–14

Franco DJ, Josephson JK, Moehn S, Pencharz PB, Ball RO (2014) Tryptophan requirement of pregnant sows1. J Anim Sci 92(10):4457–4465

Gao J, Xu K, Liu H, Liu G, Bai M, Peng C et al (2018) Impact of the gut microbiota on intestinal immunity mediated by tryptophan metabolism. Front Cell Infect Microbiol 8:13

Goldsmith G (1958) Niacin-tryptophan relationships in man and niacin requirement. Am J Clin Nutr 6(5):479–486

Gray HB, Winkler JR (2016) The rise of radicals in bioinorganic chemistry. Israel J Chem 56(9–10):640–648

Hornyák L, Dobos N, Koncz G, Karányi Z, Páll D, Szabó Z et al (2018) The role of indoleamine-2,3-dioxygenase in cancer development, diagnostics, and therapy. Front Immunol 9

Hubbard TD, Murray IA, Perdew GH (2015) Indole and tryptophan metabolism: endogenous and dietary routes to ah receptor activation. Drug Metab Dispos 43(10):1522–1535

Jockers R, Delagrange P, Dubocovich ML, Markus RP, Renault N, Tosini G et al (2016) Update on melatonin receptors: IUPHAR Review 20. Br J Pharmacol 173(18):2702–2725

Keszthelyi D, Troost F, Masclee A (2009) Understanding the role of tryptophan and serotonin metabolism in gastrointestinal function. Neurogastroenterol Motil 21:undefined

Kopylchuk H, Nykolaichuk I, Lylyk I (2020) Indexes of citrulline metabolism in rat liver under the toxic injury against the background of alimentary protein deficiency. 92:113

Lee Y-K, Lee HB, Shin D-M, Kang MJ, Yi EC, Noh S et al (2014) Heme-binding-mediated negative regulation of the tryptophan metabolic enzyme indoleamine 2,3-dioxygenase 1 (IDO1) by IDO2. Exp Mol Med 46(11):e121–e121

Leverton. (1964) The determination of tryptophan in foods. Nutr Rev 22:347–349

Liang H, Dai Z, Liu N, Ji Y, Chen J, Zhang Y et al (2018) Dietary L-tryptophan modulates the structural and functional composition of the intestinal microbiome in weaned piglets. Front Microbiol 9:1736

Liang H, Dai Z, Kou J, Sun K, Chen J, Yang Y et al (2019) Dietary l-tryptophan supplementation enhances the intestinal mucosal barrier function in weaned piglets: implication of tryptophan-metabolizing microbiota. Int J Mol Sci 20(1). https://doi.org/10.3390/ijms20010020

Lucki I (1998) The spectrum of behaviors influenced by serotonin. Biol Psychiatry 44(3):151–162

Ma B, Elkayam T, Wolfson H, Nussinov R (2003) Protein–protein interactions: structurally conserved residues distinguish between binding sites and exposed protein surfaces. Proc Natl Acad Sci 100(10):5772–5777

Maffei ME (2020) 5-hydroxytryptophan (5-HTP): natural occurrence, analysis, biosynthesis, biotechnology, physiology and toxicology. Int J Mol Sci 22(1):181

McKay MJ, Martfeld AN, De Angelis AA, Opella SJ, Greathouse DV, Koeppe RE (2018) Control of transmembrane helix dynamics by interfacial tryptophan residues. Biophys J 114(11):2617–2629

Meininger D, Zalameda L, Liu Y, Stepan LP, Borges L, McCarter JD, Sutherland CL (2011) Purification and kinetic characterization of human indoleamine 2,3-dioxygenases 1 and 2 (IDO1 and IDO2) and discovery of selective IDO1 inhibitors. Biochim Biophys Acta BBA Proteins Proteom 1814(12):1947–1954

Miller CL, Llenos IC, Dulay JR, Barillo MM, Yolken RH, Weis S (2004) Expression of the kynurenine pathway enzyme tryptophan 2,3-dioxygenase is increased in the frontal cortex of individuals with schizophrenia. Neurobiol Dis 15(3):618–629

Minhas PS, Liu L, Moon PK, Joshi AU, Dove C, Mhatre S et al (2019) Macrophage de novo NAD(+) synthesis specifies immune function in aging and inflammation. Nat Immunol 20(1):50–63

Moiseiwitsch JR (2000) The role of serotonin and neurotransmitters during craniofacial development. Crit Rev Oral Biol Med 11(2):230–239

Mossé J, Huet J (1990) Amino acid composition and nutritional score for ten cereals and six legumes or oilseeds: causes and ranges of variations according to species and to seed nitrogen content. Sci Aliment 10(1):151–173

Nichols D, Lewis K, Orjala J, Mo S, Ortenberg R, O'Connor P et al (2008) Short peptide induces an "uncultivable" microorganism to grow in vitro. Appl Environ Microbiol 74(15):4889–4897

Nongonierma AB, Fitzgerald RJ (2012) Tryptophan-containing milk protein-derived dipeptides inhibit xanthine oxidase. Peptides 37(2):263–272

Nongonierma A, Fitzgerald R (2015) Milk proteins as a source of tryptophan-containing bioactive peptides. Food Funct 6 7:undefined

Platten M, Nollen EAA, Röhrig UF, Fallarino F, Opitz CA (2019) Tryptophan metabolism as a common therapeutic target in cancer, neurodegeneration and beyond. Nat Rev Drug Discov 18(5):379–401

Posa L, De Gregorio D, Gobbi G, Comai S (2018) Targeting melatonin MT2 receptors: a novel pharmacological avenue for inflammatory and neuropathic pain. Curr Med Chem 25(32):3866–3882

Pribis P, Shukitt-Hale B (2014) Cognition: the new frontier for nuts and berries. Am J Clin Nutr 100(suppl_1):347S–352S

Rao Z, Li J, Shi B, Zeng Y, Liu Y, Sun Z et al (2021) Dietary tryptophan levels impact growth performance and intestinal microbial ecology in weaned piglets via tryptophan metabolites and intestinal antimicrobial peptides. Animals (Basel) 11(3)

Rexroth S, Poetsch A, Rögner M, Hamann A, Werner A, Osiewacz HD et al (2012) Reactive oxygen species target specific tryptophan site in the mitochondrial ATP synthase. Biochim Biophys Acta BBA Bioenerget 1817(2):381–387

Richard DM, Dawes MA, Mathias CW, Acheson A, Hill-Kapturczak N, Dougherty DM (2009) L-tryptophan: basic metabolic functions, behavioral research and therapeutic indications. Int J Tryptophan Res 2:IJTR.S2129

Roager HM, Licht TR (2018) Microbial tryptophan catabolites in health and disease. Nat Commun 9(1):3294

Rubin BY, Anderson SL, Xing L, Powell RJ, Tate WP (1991) Interferon induces tryptophanyl-tRNA synthetase expression in human fibroblasts. J Biol Chem 266(36):24245–24248

Semba RD, Ramsing R, Rahman N, Kraemer K, Bloem MW (2021) Legumes as a sustainable source of protein in human diets. Global Food Secur 28:100520

Settaluri V, Kandala C, Puppala N, Sundaram J (2012) Peanuts and their nutritional aspects—a review. 03:1644

Sudheesh C, Bhat ZR, Aaliya B, Sunooj KV (2022) Cereal proteins. In: Nutraceuticals and health care, pp 29–60

Taylor SW, Fahy E, Murray J, Capaldi RA, Ghosh SS (2003) Oxidative post-translational modification of tryptophan residues in cardiac mitochondrial proteins. J Biol Chem 278(22):19587–19590

Théate I, van Baren N, Pilotte L, Moulin P, Larrieu P, Renauld J-C et al (2015) Extensive profiling of the expression of the indoleamine 2,3-dioxygenase 1 protein in normal and tumoral human tissues. Canc Immu Res 3(2):161–172

Tomé D (2013) Digestibility issues of vegetable versus animal proteins: protein and amino acid requirements—functional aspects. Food Nutr Bull 34(2):272–274

Tricklebank MD, Pickard FJ, de Souza SW (1979) Free and bound tryptophan in human plasma during the perinatal period. Acta Paediatr Scand 68(2):199–204

Venkatesh M, Mukherjee S, Wang H, Li H, Sun K, Benechet AP et al (2014) Symbiotic bacterial metabolites regulate gastrointestinal barrier function via the xenobiotic sensor PXR and Toll-like receptor 4. Immunity 41(2):296–310

Visioli F, Strata A (2014) Milk, dairy products, and their functional effects in humans: a narrative review of recent evidence. Adv Nutr 5(2):undefined

Walther DJ, Peter J-U, Bashammakh S, Hörtnagl H, Voits M, Fink H, Bader M (2003) Synthesis of serotonin by a second tryptophan hydroxylase isoform. Science 299(5603):76

Wenefrida I, Utomo HS, Linscombe SD (2013) Mutational breeding and genetic engineering in the development of high grain protein content. J Agric Food Chem 61(48):11702–11710

Wlodarska M, Luo C, Kolde R, d'Hennezel E, Annand JW, Heim CE et al (2017) Indoleacrylic acid produced by commensal peptostreptococcus species suppresses inflammation. Cell Host Microbe 22(1):25–37.e26

Wyatt M, Greathouse KL (2021) Targeting dietary and microbial tryptophan-indole metabolism as therapeutic approaches to colon cancer. Nutrients 13(4). https://doi.org/10.3390/nu13041189

Yang J, Lee R, Schulz Z, Hsu A, Pai J, Yang S et al (2023) Mixed nuts as healthy snacks: effect on tryptophan metabolism and cardiovascular risk factors. Nutrients 15(3)

Yau W-M, Wimley WC, Gawrisch K, White SH (1998) The preference of tryptophan for membrane interfaces. Biochemistry 37(42):14713–14718

Yin J, Zhang B, Yu Z, Hu Y, Lv H, Ji X et al (2021) Ameliorative effect of dietary tryptophan on neurodegeneration and inflammation in d-galactose-induced aging mice with the potential mechanism relying on AMPK/SIRT1/PGC-1α pathway and gut microbiota. J Agric Food Chem 69(16):4732–4744

Yusufu I, Ding K, Smith K, Wankhade UD, Sahay B, Patterson GT et al (2021) A tryptophan-deficient diet induces gut microbiota dysbiosis and increases systemic inflammation in aged mice. Int J Mol Sci 22(9). https://doi.org/10.3390/ijms22095005

Zelante T, Iannitti RG, Cunha C, De Luca A, Giovannini G, Pieraccini G et al (2013) Tryptophan catabolites from microbiota engage aryl hydrocarbon receptor and balance mucosal reactivity via interleukin-22. Immunity 39(2):372–385

Chapter 2
The Function of Tryptophan Metabolism in the Gut

Xiongzhuo Tang, Wan Li, Jing Gao, Qian Jiang, Shenglan Hu, Hao Xiao, Jing Wang, Ming Qi, Leli Wang, Lang Huang, Junhao Deng, Zian Yuan, Yuezhou Yao, Feiyue Chen, Qian Xie, Mengdi Yao, Qing-Wen Gui, and Yulong Yin

X. Tang (✉)
Animal Nutritional Genome and Germplasm Innovation Research Center, College of Animal Science and Technology, Hunan Agricultural University, Changsha, Hunan, China

Yuelushan Laboratory, Changsha, China
e-mail: xiongzhuo.tang@hunau.edu.cn

W. Li (✉) · M. Qi (✉) · L. Wang (✉)
Laboratory of Animal Nutritional Physiology and Metabolic Process, Institute of Subtropical Agriculture, Chinese Academy of Science, Changsha, Hunan, China
e-mail: Liwan@isa.ac.cn; ming.qi@nih.gov; leliwang@isa.ac.cn

J. Gao (✉)
National Engineering Research Center for Oil Tea Camellia, Changsha, China
e-mail: gaojing817@hnlky.cn

Q. Jiang (✉) · J. Wang (✉) · L. Huang · J. Deng · Z. Yuan · Y. Yao · F. Chen · Q. Xie · M. Yao
Animal Nutritional Genome and Germplasm Innovation Research Center, College of Animal Science and Technology, Hunan Agricultural University, Changsha, Hunan, China
e-mail: jiangqian@hunau.edu.cn; jingwang023@hunau.edu.cn; huanglang@stu.hunau.edu.cn; dengjunhao@stu.hunau.edu.cn; 1287735078@stu.hunau.edu.cn; yaoyuezhou@stu.hunau.edu.cn; chenfeiyue@stu.hunau.edu.cn; qqqx97@stu.hunau.edu.cn; 202043105521@stu.hunau.edu.cn

S. Hu (✉) · H. Xiao (✉)
Institute of Animal Science, Guangdong Academy of Agricultural Sciences, Guangzhou, China
e-mail: hushenglan@gdaas.cn; xiaohao@gdaas.cn

Q.-W. Gui
College of Chemistry and Materials Science, Hunan, Changsha, China
e-mail: gqw1216@hunau.edu.cn

Y. Yin (✉)
Tianjin Institute of Industrial Biotechnology, Chinese Academy of Sciences, National Center of Technology Innovation for Synthetic Biology, Tianjin, China

Institute of Subtropical Agriculture, Chinese Academy of Science, Changsha, China
e-mail: yinyulong@isa.ac.cn

Y. Yin et al. (eds.), *Tryptophan in Animal Nutrition and Human Health*, https://doi.org/10.1007/978-981-97-4719-1_2

Abstract Intestinal tryptophan (Trp) metabolism generates a variety of bioactive compounds or metabolites that act both in the local gut and peripheral organs at different developmental stages, affecting the pig's health. This chapter describes the anatomic characteristics of the porcine gastrointestinal tract and its development and recapitulates the current understanding of the biological roles of Trp and its metabolites in the regulation of gut development, immunity, and gut microbial homeostasis. Additionally, how Trp metabolism-mediated nutritional intervention modulates the reproduction performance in sows, the mother-to-offspring nutritional transmission, and the effect of Trp metabolites on interorgan interactions are also discussed in this chapter.

Abbreviations

3-HKYN	3-hydroxykynurenine
5-HT	5-hydroxytryptamine
5-HT4	5-hydroxytryptamine 4
5-HTP	5-hydroxytryptophan
AHR	Aryl hydrocarbon receptor
aINS	Amygdala-anterior insula
AM	Amylose
AP	Amylopectin
ARNT	AHR nuclear translocator
ASD	Autism spectrum disorder
CaSR	Calcium-sensing receptor
CaSR	Calcium-sensing receptor
CGRP	Calcitonin gene-related peptide
CNS	Central nervous system
CRC	Colorectal cancer
CYP450	Cytochrome P450
EC	Enterochromaffin
ENS	Enteric nervous system
GCs	Granulosa cells
GF	Germ-free
GLP-1	Glucagon-like peptide 1
GPCRs	G-protein-coupled receptors
HPA	Hypothalamic Pituitary Adrenal
IA	Indoleacrylic acid
IAA	Indole-3-acetic acid
IBD	Inflammatory bowel disease
IBS	Irritable bowel syndrome
IDG	Indoxyl β-D-glucuronide
IDO	Indoleamine 2,3-dioxygenase
IDO1	Indoleamine 2,3-dioxygenase 1
IDO2	Indoleamine 2,3-dioxygenase 2

IPA	Indole-3-propionic acid
IS	Indoxyl sulfate
KA	Kynurenic acid
KD	Kidney disease
KMO	KYN 3-monooxygenase
KP	Kynurenine pathway
KYN	Kynurenine
LPS	Lipopolysaccharide
LPS	Lipopolysaccharide
MGBA	Microbiome-gut-brain axis
MT1	Melatonin receptor 1
MT2	Melatonin receptor 2
NAD+	Nicotinamide adenine dinucleotide
NAFL	Nonalcoholic fatty liver
NAFLD	Nonalcoholic fatty liver disease
NASH	Nonalcoholic steatohepatitis
NMDA	N-methyl-D-aspartate
NSPs	Nonstarch polysaccharides
pBD	Porcine β-defensin
PXR	Pregnane X receptor
QA	Quinolinic acid
ROS	Reactive oxygen species
SCFAs	Short-chain fatty acids
SP	Substance P
SpA	Spondyloarthritis
SSCs	Spermatogonial stem cells
TDO	Trp 2,3-dioxygenase
TPH	Tryptophan hydroxylase
TPH1	Trp hydroxylase 1
Trp	Tryptophan
VIP	Vasoactive intestinal polypeptide

2.1 Anatomic Characteristics of Porcine Digestive System

Pigs are monogastric omnivores, and their digestive system can be divided into the gastrointestinal tract and digestive glands ("Tracing the history of pig farming" 2022). The gastrointestinal tract consists of the mouth, esophagus, stomach, small intestine, large intestine, and anus. The digestive glands include the small digestive glands distributed along the gastrointestinal tract, as well as the large digestive glands located outside the gastrointestinal tract. These digestive glands facilitate their digestive function by secreting various active enzymes or metabolites such as salivary amylase, digestive enzymes, and bile acid to initiate nutrient absorption. The gastrointestinal tract and digestive glands often work together to accomplish the digestion, absorption, transportation, and metabolism of nutrients, providing nutrition and energy to the host (Wang 2014). Meanwhile, they also convert undigested nutritional residues into feces, which are further excreted through the anus. Nutritional digestion is a complex process combining physical, chemical, and microbial digestion (Wang 2018). Physical digestion mechanically breaks down food into smaller pieces by grinding or smashing in the mouth and gastrointestinal peristalsis. Chemical digestion is facilitated by various hydrolytic enzymes secreted in the gastrointestinal tract (Fothergill et al. 2019). Microbial digestion is initiated by the activation of the gut microbiota. Proper and effective nutrition strategies are required for pig farming at different developmental stages. This section summarizes the physiological structure of the pig digestive system and describes its major function in nutrient digestion and absorption.

2.1.1 The Gastrointestinal Tract

2.1.1.1 The Mouth and Esophagus

The mouth acts as the first site where food undergoes physical breakdown. The pig's mouth consists of tongue, epiglottis, teeth, and salivary glands. The tongue and teeth mix food with saliva through chewing or grinding, and then food passes through the esophagus, a tube that transports food from the mouth to the stomach by contraction or peristalsis. The cardiac sphincter is located at the end of the esophagus and prevents food from refluxing back into the esophagus. The epiglottis prevents food from entering the trachea and lungs. Oral digestion involves physical, chemical, and microbial processes (Sensoy 2021) and is also capable of stimulating physiological activities in peripheral organs such as the stomach, pancreas, and liver.

2.1.1.2 Stomach

The stomach is a vital part of the digestive system and plays a critical role in storing and releasing food (Fenghua et al. 2012). The stomach further mixes food delivered from the mouth and grinds it into much smaller pieces by secreting pepsin and gastric acid to increase their surface area for intestinal digestion. Pepsin, the main stomach enzyme, is secreted by gastric chief cells and digests proteins into small peptides and amino acids. Gastric acid helps break down food, and its secretion can be divided into three different phases based on the location of food along the gastrointestinal tract (Session 1991; Weiwei and Shanglian 1989). Gastric secretion starts at the cephalic phase when the food is chewed in the mouth, then switches to the gastric phase in the stomach, and finally reaches the intestinal phase when the food is delivered into the intestine.

2.1.1.3 Intestinal Structure

The intestine can be further divided into the small and large intestines. The small intestine acts as the major site for food digestion and absorption by secreting a variety of bioactive compounds, including different types of digestive enzymes and secretin (Degroote et al. 2020; Wonga 2004). Undigested and unabsorbed food enters the large intestine by the action of intestinal peristalsis, where water and electrolytes are further absorbed. Some nutritional components, such as crude fibers and ash, can be digested in the large intestine by the gut microbiota. The large intestine gland secretes intestinal fluid, lubricating the intestinal mucosa to protect against damage and forming feces toward the rectum for excretion (Dahl and Connor 2021).

2.1.2 Morphological Structure of the Small Intestine

As the pivotal site for food digestion and absorption, the small intestine is mainly composed of three parts: the duodenum, jejunum, and ileum (Qu et al. 1996). Their ability to digest and absorb nutrients is directly determined by their morphological structures, enzymatic activities, and nutritional transporters (Biviano et al. 1993). Both the weight and length of the small intestine can be considered as the main indicators for evaluating the digestion and absorption capacity of the small intestine (Marion et al. 2002; Meyer et al. 2014). It has been reported that the length of the small intestine is positively correlated with the growth and development of piglets. Piglets with a medium length of the small intestine have better growth traits (Wang et al. 2020a). There are many finger-like structures termed villi located in the intestinal epithelium and extended into the lumen of the small intestine, expanding the surface area of the intestinal walls and improving the efficiency of digestion and absorption (Zhang and Xu 2006). In addition, many intestinal crypts are found between intestinal villi in the intestinal epithelium lining of the small intestine and large intestine. Intestinal crypts contain many multipotent intestinal stem cells that undergo constant self-renewal and differentiation into multiple types of epithelial cells, including absorptive enterocytes, secretive enteroendocrine cells, goblet cells, and Paneth cells, as well as chemosensory tuft cells, to maintain normal intestinal homeostasis (Lloyd and Gabe 2008). Importantly, intestinal crypts are capable of replacing and repairing damaged or dead intestinal epithelial cells by activating regenerative capacity upon different environmental stimuli, ensuring the integrity and normal functions of the small intestine (Yang et al. 2013b).

2.1.2.1 Duodenum

The duodenum is the shortest and beginning part of the small intestine (Smith et al. 2010). Secretory components from the pancreas and gallbladder, such as pancreatic fluid and bile, are mixed with the chyme to neutralize hydrochloric acid coming from the stomach in the duodenum, which, in turn, promotes the digestion of fats, carbohydrates, and proteins (Chu and Liu 1984). Mucus and mineral salts from digestive juices increase lubrication and regulate the pH in the duodenum, while various enzymes from digestive juices are mainly involved in digesting proteins, lipids, and carbohydrates (Shiyuan et al. 2011). In addition to its digestive function, the duodenum also plays a regulatory role in gastric emptying and pancreatic release activity.

2.1.2.2 Jejunum

The jejunum, making up about two-fifths of the small intestine, is the second part of the small intestine and is surrounded by multiple blood vessels. It is located between the duodenum and ileum and contains many intestinal villi. The main function of the jejunum is to absorb large amounts of nutrients such as amino acids, fatty acids, and sugars. These absorbed nutrients can then enter the bloodstream to be transported to the organs of the body.

2.1.2.3 Ileum

The ileum is the last and longest part of the small intestine and plays a major absorptive function in vitamins, bile salts, and any nutritional products that are not absorbed by the jejunum. Unlike the existence of duodenojejunal flexure between the duodenum and jejunum, there is no clear border between the jejunum and ileum. The ileal wall is relatively thin and less permeable, and its mesentery contains higher levels of fat than the jejunum (Lim et al. 2013). In addition, the abundance of Peyer's patches, a subset of aggregated lymphoid follicles, is largely concentrated in the ileal epithelium to defend against environmental immune challenges. Undigested chyme continuously enters the large intestine through the ileocecal valve.

2.1.2.4 Digestive and Absorptive Function

The brush borders are covered by highly organized microvilli that are distributed along the small intestinal epithelium (Hartmann et al. 2022). Intestinal brush borders, located on the apical side of enterocytes, which make up 90% of intestinal crypt cells and 95% of intestinal villous cells, act as the primary and crucial site for food digestion and absorption (Adeola and King 2006). The enzymatic activities detected from brush borders can be considered as markers to evaluate the digestive maturity of weaned piglets (Henning 1985). It has been shown that there is a transiently decreased activity of brush border enzymes after weaning. Typically, the activity of brush border enzymes drops to the lowest point within 3–5 days post-weaning and then gradually increases afterward, mostly due to the ability of the intestine to adapt to dietary changes after weaning (Hampson and Kidder 1986).

Additionally, weaning also reduces the absorptive efficiency of fluids and electrolytes, resulting in impaired nutrient balance in the small intestine of piglets (Moeser et al. 2007).It is noted that the small intestine is easily exposed to a series of environmental stimuli causing impaired intestinal integrity and leading to immune and developmental dysfunction (Tang and Xiong 2022). Therefore, well-tuned intestinal epithelial barrier function plays a pivotal role in the maintenance of gut health. Intestinal epithelial barriers are generally divided into mechanical (physical) barriers, chemical barriers, immune barriers, and microbial barriers, due to their distinct functional mechanisms (Gou et al. 2022). The detailed function of the intestinal barrier is discussed in Sect. 2.2.

2.1.3 The Structure of the Large Intestine

The large intestine consists of three sections termed the cecum, colon, and rectum, and its main physiological function is to absorb water and any remaining undigested and unabsorbed nutrients coming from the small intestine, and it secretes hypertonic fluid to modulate the osmotic gradient (Williams et al. 2001; Zhang and Xu 2006). In contrast, the large intestinal epithelium is devoid of villi but retains regenerative crypt structures. In addition, the gut microbiota plays important digestive and absorptive functions in the larger intestine by secreting enzymes or metabolites. Dietary fibers are resistant to endogenous digestive enzymes in the small intestines, while they can be partially or completely fermented by the gut microbiota distributed in the large intestine to produce high amounts of volatile fatty acids and other energy substances. Then, these small energy substances and nutrients are further absorbed by the large intestine, thus contributing to the development and growth of pigs (Bindelle et al. 2007). It has been shown that the majority of carbohydrate fermentation occurs in the cecum, distal colon, and proximal colon, which accounts for 92% of the total fermentation in the large intestine (Knudsen and Hansen 1991). This implies that the main fermentation sites of undigested food by colonic microorganisms are concentrated in the anterior part of the large intestine. Interestingly, the porcine large intestine exhibits a similar fermentation pattern as the rumen in ruminants, which can effectively digest crude fiber in the diet. However, unlike the rumen, there is no anaerobic fungi in the porcine large intestine; instead, bacteria, fungi, and protozoa are the main microorganisms. Additionally, the methane production mediated by the gut flora in the porcine large intestine is lower than that in the rumen, due to its regulation by bile salts and their metabolites in the large intestine (Varel 1987).

2.1.3.1 The Cecum

The cecum is located at the beginning of the large intestine, with a cylindrical structure in the form of a blind sac, and tends to absorb salt and water and break down cellulose. In addition, the cecum receives digested chyme from the small intestine and begins to form intestinal fluid that maintains the balance of water and electrolytes in the intestine. In monogastric herbivores, the cecum is the major site for the digestion of crude fibers (Zhi-yuan and Cheng-wei 2021). However, the digestibility of crude fiber in the porcine cecum does not exhibit higher efficiency than that in other herbivores (Hongyang and Zhenjun 1998).

2.1.3.2 The Colon

The porcine colon is approximately 4.9 m in length, and its essential physiological role is to absorb remaining water, electrolytes, and nutrients to solidify the food waste. In addition, the colon secretes mucus to lubricate the food waste so that it can pass through toward the end part of the large intestine by intestinal peristalsis and muscle contractions to form feces.

A massive number of microbes colonize the colon and play important functions in metabolizing remaining carbohydrates or amino acids to generate vitamins or small bioactive compounds for the maintenance of gut health.

2.1.3.3 The Rectum

The rectum is located at the end of the colon and is the shortest part of the intestinal canal. It connects the colon to the anus by extending backward along the dorsal side of the pubic symphysis to the anus. The rectum has a thick muscular epithelial wall with folds of mucous membrane and stores feces until they are defecated from the body. Besides that, it is also capable of reabsorbing salt and water from the feces.

2.1.4 The Anus

The anus is an excretory opening located at the end of the rectum through which undigested food residues are eliminated. In addition, if food materials cannot be digested or absorbed as they pass through the digestive tract, they will be directly defecated through the anus with feces. The anus not only acts as the exit for feces but also absorbs the remaining water from the stool.

2.1.5 Digestive Glands

2.1.5.1 The Salivary Glands

There are three salivary glands termed the parotid, mandibular, and sublingual glands in pigs. The parotid glands are located near the pharynx, the submandibular glands are in front of the parotid gland, while the sublingual glands are located on the sides of the mouth. Saliva is a mixture of secretions from these three salivary glands and can lubricate the mucous membrane of the mouth, dissolve food, and facilitate swallowing actions. It also contains amylase and lysozyme to improve digestive and antibacterial activities. Salivary secretion is a complex process that is regulated by both the nervous and the humoral systems (Yang et al. 2005). Recent research reported that in addition to acetylcholine (Ach), non-cholinergic and non-adrenergic neurotransmitters such as vasoactive intestinal polypeptide (VIP), substance P (SP), and calcitonin gene-related peptide (CGRP) play important roles in salivary gland secretion by binding to their receptors (Garrett et al. 1999).

2.1.5.2 The Gastric Glands

The exocrine gastric gland consists of the oxyntic gland and pyloric gland. The oxyntic gland secretes gastric acid, which is mainly composed of pepsin, hydrochloric acid, and mucus. Hydrochloric acid is secreted by parietal cells in the gastric glands and creates an acidic environment for pepsin to digest proteins and aids in iron and calcium absorption (Guangyan 2003). After entering the small intestine, hydrochloric acid also promotes the secretion of pancreatic juice and bile, as well as gallbladder contraction. There are two types of mucus in gastric acid: soluble mucus and insoluble mucus. The soluble mucus is secreted by the gastric glands, which can lubricate food and protect the mucosa against mechanical damage. Insoluble mucus is continuously and spontaneously secreted by the gastric epithelium and covers the surface of the gastric mucosa. The production and secretion process of gastric acid are controlled by several autonomic nervous and humoral factors, including acetylcholine, goitrogens, histamine, and growth inhibitors.

2.1.5.3 The Pancreas and Liver

The pancreas is the second-largest gland in the host, composed of exocrine and endocrine glands (Fengli 2014). The porcine pancreas has similar functions as other mammals to secrete pancreatic acid and enzymes that are transported to the duodenum to break down food. The exocrine pancreas is primarily composed of pancreatic acini and ducts. Pancreatic acinar cells produce and secrete digestive enzymes, while the pancreatic ducts are responsible for the excretion of pancreatic juice[44]. The exocrine pancreatic juice contains lipase, amylase, trypsinogen, and other substances that digest fats, which are excreted into the duodenum through the pancreatic duct. However, the endocrine glands are composed of a subset of hormones produced by pancreatic islets, mainly composed of alpha cells, beta cells, gamma cells, delta cells, and epsilon cells. Alpha cells secrete glucagon, beta cells produce insulin and amylin, beta cells produce pancreatic polypeptide, delta cells secrete somatostatin, and epsilon cells generate ghrelin, all of which act together to regulate blood sugar levels (Kaneko et al. 2008). In addition to digesting proteins, fats, and carbohydrates, pancreatic fluid also modulates energy metabolism (Jianying et al. 2013; Sakai et al. 2018). The liver is the largest gland and acts as an important metabolic organ in the body. Unlike humans, the porcine liver consists of five large lobes, largely involved in detoxification, immune function, nutrient digestion, and protein synthesis. The liver is capable of synthesizing, storing, and releasing glucose and is responsible for the breakdown of proteins and lipids and the regulation of various hormone secretions (Jansen et al. 2017). Apart from its critical roles in protein, lipid, and carbohydrate metabolism, the liver also has major functions in blood reservoir, immune defense, and excretory regulation (Chang 2006; Wai et al. 2018; Xiusheng 2020).

2.1.6 Nutritional Regulation of Intestinal Development in Postweaning Piglets

As the gastrointestinal tract acts as the main site for nutrient absorption and metabolism, it is of great importance to elucidate its regulatory mechanisms associated with development. The development of the porcine gut is influenced by a variety of internal and external factors such as genetic background, nutrient metabolism, and environmental stresses. Piglets after weaning often encounter a high risk of diarrhea because of their immature gut and impaired intestinal ecology, causing vast economic losses (Sinn et al. 2017). Recently, increased evidence has suggested that nutritional intervention acts as an effective and powerful strategy to improve gut health by contributing to gut development and intestinal immune defense in piglets under conditions where the use of antibiotics is banned. Therefore, optimization of dietary nutrition and evaluation of the effects of different nutrients on gut development will be a feasible and reasonable way to increase production efficiency in pig farming.

2.1.6.1 Protein

Protein is the most important nutritional source for pigs, and its dietary level significantly affects gut development, including villus height, crypt depth, and intestinal epithelial barrier function in the small intestine of weaned piglets. It has been shown that both villus height and the ratio of villus height to crypt depth drastically increase and then decrease as the dietary protein level is gradually increased. Soy and its products are rich in protein, energy, and limiting amino acids such as lysine, threonine, and Trp, making them the most important dietary protein source for pigs. Currently, corn-soybean-based meals are the major nutritional diets in pig farming and production. Essentially, these soy products need to be fermented, enzyme-treated, or heat-treated to eliminate the anti-growth effect caused by anti-nutritional factors, allergenic proteins, and oligosaccharides present in soybeans before adding them to the pig diet. It is of great importance to develop new technologies to improve the efficiency of protein digestion and absorption from soy and its products or to explore new and high-quality protein sources that can substitute a certain proportion of soy products to reduce economic costs. Recently, the development of low -protein diets with a balanced amino acid pattern under different porcine developmental stages has become a new strategy to overcome the shortage and high-cost problems of soybean products in China.

2.1.6.2 Fat

Fat is mainly digested in the small intestine, where fat globules are emulsified into smaller pieces of lipid droplets by bile salts and lipase to increase the capacity of lipid hydrolysis. Short-chain fatty acids (SCFAs) are a group of fatty acids metabolized by the gut microbiota under fiber or polysaccharide fermentation and have been extensively studied in lipid metabolism-mediated biological processes, including cellular signaling modulation, immune function, and bacterial pathogenesis. Among these SCFAs, butyric acid can be directly absorbed and utilized in the small intestine for rapid energy replenishment, and it also stimulates repair mechanisms of the intestinal epithelium to maintain homeostasis. Because of its free and volatile nature, butyric acid is usually conjugated with sodium to form sodium butyrate for feed production. Dietary addition of sodium butyrate in weaned piglets has been reported to increase the villus height of the duodenum and jejunum and promote the proliferation of intestinal probiotics such as *lactobacilli*, both contributing to gut development.

In addition to providing energy, dietary fat promotes the absorption of fat-soluble vitamins and improves feed palatability and energy utilization. Furthermore, fat slows down the passage of chyme through the gastrointestinal tract, allowing sufficient digestion and absorption of nutrients (Long et al. 2022). Weaned piglets often fail to obtain sufficient energy from the feed to meet their growth requirements due to weaning stress and their immature gut (Baião and Lara 2005). When piglets do not get enough energy from the diet, they utilize the body fat, and the reduction of body fat is detrimental to their health. Therefore, increasing the concentration of dietary fat in piglet feed is necessary to alleviate weaning stress caused by low fat utilization efficiency. Nowadays, coconut oil or palm oil, which are rich in SCFAs, are mostly used in livestock feed production due to their high digestibility in the gut of weaned piglets. In addition, the ratio of unsaturated fatty acids to saturated fatty acids reaching approximately 3–4 is most appropriate for fat digestion in piglets.

2.1.6.3 Carbohydrates

Carbohydrates are the major energy components in pig feedstuff and act as precursors for the synthesis of various macromolecules in the body, which are beneficial for pig growth and health (Chwalibog et al. 1994; Knudsen et al. 2012). The main carbohydrate sources for piglets include whey powder, lactose, sucrose, functional oligosaccharides, nonstarch polysaccharides, and starch, all of which contribute to gut development and growth.

2.1.6.3.1 Whey Powder and Lactose

The main role of whey powder in promoting piglet growth is to provide a large amount of lactose, which can be further fermented in the digestive tract to produce lactic acid. This reduces the pH for food digestion and inhibits the growth of pathogenic bacteria. Whey powder contains high-quality whey protein with high digestibility and is devoid of anti-nutritional factors, peroxidase enzymes, and ferritin, all of which exhibit antibacterial effects. Dietary supplementation of whey powder effectively reduces weaning stress and improves the growth performance of piglets. In terms of production cost, lactose can be utilized to substitute a certain proportion of whey powder in feed production due to its advantages in palatability, digestibility, and absorptivity. Additionally, it can be converted into lactic acid after fermentation to promote the growth of probiotics for gut development.

2.1.6.3.2 Functional Oligosaccharides

Oligosaccharides are small molecular carbohydrates composed of a small number of monosaccharides and typically present as glycans by linking to lipids or amino acids (Hua et al. 2022). Although functional oligosaccharides cannot be directly digested and utilized, they can selectively act as substrates for commensal microflora, promoting the reproduction of beneficial microorganisms and inhibiting the growth of harmful ones. Oligosaccharides themselves also exhibit antigenic characteristics that initiate immune responses, thereby improving host immune function. Studies have shown that the addition of oligosaccharides promotes the growth of piglets and growing pigs and improves the reproductive performance of sows (Duan et al. 2016; Kim et al. 2008; Ru et al. 2022).

2.1.6.3.3 Nonstarch Polysaccharides

More than 80% of cereal grains consist of carbohydrates, of which nonstarch polysaccharides (NSPs) account for 10–30%. The average daily intake of piglets increases with the elevated content of NSPs in the diet. The increased body weight caused by excessive dietary supplementation of NSPs is likely a result of the increased intestinal weight (Wan et al. 2016). NSPs are capable of generating butyric acid in the gastrointestinal tract of weaned piglets to reduce the pH and inhibit the growth of pathogenic bacteria (Cheng et al. 2015a).

2.1.6.3.4 Starch

Starch serves as the primary form of carbohydrates and is the major energy source present in cereals and their derivatives for pigs (Högberg and Lindberg 2004). Starch is a polysaccharide and can be classified into two groups, amylose (AM) and amylopectin (AP), based on its chemical structures. AM consists of α-(1-4) glycosidic-bound glucose molecules forming linear polysaccharides, and its compact helical structure makes it more resistant to digestion (Bikker et al. 2006). On the other hand, AP belongs to branched polysaccharides linked by both α-(1-4) and α-(1-6) glycosidic bonds, and its highly branched structure is more easily digested than AM (Yin et al. 2004). Various factors such as morphology, granule sizes, and the AM/AP ratios can determine the physicochemical properties and energy-releasing characteristics of starches (Cummings and Stephen 2007). The lower the AM/AP ratio in the diet, the higher the starch digestibility by the action of pancreatic amylase in the small intestine (Tester et al. 2004). In addition, a lower AM/AP ratio also increases sucrase and maltase activities in the jejunum of piglets (Dona et al. 2010). Increased dietary AM content has been shown to promote the production of SCFAs, decrease intestinal pH, and activate the proliferation of bifidobacteria in the gut (Singh et al. 2010).

2.1.6.4 Amino Acids

Amino acids are vital nutrients preferentially utilized by the small intestine and are involved in a variety of important physiological and biochemical processes (Gao et al. 2020b). The utilization of amino acids by the small intestinal mucosa greatly contributes to intestinal development and the maintenance of intestinal health. Amino acids enter the intestinal epithelial cells via transporters binding to corresponding receptors, initiating extracellular signals and regulating intracellular physiological processes. Dietary proteins are hydrolyzed into small peptides and free amino acids in the digestive tract before being passively transported to the intestinal cells (Fouhse et al. 2015). Amino acids, especially functional amino acids, play key roles in regulating mucosal metabolism and the repair machinery (Gilbert et al. 2008; Tan et al. 2010; Wu et al. 1995, 2010; Yang et al. 2013a; Yao et al. 2011; Zeng et al. 2012). The daily intake of glutamate, glutamine, aspartate acts as the main source of energy for activating nutrient transport and the turnover of proteins in the small intestine.

Glutamine is a conditionally essential amino acid in catabolic conditions and provides a nitrogen source to support cell proliferation and differentiation. Dietary glutamate, aspartate, and glutamine not only provide energy for the small intestinal mucosa but also serve as precursors for protein, nucleic acid, and oxidative coenzyme II synthesis in cells, maintaining the normal morphology, structure, and function of the mucosa (Zhang et al. 2013). Glutamine can indirectly stimulate the secretion of various nutritional molecules like hormones and small peptides to promote the growth of the intestinal mucosa. In addition, glutamine also increases the polyamine concentration in intestinal epithelial cells, promoting cell differentiation and maturation and S-IgA production, thereby improving the immune function of lymphocytes and phagocytes.

Polyamines, acting as maturation factors, play important biological roles in regulating digestion, nutrient metabolism, and the normal growth of animals (Reeds et al. 2000). In the gastrointestinal tract of newborn piglets, proline oxidase, but not arginine, exhibits high activity to synthesize polyamines. Oral gavage of putrescine and proline significantly increases the proliferation of jejunal crypt cells and the activity of jejunal alkaline phosphatase, promoting the differentiation of jejunal cells in suckling piglets (Hussain et al. 2017). The absorption rates of aspartate, leucine, lysine, methionine, and proline by porcine intestinal mucosa drastically decrease within 3 hours after birth, followed by a significant increase in proline absorption, reaching the maximum value at the age of day 7. On the other hand, the absorption of aspartic acid, methionine, and lysine significantly increases with age, reaching the highest value by the age of day 28 (Kong et al. 2014). It is noted that weaning stress-induced intestinal villi atrophy and mucosal damage reduce the uptake rate of all amino acids, and then the rapid intestinal growth in postnatal piglets promotes total amino acid absorption in the intestine (Buddington et al. 2001). Arginine, glutamine, and proline are precursors for the synthesis of ammonia and play key roles in modulating cell differentiation and migration in the intestinal glandular crypts (Rao et al. 2002). Arginine improves intestinal immune function by regulating cytokine and antibody production (Tan et al. 2009).

Alpha-ketoglutarate acid directly enters the tricarboxylic acid cycle to provide energy and inhibit the degradation of glutamine and glutamate in a dose-dependent manner in the intestinal epithelium (Jiang et al. 2020). This reduces the production of alanine, ammonia, and carbon dioxide, allowing more glutamine and glutamate to be used for protein synthesis in the body.

The gut microbiota is capable of absorbing and degrading free amino acids and small peptides to produce ammonia, carbon dioxide, polyamines, nucleotides, and other nitrogen-derived substances, including nitrite, nitrate, and glutathione. It is mainly involved in the catabolism of lysine, methionine, phenylalanine, Trp, threonine, and histidine in the small intestine (Dai et al. 2012a, b, 2013).

2.1.6.5 Trace Elements

Trace elements play important roles in maintaining the normal growth and development of animals. Zinc and selenium, acting as essential trace elements, have a wide range of physiological functions. Zinc-deficient cells increase their membrane permeability and decrease their mobility. Zinc not only prevents the degradation of phospholipids but also competes with copper and iron to bind catalytic ions during radical chain reactions, which, in turn, prevent peroxidative damage on unsaturated fatty acids and sulfhydryl groups to protect the cell membrane. It has been shown that dietary addition of zinc significantly reduced the secretion of 5-hydroxytryptamine, vasoactive intestinal polypeptide, and forskolin in the intestinal mucosa of weaned piglets, thus reducing the incidence of diarrhea. Selenium is an important component of glutathione peroxidase, which protects cell membranes and mitochondrial membranes from damage by lipid peroxidation, thus playing protective and repairing roles in response to intestinal damage. Iron affects the differentiation of goblet cells, and a high iron concentration state greatly increases the number of colonic goblet cells. Iron deficiency inhibits intestinal epithelial cell development in piglets, resulting in decreased intestinal length and weight, villus height and crypt depth, as well as villus surface area. Copper is an essential trace element for pig growth, and dietary addition of copper sulfate at the weaning stage improves intestinal morphology by promoting cell proliferation and enhancing intestinal stem cell activities.

2.2 Regulation Mechanism of Tryptophan on Intestinal Development

The intestine serves as an interface between the external environment and the host (Fasano and Shea-Donohue 2005). The structural formation and functional maturation of the intestinal barrier are completed in mammals within a few weeks of birth, consisting of two developmental stages. The first stage prepares young animals for life outside the womb, while the second stage is closely related to epithelial digestion and intestinal barrier function. During this process, intestinal stem cells differentiate into mature intestinal epithelial cells, goblet cells, and intestinal immune cells in the crypt, contributing to the complete barrier and immune functions of the intestine, thus forming the first line of defense (Galenza et al. 2023). Barrier epithelial organs face the challenge of isolating the interior body from the external environment while continuously replacing cells in contact with this environment. In the intestinal tissue, development entails new cells migrating from the offspring of basal stem cells located in crypts to the top of villi, integrating into intestinal epithelium to form an intestinal barrier structure. Therefore, the migration of intestinal cells and the formation of intestinal barrier are crucial for gut development (Paerregaard et al. 2023). In the neonatal piglet model, amino acid deficiency leads to small intestine atrophy with fewer goblet cells and lower glutathione intestinal content. Trp is an essential amino acid for humans and animals (Bauchart-Thevret et al. 2009). About 99% of Trp is absorbed in the small intestine, where it can directly participate in protein deposition or produce bioactive compounds through its multiple metabolisms to regulate intestinal function and health (Li et al. 2021a). This section will introduce the impacts of Trp on intestinal development and review their regulatory mechanisms.

2.2.1 *Tryptophan Affects the Migration of Intestinal Cells*

The villus height, crypt depth, mucosal thickness, and villus surface area of the small intestine are directly related to the extent of enterocyte development, serving as representations of the digestive and absorptive function and health of the animal intestine. In young animals, intestinal development is not fully mature, characterized by slow migration of cells produced in crypts to the top of villi and a high loss rate of cell located at the top of villi. This leads to slow growth and development, resulting in small height and surface area of intestinal villi. Additionally, the small number of goblet cells and their weak mucus secretion function lead to a thin protective mucus layer, rendering the chemical barrier of the intestine incomplete in young animals. Furthermore, the arrangement and differentiation of intestinal smooth muscle cell are of significant for intestinal development. In the intestinal tract, the arrangement of smooth muscle cells is determined by molecular signals or mechanical signals (Huycke et al. 2019). Dietary protein is digested and absorbed by epithelial cells in the jejunum, with some amino acids metabolized in enterocytes to support high energy demand. Apart from their role as building blocks in protein synthesis, other amino acids play critical functions in the intestine, such as intestinal development, immune response, REDOX homeostasis, and microbiota (Coëffier et al. 2010; Vidal-Lletjós et al. 2017). There is substantial evidence that functional amino acids such as Trp, arginine, glycine, and glutamine promote intestinal development and improve intestinal inflammatory diseases (Hou et al. 2015; Ji et al. 2023). Moreover, Trp and its metabolism enhance intestinal tract activity, promote the repair of epithelial cells, and aid in the migration of newly replaced cells during enterocyte development, thereby maintaining the integrity of epithelial cells and the intestinal mucosa (Mawe and Hoffman 2013).

Partial Trp is metabolized in vivo to synthesize 5-hydroxytryptamine (5-HT) in the enterochromaffin cells (EC cells) through the serotonin pathway. Depending on the microvilli extending into the intestinal lumen and the enzymes and transporter proteins stored in the intestinal epithelium, EC cells can translate physiological and chemical signals in the intestine into endocrine signals (Bellono et al. 2017; Hata et al. 2017; Martin et al. 2017; Wang et al. 2017). 5-HT plays an important role in the development of the enteric nervous systems (ENS) and intestinal motility, secretion, inflammation, and sensory and epithelial development as a paracrine signaling molecule in the lamina propria or lumen of the intestine (Lund et al. 2018). Gut-derived 5-HT can promote intestinal motility, peristalsis, and secretion after binding to its specific receptors (5-HTRs). 5-HTRs are classified into seven families based on structure, function, and effectiveness (5-HTR1–5-HTR7). Specifically, the 5-HTR3 receptor is a ligand ion channel, and the other six are G-protein-coupled receptors (GPCRs) (Terry and Margolis 2017). There is compelling evidence that 5-HT could influence intestinal function by binding to different receptors such as 5-HTR1, 5-HTR2, 5-HTR3, 5-HTR4, and 5HTR7 (Hannon and Hoyer 2008; Pytliak et al. 2011). After binding to 5-HTR3 and 5-HTR4, 5-HT regulates motility and peristaltic reflex by activating ascending and descending interneurons and plays a role in vasodilation by locally regulating blood vessel diameter through intrinsic reflex circuits (Keating and Spencer 2019; Kendig and Grider 2015; Margolis et al. 2016; Mawe and Hoffman 2013). However, 5-HT regulates the secretion of epithelial cells like goblet cells via binding to 5-HTR2, 5-HTR3, and 5HTR4, which is neurally mediated or through the paracrine pathway acting on nearby enterocytes (Raybould et al. 2003; Tuo and Isenberg 2003). Meanwhile, binding to 5-HTR4 activates neurogenesis and enteric protection through neuronal 5-HT (Fiorica-Howells et al. 2000; Liu et al. 2009; Matsuyoshi et al. 2010). Additionally, 5-HT has an impact on intestinal mucosal growth by interacting with serotonergic neurons and projecting submucosal cholinergic neurons after binding to 5-HT2A receptors (Gross et al. 2012; Tackett et al. 2017). A large number of research have shown that 5-HT regulates the migration of vascular smooth muscle cells and mast cells through the CaSR/Rac1/PLC-γ1 signaling pathway by upregulating the protein expression of calcium-sensing receptors (CaSR), total Ras-related C3 botulinum toxin substrate 1 (total Rac1), Rho family member 1 of GTP-binding protein (GTP-Rac1), and phosphorylated phospholipase Cγ1 (p-PLC-γ1) (Gu et al. 2021a). Meanwhile, the Trp metabolite 5-HT could promote intestinal development and enhance the intestinal barrier by regulating the reconstitution of villus and vascular to meet the nutritional requirement (Liu et al. 2021c). In conclusion, Trp-metabolized 5-HT has an important influence on the development of intestinal structure and function by mediating the motility, secretion, and function of intestinal cells.

2.2.2 *Tryptophan Regulates the Formation of the Intestinal Barrier*

Intestinal development is closely related to the integrity of the intestinal barrier. The intestinal barrier is a highly complex system composed of many components, including the mucus layer, the epithelial layer, the lamina propria, and components such as microbiota and their secretions (Scaldaferri et al. 2012). The development and integrity of the intestinal barrier are crucial for maintaining animal health, alleviating intestinal damage, and preventing various diseases. Meanwhile, it also significantly influences intestinal development. The mucus layer covering the surface of the intestinal tract serves as the first line of defense against mechanical, chemical, and biological insults, preventing antigens, toxins, and bacteria from directly reaching epithelial cells. This mucus barrier, largely influenced by dietary habits, is composed of mucin complexes secreted by goblet cells (Fang et al. 2021). Another important component of the intestinal barrier is the intercellular junctional complex, crucial for maintaining barrier integrity. Intestinal epithelial cells are joined by tight junctions and adherens junctions, forming a complex cell-cell adhesion system (Chelakkot et al. 2018). Additionally, tight junctions can form both an absolute barrier and a paracellular ion, solutes, or water channel. The mucosal barrier, composed of an absolute mechanical barrier and a mucus barrier, effectively prevents invading organisms from entering the body, promotes the antigen shift of bacteria, and activates the inflammatory reaction of the intestinal mucosa (Horowitz et al. 2023). Trp has been reported to influence the taste receptors T1R1 and T1R3 expressed in cholecystokinin (CCK)-secreting cells, activating the expression and secretion of intestinal hormones (Daly et al. 2013). In addition, Trp participates in the development of the mucus barrier by regulating amino acid-sensing GPCRs such as the CaSR and GPR142, inducing the releases of hormones such as glucagon-like peptide 1 (GLP1) (Gribble and Reimann 2016).

Most Trp is absorbed and metabolized in the intestine, impacting intestinal protein deposition. Furthermore, gut-derived 5-HT and indole can act on the intestinal barrier to influence the integrity of barrier structure and function, regulating the intestinal development of newborn animals. Recently, some evidence suggested that Trp has an impact on the intestinal barrier through Trp-derived gut bacteria products, such as indole. For example, Li et al. found that indole-3-propionic acid (IPA), a Trp derivative, could upregulate the expression of claudin-1, occludin, and ZO-1, increase transepithelial electrical resistance, decrease paracellular permeability, and adjust the intestinal mechanical barrier. Importantly, IPA also strengthened the mucosal barrier by increasing the expression of mucins (MUC2 and MUC4) and the secretion products of goblet cells (TFF3 and RELMβ). Additionally, IPA reduced the expression of inflammatory factors induced by lipopolysaccharide to maintain the intestinal barrier (Liang et al. 2019).

The intestine is one of the most important immune organs and contains more than half of the immune cells in the body. The development of intestinal immunity is closely related to animal health. Intestinal homeostasis depends on the interaction of enteric microorganisms, nutrient substances, and the host immune system. Trp-metabolized kynurenine (KYN) and indoles are regarded as activators of the aryl hydrocarbon receptor (AHR), GPCR, and pregnane X receptor signaling, participating in the regulation of the host immune response. Meanwhile, the supplementation of Trp in the diet has a positive effective on the development of immune cells and the composition of enteric microorganisms, promoting innate immune responses in the intestine (Liang et al. 2019; Liu et al. 2019). It has been reported that dietary Trp can raise the level of secretory IgA (sIgA) as well as mRNA levels for porcine β-defensin (pBD) 2 and 3 in the intestinal of piglets (Friedman 2018). Additionally, the supplementation of Trp plays a role in anti-inflammatory response by meeting the demand for specific amino acids by improving the perception and utilization of amino acids (Elizei et al. 2017; Metghalchi et al. 2015; Wang et al. 2006; Zheng et al. 2019). In conclusion, dietary Trp can promote the intestinal mucosal barrier and maintain commensal balance, resulting in higher intestinal immunity and the development of the intestinal mucosal immune system.

The composition and homeostasis of microbes are closely related to intestinal digestion and absorption function, as well as the development of the microbial intestinal barrier. Meanwhile, it has been verified that dietary protein has a significant impact on the composition of intestinal microbiota. For weaned piglets, dietary Trp supplementation can increase the α-diversity and abundance of SCFA-producing bacteria such as *Prevotella* and rose bacteria while decreasing the abundance of

Clostridium and *Enterobacter*. The alteration of intestinal microbial structure increases the level of SCFA, contributing to the energy metabolism of epithelial cells and the intestinal barrier function (Liang et al. 2018). All of this indicates that Trp has an important influence on intestinal development and improvement by mediating the composition and structure of intestinal microorganisms in piglets.

2.3 Regulatory Impact of Tryptophan on Intestinal Immune Response

The gut microbiota and its metabolites play crucial roles in preserving intestinal health. In recent years, the biological connections between the gut microbiota, microbial metabolites, and intestinal immune function have become evident by integrating analyses of the intestinal microbiome, transcriptome, and metabolome. The gut microbiota impacts the duration and process of the immune system response. Likewise, the immune system also determines the distribution and complexity of the gut microbiota. Recent studies have highlighted the significant effects of dietary nutrients on the distribution and makeup of the gut microbiota, as well as on the relationship between the gut microbiota and immunological reactions (Thorburn et al. 2014). L-tryptophan (Opitz et al. 2011) is an essential amino acid in mammals that plays a critical role in the maintenance of gut immune homeostasis, and its abnormal metabolism is often connected to various immune disorders.

Altering the gut microbial structure has been shown to influence the levels of Trp and its metabolites (Clarke et al. 2014). This chapter briefly summarizes the current understanding of the relationship between Trp metabolism and the gut microbiota, the role of Trp in host-microbiota interactions, and the impact of Trp metabolism on gut immunological balance. Comprehending these connections could offer new strategies for treating different intestinal illnesses linked to the microbiota and Trp metabolism.

2.3.1 Regulatory Impact of the Gut Microbiota on Intestinal Immune Response

The microbiome is composed of bacteria and their combined genetic material, both of which affect the host's metabolic characteristics and immune system (Gordon 2012). Bidirectional interactions between the gut microbiota and the host immune system occur throughout the host's lifespan. The microbiota affects the development of the immune system, and in turn, the immune system modulates the composition and diversity of the gut microbiota. Moreover, the gut microbiota plays a role in immune cell recruitment and triggers inflammation. The interaction between the gut microbiota and enterocytes impacts the gut environment and significantly affects intestinal immunological balance, as demonstrated by Adhyaru et al. in 2018, with long-lasting effects (Adhyaru and Jacobson 2018). Aberrant gut microbiota and heightened gut permeability have been shown to be associated with the development of various diseases, such as autoimmune diseases and neurodegenerative disorders (Anderson et al. 2016).

The gut microbiota interacts with the host immune system through various chemicals and signaling pathways, affecting both local and peripheral organs such as the gut, liver, lung, and brain (Anders et al. 2013). Additionally, host-gut microbiota interactions also affect metabolic processes and long-term health outcomes (Blumberg and Powrie 2012; Elbeldi-Ferchiou et al. 2011). The gut immune system modulates the formation of the bacterial layer on the intestinal epithelium to reduce the risk of negative effects and organize an intact intestinal barrier (Blumberg and Powrie 2012). Therefore, it is of great importance to maintain the well-tuned relationship between the gut microbiota and gut immune system (Hamard et al. 2007).

2.3.2 Interactions Between the Host Immune System and the Microbiome Related to Tryptophan Metabolism

2.3.2.1 Tryptophan

Tryptophan (Trp) plays a pivotal role in regulating inflammatory responses in mammals by acting as an essential component of dietary protein (Marsland 2016). Research has consistently demonstrated that mice subjected to a low-Trp diet exhibit heightened susceptibility to chemically induced inflammation, while mice given Trp-rich diets show reduced inflammation and milder symptoms of conditions such as colitis (Hamard et al. 2007). For instance, investigations conducted on both mice and piglets have underscored the beneficial effects of Trp-rich diets in alleviating inflammation and mitigating the severity of diseases like colitis, in stark contrast to Trp-deficient diets (Kim et al. 2010). Moreover, mice subjected to Trp-deficient diets tend to manifest exacerbated inflammation in their central nervous system, a manifestation that is notably ameliorated upon Trp supplementation. These findings strongly suggest that Trp exerts significant regulatory effects on the immune system, potentially through the generation of Trp metabolites. Such insights underscore the critical role of Trp and its associated regulatory pathway in modulating inflammatory responses across mammalian species.

2.3.2.2 Aryl Hydrocarbon Receptor Signaling

Aryl hydrocarbon receptor (AHR) is a cytosolic transcription factor activated by ligands that plays a crucial role in regulating host immunity. It participates in modulating adaptive immunity and maintaining intestinal immune homeostasis (Hubbard et al. 2015a). It has been reported that AHR-null mice exhibit severe symptoms and mortality in dextran sodium sulfate (DSS)-induced intestinal damage models, as demonstrated by Arsenescu et al. in 2011 (Arsenescu et al. 2011). In addition, AHR activation induced by 2,3,7,8-tetrachlorodibenzo-p-dioxin (TCDD) has been shown to reduce lethality and symptom severity, as shown by Takamura et al. in 2010 (Takamura et al. 2010). AHR-deficient mice are vulnerable to intestinal stressors, highlighting the essential function of AHR in preserving gut immunological and barrier functioning (Sutter et al. 2011). Moreover, AHR-deficient mice also exhibit increased vulnerability to *Listeria monocytogenes* and *Citrobacter rodentium* infection (Qiu et al. 2012). Basically, there are two mechanisms of AHR-mediated promotional effects on the maintenance of immunological homeostasis. Firstly, AHR plays an antibacterial role in the gut by inducing IL-22 transcription in an AHR-dependent manner. Lee et al. reported that AHR activates innate lymphoid cell 3 (ILC3) expression to stimulate IL-22 transcription (Lee et al. 2012). In the gut, IL-22 also regulates the release of antimicrobial peptides and maintains the balance between gut immune reactions and the gut microbiota by modulating the composition and diversity of gut microbes, as stated by Zelante et al. (2013). Secondly, the anti-inflammatory function of AHR through its regulation of the formation of intraepithelial lymphocytes and innate lymphoid cells has also become evident (Zelante et al. 2013; Zelante et al. 2014). These immune cells prevent the host from pathogenic microorganism infection and further contribute to the maintenance of gut homeostasis (Hubbard et al. 2015b).

Microbial metabolites acting as ligands of AHR play a crucial role in host immunity by shielding the mucosa from inflammation (Rooks and Garrett 2016). Overexposure to AHR ligands has negative impacts on intestinal immunity, which can be mitigated by consuming more dietary AHR ligands (Schiering et al. 2017). AHR can directly target and activate specific genes such as interleukin-6, interleukin-22, prostaglandin G/H synthase 2 (PTGS2), vascular endothelial growth factor A (VEGFA), and cytochrome P450 1A1 when exposed to AHR ligands during inflammation. Additionally, the intake of AHR ligands can counteract the AHR-mediated control of intestinal balance (Hubbard et al. 2015b).

Consuming a meal enriched with Trp can activate AHR and consequently raise colonic IL-22 expression. Dietary supplementation with Trp protects the epithelial layer from acute colitis and intestinal inflammation through activation of AHR signaling (Hashimoto et al. 2012). Dietary Trp-mediated AHR activation has been shown to improve colitis symptoms and severity caused by DSS damage, as reported by Islam et al. in 2017 (Islam et al. 2017). The regulatory effects of dietary Trp are largely determined by its metabolites, which function as AHR ligands, rather than Trp itself (Opitz et al. 2011). Various endogenous and bacterial-derived Trp metabolites have been shown to function as ligands to bind AHR and modulate intestinal immune function (Zelante et al. 2013). For instance, KYN, KA, xanthurenic acid (XA), and cinnabarinic acid directly act as AHR ligands and are capable of activating AHR-dependent target gene expression (Romani et al. 2014). AHR activation regulates the transcription of IL-22 in human and murine CD4+ T-cells, as demonstrated by Lowe et al. in 2014 (Lowe et al. 2014). Additionally, various bacterial-derived Trp metabolites such as indole, indole acetic acid, skatole, indole propionic acid, and tryptamine have been also confirmed as AHR ligands (Bittinger et al. 2003).

AHR and IDO1 play important roles in the breakdown of microbial Trp and host-produced Trp metabolites to regulate AHR-dependent T-cell immunological balance in the mucosal layer. IDO1 is activated in response to pro-inflammatory cytokines, leading to the production of KYNs. KYNs act as ligands for the AHR and control immunological balance, promoting the development of regulatory T cells and preventing excessive inflammation in mice (Bessede et al. 2014). Trp metabolism-derived AHR ligands may be involved in establishing the symbiotic relationship between the host and gut bacteria (Hubbard et al. 2015a). IDO1 deficiency results in an increase in commensal-driven AHR/IL-22 activity, while AHR activation also regulates IDO1 levels (Bessede et al. 2014). The reciprocal relationship between IDO1 and AHR is essential for promoting the coevolution of commensal fungi with the mammalian immune system and gut microbiota. This interaction benefits the host's survival and the establishment of fungal commensalism, particularly in the presence of intense inflammation, and helps maintain balanced immune responses.

The synthesis of AHR ligands and subsequent AHR activation are controlled by both exogenous and endogenous factors, including dietary conditions, bacterial structures, and endogenous enzymatic activities (Hubbard et al. 2015b). Trp metabolites generated by gut bacteria can signal the host through the AHR to protect against mucosal inflammation. *Lactobacilli* species metabolize Trp to generate AHR ligands, like indole-3-aldehyde (IAld), which stimulates innate lymphoid cells, leading to enhanced mucosal defense against pathogenic *Candida albicans* infection. IL-22 production promotes the expression of antimicrobial peptides in the gut epithelium to reduce pathogenic bacterial infection by sequestering metal ions (Rooks and Garrett 2016; Zelante et al. 2013). The bidirectional relationship between AHR and the gut microbiota has been well documented, and this connection largely impacts host metabolism (Korecka et al. 2016). Microbial-generated metabolites like SCFAs and Trp metabolites can stimulate the AHR and its target genes in the intestine or liver. Consequently, AHR signaling also impacts the microbial structure in the intestine. Therefore, AHR plays a crucial role in regulating intercommunication between the host and the gut microbiota, which in turn modulates host metabolism and the immune system.

2.3.3 Factors Affecting the Interactions Between the Host Immune System and the Microbiome Related to Tryptophan Metabolism

2.3.3.1 Diseases

The significant effects of the gut microbiota on host physiology and pathology have been thoroughly researched. The progression and development of various immunological diseases, including pediatric spondyloarthritis, irritable bowel syndrome, inflammatory bowel disease, and colon cancer, are linked to the gut microbiota. Metabolites of Trp have been reported to be involved in regulating these immunological disorders. Importantly, the gut microbiota regulates Trp metabolite levels, which in turn significantly alter the host intestinal immune homeostasis (Uronis et al. 2009).

Spondyloarthritis

Spondyloarthritis affects around 1% of the population in the United States. According to Vereecke et al. 50% of people with SpA may experience intestinal inflammation without showing obvious gastrointestinal symptoms, a condition known as subclinical inflammation (Vereecke and Elewaut 2017). The microbiota has been proposed as a key component impacting immunological reactivity in SpA patients, due to the tight association between SpA and intestinal inflammation. Stoll et al. found taxonomic variations of bacteria in the feces of juvenile or adult SpA patients when compared to healthy persons (Stoll et al. 2014). Research examined the impact of the gut microbiota on SpA patients and found that these patients exhibit reduced gut microbial diversity, resulting in decreased Trp metabolism and lower levels of Trp metabolites (Lawrence et al. 2008). Similarly, rheumatoid arthritis patients also exhibited low levels of Trp metabolites (Kang et al. 2015). Alterations in the gut microbiota have been linked to changes in fecal metabolomics and abnormalities in Trp metabolism in pediatric spondyloarthritis (Liu et al. 2010; Stoll et al. 2014). In this study, 21 distinct Trp metabolites from different Trp metabolic pathways were identified to be decreased in the feces of SpA patients, including three bacteria-derived Trp metabolites, such as indole-3-acetate, methyl indole-3-acetate, and indole-3-acetaldehyde; eight KP-generated metabolites like 3-hydroxy-L-kynurenine, anthranilate, glutaryl-CoA, and kynurenate; and six metabolites from the 5-hydroxytryptophan pathway. Moreover, the 16S sequencing data confirmed the reduced Trp metabolism level in SpA patients when compared to controls (Stoll et al. 2016). The gut microbiota may serve as a pro-inflammatory factor in patients with enthesitis-related arthritis, a type of SpA, ascribe to changes in Trp catabolism. Many Trp metabolites are linked to intestinal immune responses and inflammation, but further research is needed to determine which ones are primarily responsible for producing subclinical inflammation in SpA.

Irritable Bowel Syndrome (IBS)

IBS has been diagnosed as one of the most common chronic GI disorders, with a prevalence of ~15% in Western populations. The intricate pathophysiology of IBS involves various cell types including intestinal epithelial cells, neuronal cells, and immune cells (Xavier and Podolsky 2007). Recent studies have increasingly highlighted disruptions in gut microbiota diversity and composition as key factors in the etiology of IBS. Notably, alterations in the balance of intestinal microflora have emerged as pivotal contributors to the symptomatology of IBS, characterized by an elevated Firmicutes to Bacteroidetes ratio, augmented abundance of *Streptococcus* and *Ruminococcus* species, and decreased populations of *Lactobacillus* and *Bifidobacterium*.

Furthermore, a compelling association between IBS and enhanced Trp metabolism via the kynurenine pathway (KP) has been unveiled. The positive correlation between the KYN to Trp ratio and symptom severity in IBS implicates the involvement of IFN-γ activation and subsequent IDO1-mediated Trp oxidation as potential pathogenic mechanisms. Dysfunction within the serotonergic system also emerges as a significant contributor to the pathophysiology of IBS, as evidenced by studies demonstrating that acute Trp increase treatment exacerbates gastrointestinal symptoms in IBS patients compared to acute Trp depletion treatment (Jenkins et al. 2016). Notably, both gut and brain serotonin concentrations affect the severity of IBS symptoms.

Additionally, microbial-origin stimuli and the gut serotonergic system are believed to play pivotal roles in shaping the symptomatology of IBS. Addressing gastrointestinal symptoms through serotonergic modulation in IBS patients necessitates a comprehensive investigation into the in vivo effects of local concentrations and proportions of various colonic SCFAs on mucosal serotonin homeostasis. Balancing mucosal serotonin homeostasis through manipulation of the complex microbial community, given the concentration-dependent effects of SCFAs on TPH1 expression, poses a considerable challenge. Moreover, a more nuanced understanding of the interplay between SCFAs and the gut serotonergic system is imperative.

Inflammatory Bowel Disease (IBD)

IBD is a chronic inflammatory disease consisting of Crohn's disease and ulcerative colitis. Dysregulated mucosal immune system, imbalanced gut microbial ecology, and disrupted mucosal barrier are often associated with the development of IBD (Franceschi et al. 2000; Shanley et al. 2009). Among these effectors, the gut microbial community has gained much attention due to its regulatory roles in maintaining gut health (Levin and van den Brink 2014). The development of IBD is linked to changes in the structure of the gut microbiota. It is still debated whether these changes are the cause or a consequence of inflammation in IBD (Gkouskou et al. 2014). Various data suggest that the gut microbiota may initiate inflammation in IBD because antibiotic treatment can significantly reduce the process of intestinal inflammation (Sartor 2004). Altered interactions between the gut microbiota and mucosal immune system are believed to lead to heightened intestinal permeability and inflammation in IBD (Hold 2016).

IBD patients experience decreased bacterial diversity, leading to heightened intestinal permeability and greater infiltration of intestinal bacteria. This can trigger immune reactions and result in systemic inflammation, weakening the protective linings in the intestines. The overgrowth of certain bacteria that break down mucus in IBD may be linked to changes in sugar molecules in the mucus, as proposed by Png et al. in 2010 (Png et al. 2010). Various bacteria have been shown to be linked to IBD. In IBD patients, there is a drop in Bacteroidetes and Bacteroides levels, and an increase in *Desulfovibrio* and Bilophila levels. The invasive capability of *Fusobacterium* is positively correlated with the severity of IBD, indicating that the composition of *Fusobacterium* could impact IBD pathogenesis. In contrast, certain gut bacterial species have been reported to exhibit protective effects against IBD (Kostic et al. 2014). For instance, *Bifidobacterium*, *Lactobacillus*, and

Faecelibacterium prevent mucosal inflammation by reducing inflammatory cytokines or promoting the production of IL-10, an anti-inflammatory cytokine in the gut (Llopis et al. 2009). Changes in Trp metabolism are often linked to IBD. IBD patients show elevated plasma levels of KYN and KA, along with lower quantities of plasma Trp. In addition, increased IDO1 activity has been observed in the peripheral blood and colon cells of IBD patients (Greenhill 2014). Elevated expression of various pro-inflammatory cytokines such as IFN- γ, IL-1, and IL-6 in IBD patients are believed to trigger Trp catabolism, leading to a decrease in plasma Trp levels and an increase in Trp catabolite levels. Elevated levels of 5-HT have been observed in the inflamed mucosa of patients with Crohn's disease (Lee et al. 2012) and in animal models with IBD, indicating the significant involvement of 5-HT in promoting intestinal inflammation in IBD (Lowe et al. 2014). Bacteria-mediated Trp metabolism plays an important role in the development of IBD. A considerable drop in microbial Trp metabolites including indole acetate and indole propionate, two metabolites that have been shown to have anti-inflammatory properties in the intestines, has been observed in dogs with IBD. Similarly, the expression levels of indole-3-acetic acid are decreased in feces from IBD patients, indicating that microbial Trp metabolites may play a critical role in the development of IBD (Lamas et al. 2017). Additionally, there is a notable decrease in the presence of bacteria capable of breaking down terminal fucose residues from intestinal mucins using α-L-fucosidases. This decrease is associated with lower levels of indole acrylic acid and indole-3-propionic acid synthesized from tryptophan (Sigurdsson et al. 2023). Reestablishing bacterial Trp metabolism in the gut to enhance the production of indole acrylic acid to facilitate their anti-inflammatory function may be a potential treatment for patients with IBD.

Colorectal Cancer

Colorectal cancer is the third most common cause of death among cancer patients. Research has investigated the impact of the gut microbiota on colorectal cancer progression. Genetically modified animals had reduced tumor masses compared to conventionally bred animals. Antibiotic-induced reduction of gut microbiota has been demonstrated to decrease the incidence of colorectal cancer (Raman et al. 2013). Changes in gut microbiota composition have been shown to reduce temporal stability but increase the diversity of Clostridium leptin and Clostridium coccoides subgroups compared to control (Grivennikov et al. 2012). The metabolism through the KP has been identified as a powerful target for immunotherapy. KP is triggered in various types of tumors; IDO1 triggers KP in the tumor as well as in infiltrating myeloid-derived cells in colorectal cancer. IDO1 seems to be constantly active in cancer patients and is frequently upregulated in colorectal cancer, in which decreased levels of Trp and elevated levels of KP metabolites have been observed (Liu et al. 2010). Furthermore, IDO1 is crucial for regulating the composition of the microbiota, and activation of IDO1 inhibits microbial growth. Thus, it is likely that the increased expression of IDO1 in colorectal cancer may lead to reduced temporal stability of the gut microbiota. Reducing the growth rate of gut bacteria through IDO1 activation may serve as a self-regulating defensive mechanism against colorectal cancer, as gut microbiota depletion increases the risk of colorectal cancer. Research has demonstrated that inhibiting IDO1 activity can boost the immune response of tumor-bearing animals against tumors. Therefore, the impact of IDO1 on the gut microbiota should be considered. Selective inhibition of IDO1 is suggested to enhance cellular immunity and has therapeutic potential in cancer, particularly in colorectal cancer.

IDO1 acts as a mediator in the communication between the host and the gut microbiota, impacting the development of gastrointestinal cancer. Activation of pattern recognition receptors (PRRs), particularly TLR4 and TLR9, might promote the expression of IDO1 during host-microbe interactions (Santhanam et al. 2016). LPS activate TLR4 and subsequently IDO1, leading to a decrease in the production of LPS and toxins by regulating gut microorganisms, potentially lowering the likelihood of gastrointestinal cancer. Moreover, developing new antagonists for PRRs to decrease the expression of IDO1 could be advantageous in combating malignancies. Microbial Trp metabolites like indoles have been shown to be often associated with colorectal cancer, likely due to the activation of AHR. The regulatory function of the gut microbiota and Trp metabolism in the pathogenesis of colorectal cancer may provide potential to develop new strategies to treat colorectal cancer.

2.3.3.2 Stress

The bidirectional interaction between stress responses and gut microbiota has been increasing recognized in recent years (Kleiman et al. 2019). Physical and psychological stresses are capable of changing the composition and diversity of gut microbiota in both rats and primates (Bailey 2014). Moreover, altering the gut microbiota has been demonstrated to cause behavioral changes such as stress, depression, and anxiety. Recolonization of fecal bacteria from pathogen-free mice to the gut of germ-free (GF) mice has been reported to partially alleviate stress responses (Sudo et al. 2004). Stress-induced biological function is largely linked to Trp metabolism. Stress-related hormones and neuropeptides affect Trp absorption and serotonin production and release in the brain. Serotonin release caused by stress is associated with increased TPH1 expression. Moreover, the expression and activity of TDO have also been induced by stress-related glucocorticoid hormones (Ruddick et al. 2006). Immune stress induces the release of IFN-γ to activate IDO expression and promote NAD+ synthesis in macrophages using quinolinic acid (QA) as the substrate (Ruddick et al. 2006). Disrupted Trp metabolism and IDO1 activity may cause stress-related diseases. Likewise, stress-induced cortisol can elevate IDO1 activity and gut permeability.

The gut microbiota affects 5-HT production through modulation of IDO expression in response to stress (Forsythe et al. 2010). Excessive intestinal stress disrupts intestinal barrier function, increases gut permeability, and activates the expression of pro-inflammatory cytokines, leading to changes in Trp metabolism and IDO activity. Pro-inflammatory cytokines act together with 5-HT to regulate the secretion of corticotropin-releasing hormone and vasopressin in response to stress. It has been shown that probiotics can improve gut barrier function and serotonin synthesis in response to stress. Probiotic treatment is considered a potential therapeutic application to treat stress-induced gastrointestinal disorders (Bhattarai et al. 2017). Further research is required to fully comprehend how stress impacts the physiological relationships between the gut microbiota and Trp metabolism.

2.3.3.3 Probiotics

Probiotics are live microorganisms beneficial to host health. Research shows that dietary supplementation with probiotics alters gut ecology and improves gut immune function (Nagata et al. 2016). Probiotics can specifically impact the metabolism of Trp and its concentration. Plasma Trp concentration and KA levels were significantly increased in rats treated with bifidobacteria compared to the control group. Bifidobacteria treatment also reduced levels of pro-inflammatory cytokines like IFN-γ, IL-6, and TNF-α in rats after mitogen stimulation. These findings suggest that bifidobacteria attenuate pro-inflammatory immune responses by raising plasma Trp and KA levels (Desbonnet et al. 2008). Addition of *L. reuteri* improved the formation of indole-3-aldehyde (IAld) by activating Trp metabolites (Opitz et al. 2011; Zelante et al. 2013). IAld activates AHR in gut-resident T cells and ILCs to promote IL-22 production and protect against colonic inflammation. In addition, *Lactobacillus fermentum* VRI-003 treatment increased the IFN-γ response to stimulant IDO1 expression (Cox et al. 2010).

There are several hypothesis about how probiotic treatment impacts the availability of Trp and its metabolism. Firstly, probiotics affect the composition and diversity of the gut microbiota, which in turn directly influences Trp metabolism and immune-responsive pathways (Opitz et al. 2011; Zelante et al. 2013). Probiotic treatment in vulnerable individuals can alter the gut microbiota and enhance the production of AHR ligands through Trp metabolism, potentially preventing intestinal inflammation in the host (Etienne-Mesmin et al. 2017). Introduction of three commensal *Lactobacillus* strains with strong Trp metabolic capabilities in Card9-/- mice led to the recovery of intestinal IL-22 production and reversed their vulnerability to colitis (Lamas et al. 2017). So far, only a limited number of probiotics have been studied for their capacity to influence intestinal immune function through Trp availability and metabolism. Additionally, the essential functional genes in these probiotics have not been identified. Therefore, a variety of probiotics need to be explored in the future. Developing new probiotics that target genes related to Trp metabolism through genetic engineering might be an alternative way to investigate in the future.

2.3.3.4 Aging

Aging is intricately intertwined with alterations in the gut microbiota and is often accompanied by physiological changes in the gastrointestinal (GI) tract and a decline in immune system function (Salazar et al. 2017). These changes collectively contribute to an increased susceptibility to infection, malnutrition, and other functional deficiencies among elderly individuals (Salazar et al. 2017). Notably, aging is typified by reduced bacterial diversity, diminished beneficial microorganisms, altered bacterial species, and increased facultative anaerobic bacteria within the gut microbiota, indicating its relationship with microbial changes (Cheng et al. 2015b).

Furthermore, alterations in microbial composition have been implicated in immune senescence and inflammation among older individuals (Salazar et al. 2014). Notably, aging significantly impacts Trp metabolism, with Trp levels playing important roles in immune tolerance and gut microbial maintenance (Rampelli et al. 2013). Analysis of *Kyoto Encyclopedia of Genes and Genomes* (KEGG) orthologous genes of gut microbes from young to old individuals has revealed age-related upregulation of genes involved in the Trp metabolism pathway, consistent with the observed reduction in serum Trp concentrations among centenarians (Rampelli et al. 2013). The increased Trp consumption by the gut microbiota is hypothesized to affect Trp bioavailability to the host, potentially contributing to heightened immune activation, particularly in individuals with inflammatory diseases (Collino et al. 2013, Gupta et al. 2012).

Moreover, dietary Trp restriction has been shown to be associated with delayed aging and extended lifespans in animal models (van Beek et al. 2017). Utilizing the accelerated aging Ercc1-/Δ7 mouse model, studies have elucidated the intricate relationship among aging, gut microbiota, and Trp metabolism (Gurkar and Niedernhofer 2015). Dietary Trp restriction improved the diversity of gut microbiota, restoring the gut microbial composition from the older mice to that of younger counterparts (van Beek et al. 2017). Additionally, dietary Trp restriction impedes B-cell development in the bone marrow, with decreased abundances of *Alistipes* and *Akkermansia* spp., two bacterial species that are positively correlated with the reduction of B-cell precursors.

In summary, dietary Trp restriction emerges as a potent intervention strategy to modulate the function of gut microbiota and gut immunity as well as the aging process. However, further research is needed to elucidate the complex interplay between dietary Trp metabolism, gut microbiota, and B-cell development during aging. Understanding the mechanisms of how specific bacterial species determine the number of B-cell precursors and elucidating the key metabolites and signals mediated by this connection during the aging process await further investigation.

2.4 Mechanisms of Tryptophan Regulation of Intestinal Microbes

The animal gut is a complex and diverse habitat for microbial communities, mainly consisting of bacteria, viruses, archaea, and eukaryotes (Rajilic-Stojanovic and de Vos 2014). Bacteria make up the bulk of the community and are divided at the phylum level into Firmicutes, Proteobacteria, Actinobacteria, and Bacteroidetes (Liu et al. 2022). The composition of gut microbes is influenced by many factors, such as developmental stage, gut site, and exogenous intervention, and is constantly changing from infancy to old age (Milani et al. 2017). In addition, changes in gut microbial composition in early infancy are strongly associated with the development of disease later in life (Ihekweazu and Versalovic 2018). Studies of animal gut microbes have found that the composition and structure of microbes in the small and large intestine of pigs are very different, with compartmentalized metabolism of nutrients (Zhu et al. 2014). The microbial composition of the small intestine lumen and intestinal wall differs from one ecological site to another, with differences in the metabolism of amino acids (Yang et al. 2014). On the other hand, the microbial composition and function of the mammalian gastrointestinal tract are rich and diverse and are closely related to the health of the host. Recent studies have shown that the effects of serotonin on the body may be influenced by gut microbes and that serotonin can also regulate gut microbial composition and function. Trp is the only precursor of serotonin, and studies have shown that Trp metabolism is a key metabolic pathway in establishing the microbe-gut-brain axis (Gao et al. 2020a). Serotonin, also known as 5-HT, is a monoamine neurotransmitter responsible for the regulation of mood, energy, memory, and other functions (Israelyan and Margolis 2018). 5-HT is produced in the brain by TPH2 and in the intestine by TPH1 in intestinal chromaffin cells (Guo et al. 2021). At a young age, 5-HT is used as a physiologically active substance to promote the development of target organs such as the brain, intestine, and internal organs of the animal organism (Ugrumov 2010a). 5-HT also acts as a signaling molecule to regulate the function of the intestinal tract by activating 5-HT receptors that transmit signals to neurons, which in turn regulate a range of physiological activities such as secretion, peristalsis, vasodilation, and nutrient absorption in the intestine (Mohammad-Zadeh et al. 2008). 5-HT produces melatonin through N-acetyltransferase and hydroxyindole oxygen methyltransferase, which is secreted by the pineal gland in the brain and regulates sleep patterns and rhythms (Kaczor and Skalski 2016). It is also a powerful antioxidant, protecting the reproductive system and heart muscle cells and inhibiting cancer (Samanta 2022). It also plays a role in the improvement of autoimmune diseases (Ren et al. 2017).

2.4.1 The Gut-Brain Axis and the Microbial Factors that Affect It

The gut-brain axis is a system of bidirectional cross-border communication between the gut and the brain mediated by intestinal microbes, also known as the microbiota-gut-brain axis (MGBA) (Guo et al. 2021). The sensory signals transmitted by the gut change reflex activity, sensation, and emotion through the vagus-spinal cord-central nervous system axis. At the same time, the gut also affects the brain through the manipulation of neurotransmitters and neuroactive molecules. Conversely, the neural activities of the brain can also affect intestinal secretion, movement, permeability, and immune function (Romani et al. 2014). Gut microbes are important influencers of the gut-brain axis, and studies have found a high correlation between IBS and depression (Lee et al. 2017). In mice with depression, the number of bifidobacteria in the gut is reduced (Desbonnet et al. 2010). In addition, altered intestinal microflora in mice led to changes in Trp metabolism, reduced bile acid circulation, and consequently, decreased intestinal motility (Golubeva et al. 2017). Lipopolysaccharide (LPS) and peptidoglycan (PGN) production by gut microbes mediate the development of corresponding immune responses in the gut and brain by binding to PRRs, including Toll-like receptors and nucleotide-binding oligomerization domain (NOD)-like receptors (Arentsen et al. 2017). In addition to being present in the blood, LPS and PGN have been found to cross the blood-brain barrier into the brain and affect 5-HT signaling by activating NOD receptors on the surface of periventricular organs, choroidal plexuses, meningeal cells, astrocytes, stretch cells, and endothelial cells (Pusceddu et al. 2019; Vargas-Caraveo et al. 2017). This also affects brain function and mental status by activating intestinal PRRs to produce pro-inflammatory factors that enter the bloodstream (Yarlagadda et al. 2009). In addition, gut microbes produce or consume large amounts of neurotransmitters (e.g., dopamine, norepinephrine, serotonin, or γ-aminobutyric acid) that affect the host's neurological response (Strandwitz 2018).

2.4.2 Influence of the Main Intestinal Microorganisms on MGBA

2.4.2.1 Probiotics

Probiotics in animals are a naturally occurring group of active microorganisms that, when consumed in sufficient quantities, can have beneficial health effects on their hosts (Cryan et al. 2019). Although their exact mechanism of action remains to be fully understood, the use of probiotics is beneficial in regulating the gastrointestinal microbiota (Yelin et al. 2019). After dogs with inflammatory enteritis were treated with probiotics, the results showed a reduction in the inflammatory bowel disease activity index and $CD3^+$ cells, an increased number of $FoxP3^+$ and TGF-β^+ cells, and normalized intestinal dysregulation due to enteritis (Rossi et al. 2014). Studies have found that *Probiotic-4* can ameliorate cognitive dysfunction in aged mice by regulating intestinal dysbiosis and MGBA deficiency (Yang et al. 2020). Moreover, research has shown that probiotics can modulate MGBA; improve metabolic, immune, and endocrine functions; reduce stress behavior; and enhance cognitive and memory abilities in animals (Cryan et al. 2019). Fleming et al. (2019) demonstrated that the use of dietary probiotics in piglets early in life modulated the concentration of SCFAs and hippocampal serotonin in the colon, blood, and brain and promoted memory recognition. These studies suggest that probiotics can have a positive effect on energy metabolism, intestinal barrier function, immune function, and emotional-cognitive behavior in livestock.

2.4.2.2 SCFAs

SCFAs are derived metabolites of intestinal microorganisms, produced during the anaerobic fermentation of gastrointestinal dietary fiber bacteria, consisting mainly of acetic, propionic, and butyric acids, which have neuroactive effects (Peng et al. 2021). As extremely important signaling molecules in the bidirectional communication system of the MGBA, SCFAs are directly or indirectly involved in the regulation of various MGBA pathways (Peng et al. 2021). SCFAs also play a role in immune regulation in two main ways: (1) SCFAs bind to a variety of GPCRs (Sittipo et al. 2022), of which GPR41 is expressed in the central nervous system (CNS), allowing signals from SCFAs to reach the CNS directly (Nohr et al. 2013). GPR43 is expressed in white adipose tissue, allowing SCFAs to act as hormonal molecules that stimulate energy expenditure in skeletal muscle and the liver (Kimura et al. 2013). (2) SCFAs exert immunosuppressive functions by inhibiting histone deacetylase (HDAC) and stimulating the hypothalamic-pituitary-adrenal (HPA) axis as potential epigenetic modifiers (Farzi et al. 2018). Butyric acid reduces intestinal inflammation by inhibiting HDAC expression of FoxP3^{+} and promoting T-lymphocyte secretion (Zhou et al. 2018). Under stress, the HPA axis is hyperactive, cortisol concentrations are elevated, the level of SCFAs in the intestine is significantly reduced, and chicken feed intake decreases (Meimandipour et al. 2011). Butyric acid increases the proportion of cholinergic neurons in the enteric plexus by inhibiting HDAC energy (Soret et al. 2010). This suggests a possible role for SCFAs in maintaining physiological homeostasis in the ENS. In mammals, SCFAs can influence the proliferation and differentiation of intestinal epithelial cells and affect the MGBA by maintaining the integrity of the intestinal and blood-cerebrospinal fluid barriers (Peng et al. 2007). The above studies suggest that SCFAs are closely related to the MGBA and are involved in influencing immune function, brain cognition, and energy metabolism in livestock organisms, thus regulating feeding behavior and livestock health.

2.4.3 Tryptophan and Its Main Metabolite 5-HT In Vivo

Trp is not only an essential amino acid required for animal growth and development but also a key molecule in the MGBA as it serves as an important precursor for the neurotransmitter 5-HT and many other metabolites that contribute to neuroendocrine signaling (Ruddick et al. 2006) within the CNS and ENS. The role of Trp in the CNS is largely influenced by the KYN metabolic pathway (Kennedy et al. 2017). Trp can be converted to indole by a combination of intestinal microbes and tryptophanase. Indole can increase the expression of genes related to the mucosal barrier and mucin production, while also enhancing the expression of anti-inflammatory cytokine L 10 and decreasing the expression of pro-inflammatory cytokine L-8 (Bansal et al. 2010), thereby improving the intestinal epithelial barrier and alleviating enteritis in livestock.

5-HT released from intestinal chromaffin cells plays a key role in immune regulation and can promote the secretion of lymphocytokines and monocytokines (Dürk et al. 2005), which are involved in the regulation of CNS function. 5-HT derived from enterochromaffin cells also alters afferent vagal activity and affects GBA signaling. In animals, approximately 98% of 5-HT is produced by intestinal chromaffin cells (Foster et al. 2017). However, gut microbes have been found to affect host gut 5-HT function (Hata et al. 2017). 5-HT does not cross the blood-brain barrier, but 5-HTP, produced in the blood by the enzyme TPH, can cross the blood-brain barrier. After crossing, it is rapidly converted to 5-HT, which acts centrally (Geng et al. 2023). The ability of Trp to synthesize 5-HT is widely dependent on mammalian tissues and several microorganisms (Bailey et al. 2011). 5-HT is involved in the regulation of appetite and stress responses in animals. It has been found that supplementation of Trp in the diet enhances 5-HT activity in the hypothalamus of livestock, reducing the intensity of stress responses in animals under stress and altering their social behavior, especially aggressive behavior (Stracke et al. 2017). Trp also affects the appetite of pigs by acting on the hypothalamic feeding center through 5-HT. It has been found that insufficient Trp leads to a decrease in pig feeding, while high Trp levels inhibit pig feeding (Zhu et al. 2020).

2.4.4 Tryptophan Metabolism in Intestinal Bacteria, Fungi, and Protozoa

In the animal gut, there is a rich and diverse microbial community, including bacteria, fungi, protozoa, viruses, archaea, etc. (Wang et al. 2022). Apart from viruses and archaea, most bacteria, fungi, and protozoa can metabolize Trp through different pathways, producing bioactive molecules that affect the health of the host (Choera et al. 2017; McGettrick et al. 2016).

2.4.4.1 Tryptophan Metabolism in Bacteria

Intestinal bacteria can convert Trp into a variety of bioactive molecules to maintain a healthy homeostasis in the gut (Wang et al. 2022). They metabolize excess dietary Trp and produce metabolites such as IAA, indole sulfate (ISA), IAAld, IA, IAld, and tryptamine, most of which are AHR agonists (Dong et al. 2020). *Lactobacillus royi* provides indole derivatives of L-tryptophan that activate the AHR, leading to decreased Tpok gene expression and recoding of $CD4^+$ epithelial lymphocyte ($CD4^+$ IELs) cells to double-positive epithelial lymphocyte (DP IELs) cells, and other indole derivatives may have similar effects (Cervantes-Barragan et al. 2017). Bacterially mediated Trp metabolites contain indole bioactivities such as ISA and the antioxidant indole-3-propionic acid (IPYA) (Wang et al. 2022), where the production of IYPA is entirely dependent on the presence of intestinal flora and can promote colonization by Clostridium perfringens (Wikoff et al. 2009). Bacterial-derived metabolites can directly impact host physiology through specific pharmacological targets, with tryptamine, produced by bacterial metabolism of Trp, acting as a ligand for the intestinal epithelial GPCR 5-HT receptor to regulate gastrointestinal motility (Cryan et al. 2018).

2.4.4.2 Tryptophan Metabolism in Fungi

Both bacteria and fungi play important roles in the physiological development of the host, but the role played by fungi is less well-known and requires further study (van Tilburg Bernardes et al. 2020). Because fungi constitute less than 1% of the gut microbiota, their role and importance have long been overlooked (Arumugam et al. 2011; Underhill and Iliev 2014). Intestinal fungal colonization induces the expansion of germinal center (GC)-dependent B cells in extraintestinal lymphoid tissue and the production of systemic antibodies that protect disseminated Candida albicans or Candida ear infections (Doron et al. 2021). Related studies have identified the transcription factor UME6 as a major regulator of filamentous structure, which inhibits bacterial intestinal colonization by activating the mycelium-specific pro-inflammatory secretory protease SAP6 and the expression of the mycelial cell surface adhesin Hyr1 (Witchley et al. 2019). Similar to bacteria, many fungi are involved in the biosynthesis of IAA, utilizing the indoleacetamide and indolepyruvate pathways (Wang et al. 2022).

2.4.5 Tryptophan Metabolism in Protozoa

Protozoa are prevalent in the animal gut and are often considered parasites harmful to their hosts. However, gut eukaryotes are symbiotic microorganisms that interact without causing harm, and some protozoa even contribute to the stability of the intestinal micro-ecosystem (Lukes et al. 2015). Cryptosporidium, flagellates, endamoeba, budding protozoa, and helminths are commonly found in the intestine (Wang et al. 2022). These amoebae have acquired bacterial genes to produce bacteriostatic activity against cresol, indicating that gut eukaryotes can benefit mammalian hosts (Lukes et al. 2015; Reynolds et al. 2015). Flagellates can also generate large amounts of indole by activating tryptophanase (Tna A) activity, while *Trypanosoma brucei* can enhance Trp metabolism, leading to increased production of IPYA and indole lactate (Nývltová et al. 2017). Intestinal parasites and certain bacterial microbiota are believed to possess potent immunomodulatory effects. Some studies indicate that helminth infection alters the composition of the bacterial gut microbiota, and the presence and composition of the bacterial microbiota affect the settlement and persistence of worms in mammalian hosts (Reynolds et al. 2015). Some parasites are of interest due to their beneficial immunomodulatory effects, which are also associated with disruptions in Trp metabolism and increased tryptophan consumption (Zhang et al. 2021).

2.4.6 Potential Mechanisms of Action of Intestinal Flora in the Regulation of Tryptophan Metabolism

2.4.6.1 Intestinal Flora Regulate the Canine Urinary Ammonia Pathway by Influencing IDO

Although the mechanism by which gut flora affect the Trp-KYN metabolic pathway is not fully understood, they primarily regulate the expression and activity of IDO in KYN metabolism through multiple pathways. Intestinal flora can influence the activity of enzymes in the KP, and IDO expression in the gut of GF mice is deficient compared to mice with normal intestinal flora (Rhee et al. 2012). Harrington et al. (Harrington et al. 2008) showed that the upregulation of the IFN-γ gene in IFN-γ knockout and GF mice depended on normal commensal flora. Since IFN-γ is a potent inducer of IDO, IDO expression increased when the mouse intestine was colonized with flora. Clarke found that GF mice had a reduced plasma urine/Trp content (KYN/Trp) ratio. However, removing GF mice from the GF facility after 3 weeks of weaning and then housing them with conventional mice until 8–9 weeks of age to achieve colonization with normal flora revealed that the KYN/Trp ratio in the colonized mice could be restored to normal levels (Clarke et al. 2013). Another study found that plasma KYN concentrations and KYN/Trp ratios increased to a similar extent on days 4 and 30 after colonization (El Aidy et al. 2012). However, not all bacteria promote IDO activity. Studies examining BioBreeding diabetes-prone (BBDP) rats fed *Lactobacillus johnsonii* found that *Lactobacillus johnsonii* inhibited IDO activity, while serum KYN concentrations were reduced in BBDP rats. This suggests that *Lactobacillus johnsonii* may allow H_2O_2 concentrations in the ileum to increase, which could be responsible for the inhibition of IDO activity (Valladares et al. 2013). This is consistent with Gao et al. (Gao et al. 2018), who reported that the main reason for the decrease in IDO activity induced by *L. johnsonii* colonization was the increased production of H_2O_2 in the ileal lumen. H_2O_2 acts as a signaling molecule, inhibiting IDO activity by affecting host-bacteriophage symbiotic interactions. Another study (Freewan et al. 2013) found that H_2O_2 could induce protein oxidation and inhibit IDO activity by activating peroxidase. Moloney et al. (2017) found that the microbiota regulates the expression of miRNAs associated with KP metabolism by comparing miRNAs in the hippocampus of conventional, GF, and colony-transplanted mice. They demonstrated that intestinal flora can regulate the expression of KP genes in the hippocampus. REGA3 is an antimicrobial protein in the gastrointestinal tract that can influence the composition of the intestinal flora, thereby increasing the number of lactobacilli. Some studies have reported that *Lactobacillus* increased by REGA3 may upregulate IDO in the intestinal epithelium through the production of ornithine to produce KYN (Yan et al.).

2.4.6.2 Intestinal Flora Regulate the 5-HT Pathway by Affecting Tryptophan Hydroxylase

Compared to specific pathogen-free (SPF) mice, GF mice have morphologically larger intestinal enterochromaffin cells, suggesting that intestinal microbes may influence the development and function of enterochromaffin cells (Uribe et al. 1994). Sjδgre et al. (Sjogren et al. 2012) found significantly increased levels of Trp (a substrate for TPH1) in the feces and serum of GF mice, despite insufficient levels of 5-HT in the colon and serum. This suggests that disruption of host TPH1 expression led to the accumulation of Trp. Yano et al. (Jessica M Yano et al. 2015) further found that oral administration of 5-HTP, a product of TPH1, adequately alleviated the deficit in colonic and serum 5-HT expression in GF mice. Meanwhile, supplementation with Trp did not restore 5-HT expression. This suggests that the microbiota promotes 5-HT synthesis by increasing TPH1 expression in colonic enterochromaffin cells. To further investigate the mechanism by which intestinal flora affects 5-HT production, researchers increased the intraluminal concentration of deoxycholic acid in the colon of GF mice to match that of SPF mice. They found that this led to an increase in 5-HT levels in the colon and serum. Therefore, they concluded that intestinal microbes promote 5-HT production by converting bile acids secreted by the liver into deoxycholic acid. In addition, studies have reported on 5-HT in the gut of TPH1-deficient ($TPH1^{-/-}$) mice (Manocha and Khan 2012). Kwon et al. (Kwon et al. 2019) observed significant differences in gut microbial composition between $TPH1^{-/-}$ and $TPH1^{+/-}$ mice from the same litter. They found that 5-HT directly regulates the growth of commensal bacteria in a concentration-dependent and species-specific manner in vitro using an isolated commensal system. Another study (Reigstad et al. 2015) compared TPH1 RNA and protein expression in the proximal colonic segments of GF mice, conventionally reared mice, and mice colonized with human intestinal flora (HM) using RT-PCR and Western blotting (WB). They found that both TPH1 mRNA and protein expression were significantly higher in HM mice than in GF mice, while 5-HT levels were significantly lower in the colon of GF mice than in HM mice. To determine whether TPH1 mRNA and protein expression were specific to the human intestinal flora, TPH1 mRNA and protein expression were also assessed in conventionally reared mice, which were found to have increased TPH1 mRNA expression like that of HM mice. These findings suggest that microbiota deficiency is associated with reduced colonic TPH1 expression and that both human and mouse intestinal microbiota can influence the 5-HT biosynthetic pathway. This study also found that SCFAs (butyric and acetic acids) produced by distal gut microbes significantly increased TPH1 RNA expression in an in vitro model of human intestinal chromaffin cells. This suggests that gut microbes affect intestinal chromaffin cells and TPH1 via SCFAs, which in turn promote 5-HT production in the colon.

Intestinal flora may also influence the production of 5-HT through Trp metabolism via the serotonin transporter (SERT). SERT promotes the degradation of 5-HT by intracellular monoamine oxidase and maintains it at normal levels in vivo.

In SERT mice plasma, 5-HT is deficient (Greig et al. 2019). Singhal et al. (2019) assessed the fecal and cecum microbial composition of SERT$^{+/+}$ and SERT$^{-/-}$ mice and found significant changes in the microbiome community structure in SERT$^{-/-}$ mice compared to SERT$^{+/+}$ mice. SERT$^{+/+}$ mice had a more coexisting network of intestinal microbial flora, whereas the microbiome of SERT$^{-/-}$ mice consisted of discontinuous sub-networks, suggesting intestinal bacteria of SERT$^{-/-}$ mice. Golubeva et al. (2017) investigated the availability of 5-HT in the colon and ileal tissues of a BTBR mouse model of autism spectrum disorder and C57BL/6 mice. Compared to C57BL/6 mice, BTBR mice showed a 50% reduction in 5-HT levels in colon and ileal tissues, a downregulation of TPH1 expression, and an upregulation of SERT gene expression. Since the activity of TPH determines the amount of 5-HT produced by Trp, while the SERT gene controls the uptake and breakdown of 5-HT, the test results indicate a decrease in 5-HT synthesis and an increase in decomposition, leading to the reduced bioavailability of 5-HT in the colon and ileum of BTBR mice. At the same time, they found that the concentration and availability of 5-HT in the colon and ileal tissues were strongly associated with the relative abundance of the *Clostridium* genus and also a strong correlation between the *Bacillus* genus and the expression levels of TPH1 and SERT genes. This confirms the importance of *Clostridium* bacteria in 5-HT production by Trp. Furthermore, Hata et al. (2017) found that 5-HT levels were significantly lower in the cecum and colon lumen of GF mice than in SPF mice and that in SPF mice, most 5-HT was in the unconjugated free form, whereas in GF mice, approximately 50% of 5-HT was present in the glucuronide-conjugated form, suggesting that the process of de-conjugation of glucuronide-conjugated 5-HT by intestinal flora may contribute to the promotion of free intestinal luminal 5-HT production. In addition, it has been found that intestinal flora can release 5-HT and activate 5-HT receptors by modulating neurons in the intermuscular plexus (De Vadder et al. 2018).

2.4.7 Intestinal Flora Directly Metabolize Tryptophan and Alter Tryptophan Availability in Their Hosts

The intestinal flora can metabolize Trp directly into metabolites such as tryptamine, fecal odorant, IPA, IAA, IAld, and ILA (Schiering et al. 2017). It was found that the most downstream aromatic L-amino acid decarboxylase after the non-ribosomal peptide synthetase gene cluster in *Bacillus atrophaeus C89* is associated with Trp decarboxylated formation of tryptamine (Yuwen et al. 2013). Also, *nematophagous pathogenic bacteria* can metabolize Trp to produce tryptamine (Qi et al. 2019). In addition, Williams et al. (2014) found that *Clostridium sporogenes* decarboxylates Trp by a specific Trp decarboxylase. Wikoff et al. (2009) found that *Clostridium perfringens* could produce IPA in culture. GF mice were colonized with *Clostridium perfringens* strain ATCC15579, and blood samples were taken at regular intervals after colonization. It was found that IPA was not detected in samples taken shortly after the introduction of the microorganism. This suggests that *Clostridium perfringens*, which can produce IPA in vivo, can introduce IPA into the blood of the host after colonization of the gastrointestinal tract. Other studies have found that the genetic engineering of *Clostridium perfringens* to block the degradation of Trp reduced the level of IPA in the blood of mice (Dodd et al. 2017b), and feeding *Lactobacillus royi* to mice could convert Trp to IAA (Cervantes-Barragan et al. 2017). Wilck et al. (2017) analyzed the feces of mice fed a normal salt diet and a high salt diet by 16Sr DNA gene sequencing and found that the high salt diet altered the fecal microbiome and depleted *Lactobacillus*, significantly reducing the levels of IAA and ILA in the feces, while the levels of IAld remained unchanged. Further isolation of the feces revealed that the mentioned *Lactobacillus* was *Lactobacillus rhamnosus*. In the case of mice fed with *Lactobacillus rhamnosus* isolates, ILA, IAA, and IAld were detected in the feces, whereas *Lactobacillus rhamnosus* was not found to produce ILA, IAA, and IAld in the feces of GF mice. It has been shown that the intestinal microorganisms *Lactobacillus* spp., *Prevotella* spp., *Clostridium* spp., *Actinomyces* spp., and *Macrocystis* spp. can synthesize fecal odorants through decarboxylation of the closest precursor to fecal odorants, IAA (Cook et al. 2007). Whitehead et al. (Whitehead et al. 2008) observed the production of fecal odorant and IAA within 48 h in conventional media supplemented with Clostridium drakei and Clostridium faecalis, indicating that both bacteria can degrade Trp to form IAA and fecal odorant.

2.4.8 Summary

The gut microbiota has an important influence on the KYN, 5-HT, and microbial metabolic pathways of Trp. A growing number of studies (Bosi et al. 2020; Comai et al. 2020) demonstrated that Trp metabolism is associated with diseases such as IBD, IBS, neurodegenerative diseases, and psychiatric disorders (anxiety, depression, autism, etc.). Gut flora plays an important role in these diseases, suggesting that Trp metabolism may mediate part of the role of gut microbes. Trp metabolites are directly or indirectly controlled by gut flora, suggesting that a variety of research approaches can be used to help determine the impact of gut flora on various diseases, including the use of GF animals deficient in flora, probiotic supplementation, antibiotic administration, and fecal transplantation studies. Modulation of Trp metabolism by probiotics affecting the composition of the gut flora may also be a potential and effective therapeutic approach. However, due to the complexity of gut microbiota-host interactions, the complexity of diseases and models, and the fact that most current studies are basic, further research is still needed on the influence of gut flora on host health through Trp metabolism.

2.5 Regulation Mechanism of Tryptophan and Its Metabolites on Reproduction and Lactation Function

2.5.1 *Effect of Tryptophan on Reproduction and Lactation*

Tryptophan, as a functional amino acid necessary in animals, is involved in the production of animal proteins. Its metabolites, including 5-HT, melatonin, NAD, nicotinamide adenine dinucleotide phosphate (NADP), and niacin, play roles in animal nutrition, development, reproduction, lactation, and antioxidant activity. Trp was identified as the third limiting amino acid for lamb growth in corn-based diets by Van et al. (2008). Animal growth performance can be improved by Trp. In animals, Trp deficiency can result in decreased feed intake, decreased nitrogen deposition rate, stunted growth, and poor resistance. In addition, Trp improves nitrogen deposition and daily weight gain while decreasing urinary nitrogen excretion. Ma et al. found that a daily dose of 2.0 g of Trp improved nitrogen retention, increased daily gain by 914 g, and significantly accelerated cashmere growth in the Inner Mongolian cashmere goat (Ma et al. 2010).

Sows that are fed with lysine, methionine, threonine, and Trp have been shown to have larger uteri and ovaries, which may be related to increased litter size after birth (Li et al. 2011). However, it is not known whether supplementation with Trp alone would have the same effect on the uterus and ovaries. Li et al. found that 0.21% Trp supplementation increased litter size and stillbirth size but had no effect on birth size. Piglet weight at weaning has a significant influence on post-weaning weight gain, and sow milk production is a major determinant of piglet growth (Paulicks et al. 2006). During peak lactation, lactating sows produce milk containing at least 50% of the nitrogen and calories provided to the piglets. Only a few studies have been carried out on the effect of Trp on the lactation performance of sows, and the mechanism is not known. The milk yield of lactating sows was altered by dietary supplementation of 0.15%, 0.18%, 2.4%, and 3.0% Trp (6840, 8192, 8362, 9035, 8618, and 8140 g/d), but there was no significant change in milk fat, protein, or lactose content (Charney et al. 1982). The increase in milk fat, protein, and lactose synthesis may be due to increased milk production rather than the influence of Trp. Milk production is determined by the genetic ability of the sow and feed intake but is also influenced by litter size and the ability of the piglets to nurse. Lewis and Speer (1974) found that 0.091% Trp supplementation had a significant effect on milk yield during the first 3 weeks of lactation.

Both initiation and maintenance of lactation are regulated and sustained by the neurohormonal system. The hormone that maintains lactation in monogastric animals is prolactin (PRL), while the hormone that maintains lactation in ruminants is growth hormone. During pregnancy, estrogen and progesterone levels rise to promote mammary gland development, and these two hormones decline rapidly before parturition, while prolactin (or growth hormone) rises to initiate and maintain lactation. Studies have shown that the addition of 20 g of 5-HTP to the basal diet can increase plasma levels of PRL and GH in dairy cows and improve lactation (Deng et al. 2020). Charney et al. found that blood levels of PRL and GH were significantly elevated in healthy subjects following intravenous administration of Trp (Charney et al. 1982). Trp can increase blood-free Trp levels and increase the amount of Trp that passes through the blood-brain barrier, thus increasing the level of 5-HT in the brain and promoting the secretion of hormones such as PRL and GH, which act on the mammary tissue, promoting the secretion of breast milk (Kasuya et al. 2010).

2.5.2 *Effect of Tryptophan on Plasma Hormone Levels*

Intravenous administration of 38.5 mg/kg BW Trp significantly enhanced both cerebrospinal fluid 5-HT and plasma GH levels in cows (Kasuya et al. 2010). Murakami et al. discovered a considerable increase in plasma GH levels after administering 5-HT to the ventricles of mice's brain (Yoshio et al. 1986). Trp, as a precursor substance of 5-HT, can raise the amount of 5-HT in the brain, and 5-HT, as an important neurotransmitter in the CNS, can stimulate the release of PRL and GH in the hypothalamus and pituitary gland. Furthermore, Kasuya's research found that melatonin, a Trp metabolite, stimulates GH secretion via the hypothalamus (Kasuya et al. 2008). Trp, as a precursor of melatonin, undergoes hydroxylation, decarboxylation, acetylation, and methylation to create melatonin, which increases GH production.

2.5.3 *Effects of Tryptophan Metabolites on Reproductive Lactation*

2.5.3.1 Role of Melatonin in Animal Reproduction

Melatonin, which has distinct circadian and seasonal rhythms, is synthesized from Trp, a molecule that serves as a precursor. Therefore, melatonin secretion is intimately connected to animal reproductive activity and is crucial for controlling seasonal reproduction. Melatonin's regulation of the reproductive system has been found to have different effects on different animal species. Melatonin inhibits the reproductive system of humans and long-day animals such as cows, rats, and birds while enhancing the reproductive system of short-day animals such as sheep and deer. Melatonin has no effect on light-insensitive species. Young et al. found that reducing light exposure in hamsters living in a normal environment shrunk the testicles of male hamsters and significantly decreased the contents of the luteinizing hormone (LH), PRL, and follicle-stimulating hormone of female hamsters. Injecting melatonin under normal light conditions also caused a decrease in reproductive function, and the effect of the melatonin injection was like that of short daylight. Subcutaneous melatonin treatment produced similar results in mice. It was found that after melatonin implantation in sheep, the time of first estrus in the melatonin group was 30 days earlier than in the control group, and the estrus rate was 40% higher in the melatonin group (Young and Nelson 2001). Therefore, melatonin may inhibit the reproductive function of long-day breeders and enhance the reproductive activity of short-day breeders. On the one hand, melatonin regulates the release of endogenous melatonin in seasonal breeding animals through the regulation of the hypothalamic-pituitary-gonadal axis and the change of sunshine duration and acts on the hypothalamic-pituitary-gonadal axis to promote the secretion of endogenous hormones and the development of the reproductive system (Frungieri et al. 2005). On the other hand, studies have found that there are also many melatonin receptors in follicular fluid and ovarian tissue, indicating that melatonin can directly interact with the ovary to regulate its physiological function by binding to specific receptors. Melatonin can stimulate progesterone secretion by ovarian granulosa cells, increase LH receptor expression, and decrease GnRH receptor expression in luteal granulosa cells (Picinato et al. 2008). A large number of studies have shown that melatonin can promote follicle development, oocyte maturation, and early embryo development, playing an important role in the seasonal reproductive activities of animals (Abecia et al. 2002; Shi et al. 2009).

2.5.3.2 Effects of Melatonin on Male Reproductive System

Melatonin can significantly improve the state of the male reproductive system. Besides melatonin, melatonin receptor 1 (MT1) and melatonin receptor 2 (MT2) can mediate other hormonal signals and upregulate testosterone levels (Yuan et al. 2019). In embryonic and juvenile stages, melatonin can promote the secretion of male hormones and the development and maturation of the reproductive system by binding to MT1 and MT2 in testicular tissue (Johnston et al. 2006). In adulthood, melatonin can promote the secretion of male hormones and the formation of sperm and reproductive activity in seasonally estrus animals (Tabecka-Lonczynska et al. 2017).

In addition, melatonin's anti-inflammatory and antioxidant properties are beneficial to the male reproductive system. Melatonin can improve the survival rate of cryopreserved spermatogonial stem cells (SSCs) and reduce reactive oxygen species (ROS) generated in cells during freeze-thaw cycles. Melatonin also has a positive preventive and therapeutic effect on male reproductive problems. Gram-negative bacterial orchitis decreases testosterone synthesis and spermatogenesis, where lipopolysaccharide (LPS) in the cell wall binds to Toll-like receptor 4 and activates an immune response. Melatonin treatment can increase the phagocytic activity of macrophages in the testis and the release of testosterone in Leydig cells, while blocking the p38 MAPK pathway, which inhibits the secretion of the pro-inflammatory cytokine TM, thereby reducing the inflammatory response (Deng et al. 2020).

2.5.3.3 Effects of Melatonin on Female Reproductive System

Melatonin regulates the secretion of reproductive hormones such as GRH and LH through the hypothalamic-pituitary-gonadal axis in livestock and poultry. Studies have shown that exogenous melatonin increased transcript levels of GnRH-1, GnRH-3, and GnRH receptors in the brain, which is influenced by external light conditions. Removal of the ovaries in female mice can increase the amount of melatonin secreted by the pineal gland. However, this was not found in male mice, indicating that GnRH may be involved in the control of melatonin release in the pineal gland of adult female rats (Ishizuka et al. 2000).

During the ovulation process of female animals, mitochondria can produce a significant amount of melatonin, which can reduce excess ROS, while promoting oocyte maturation and reproductive performance of livestock and poultry. Tian et al. found that the concentration of melatonin in bovine follicular fluid is approximately 10–11 M used radioimmunoassay (Tian et al. 2014). Exogenous melatonin increased oocyte maturation by upregulating oocyte maturation genes (GDF9, MARF1, and DNMT1a) and oocyte-related genes (PTX3 and HAS12). Tao et al. established a hypoxia model of porcine granulosa cells (GCs) and found that melatonin significantly reduced ROS levels and increased the expression of antioxidant enzymes like superoxide dismutase (SOD1) and GC activity while inhibiting caspase-3 activation in GCs and reduced cell apoptosis (Tao et al. 2021). Furthermore, melatonin can reduce IL-10TNF and IL-6 levels in inflammatory conditions, as well as reduce intrauterine inflammation and autophagy caused by pregnancy complications (Sagrillo-Fagundes et al. 2018). In pregnant mice, Lee et al. showed similar results with exogenous melatonin (Lee et al. 2019). Melatonin can promote the differentiation of spermatogonial stem cells into functional sperm by increasing the expression of the postmeiotic marker genes Dnmt3a and Bcl-2 while decreasing the expression of Bax (Deng et al. 2016).

2.5.3.4 Melatonin and Lactation

Because the body cannot produce endogenous melatonin for a period after birth, it is critical to alter its biological cycle and promote immune system development by consuming melatonin in breast milk. As a result, the level of melatonin in breast milk during lactation is critical for puppy development. Linear expression vectors of the melatonin-synthesizing rate-limiting enzymes AANAT and HIOMT in sheep were found to significantly increase the melatonin content of breast milk. The results showed no significant difference among growth performance, many physiological indices, and milk composition between the transgenic sheep model and the control group. Melatonin and lactose levels of the transgenic sheep were increased, while somatic cell count was decreased. The reason is the increase of melatonin content in mammary gland, which exerts its immunomodulatory function and reduces the somatic cell count in milk. Illberova et al. studied the changes in melatonin levels in the blood and milk of women 3–7 weeks postpartum and found that the melatonin content in milk was much higher at night than during the day, which is consistent with serum melatonin levels (Illnerová et al. 1993). Eriksson et al. investigated the diurnal variation of melatonin levels in the serum and milk of Elxia cows at different stages of lactation. The results showed that milk from cows in early lactation contained less melatonin, which was not related to changes in blood melatonin levels, but milk from cows in late lactation contained more melatonin at night, which is changed in a similar way to plasma (Eriksson et al. 1998). Castro et al. studied the effect of light cycle on melatonin levels in plasma and milk of dairy cows and found that milk melatonin levels would increase with the increase in plasma melatonin levels at night, but there was a time lag (Castro et al. 2012). The results suggest that melatonin, as a precursor of Trp, can enter the mammary gland via active blood transport and be released into milk. Blood melatonin levels are directly related to milk melatonin levels.

In conclusion, the effects of melatonin on lactation may vary between species and physiological situations. Melatonin reduces prolactin secretion in mammals, including humans. PRL hormones are important regulators of milk production and secretion in mammary cells. Melatonin levels can increase and block PRL secretion, reducing milk production and secretion. This is one reason why some mammals produce less milk at night or in the dark. The effects of melatonin on lactation are complex and influenced by a variety of circumstances. Melatonin can suppress milk production and secretion under normal conditions, but its role may be modified by other regulatory variables in specific physiological situations.

2.5.3.5 Mechanism of Influence of Serotonin on Lactation

Supplementation with Trp in sow diets can increase levels of 5-HTP, which is a Trp metabolite critical for maintaining milk calcium content and mammary gland homeostasis (Weaver et al. 2016). Serotonin, a recognized regulator of mammary homeostasis, is generated from L-Trp in a two-step process that begins with the rate-limiting enzyme Trp-hydroxylase, followed by the conversion of 5-HT to 5-HTP by aromatic amino acid decarboxylase (Hernandez et al. 2008, 2009; Matsuda et al. 2004; Stull et al. 2007; Wang et al. 2002).

Weaver et al (2016) found that supplementing with 0.12% serotonin could increase the calcium status during lactation, and they confirmed that the circulating serotonin level on the first day of lactation was a good indicator of calcium status in dairy cows. Most mammals have difficulty maintaining a certain blood calcium level at the beginning of lactation. The increased need for calcium during pregnancy and lactation places a significant strain on calcium homeostasis. The ability to control circulating calcium concentrations and mobilize bone tissue during early lactation is critical not only for lactation but also for the health of the sow. Since bone resorption is the primary source of calcium in mammalian milk, additional research is needed to investigate the effect of Trp supplementation on bone turnover in lactating sows (Kovacs and Fuleihan 2006).

It has been found that FAS and LS and β-casein secretion increased with the addition of Trp to the culture medium, suggesting that the addition of Trp could enhance the lactation ability of primary mammary gland epithelial cells. Trp supplementation increased the lactose concentration of sows' milk in vivo on days 7 and 14 of lactation. The increasing milk lactose concentration caused by Trp supplementation could be because of its ability to enhance LS in the mammary epithelial cells. However, Trp supplementation did not affect milk fat and protein in vivo but increased FAS and β-casein in vitro, which could be due to other animal factors (Miao et al. 2019). This is consistent with the result of Wang et al. (2015a) that L-Trp enhances tight junction protein expression in porcine intestinal epithelial cells. The enhanced β-casein was not 5-HT-mediated because THI had no effect on the secretion of β-casein. This is consistent with a previous study in mice (Matsuda et al. 2004), which reported that 5-HT inhibited milk protein RNA abundance in PRL-stimulated mouse mammary epithelial cells. Correspondingly, treatment of mammary gland explants and mouse mammary epithelial cells with metergoline, a 5-HT receptor antagonist, leads to increased milk protein gene expression. Conversely, pre-treatment of cells with THI significantly reduced Trp-enhanced FAS and LS synthesis in PMECs, suggesting that 5-HT mediated the enhanced FAS and LS synthesis by Trp. In vitro study with mouse mammary epithelial cells showed that supplemented Trp in culture medium significantly increased the intracellular level of calcium, which was significantly reduced by pre-treatment with THI. It is implied that the higher concentration of calcium found in the milk of Trp-supplemented sows was related to 5-HT metabolism. This is consistent with the study of mammary gland-specific knockout THI in mammalian mice, which found that the level of calcium transporter gene was lower than that of the control group (Laporta et al. 2014a).

It has been demonstrated that 5-HT regulates the induction of PTHrP, which is synthesized only within the mammary gland and is responsible for calcium mobilization from the bone in lactating mammal (Hernandez et al. 2012). Manipulation of 5-HT-induced PTHrP synthesis near the end of pregnancy could be critical in preventing the onset of hypocalcemia during the early lactation period (Laporta et al. 2013). The calcium-sensing receptor (CaSR), which is greatly expressed in the mammary gland during lactation, plays a crucial role in maintaining maternal calcium concentrations (VanHouten and Wysolmerski 2013). In rodent models, it has been reported that 5-HT is essential for calcium liberation from bone during lactation to sustain maternal calcium homeostasis (Hernandez et al. 2012; Laporta et al. 2014a, b). Laporta et al. (2013) have shown that in the transition from pregnancy to lactation, supplementation of the serotonin precursor, 5-HTP, increased postpartum serum serotonin, PTHrP, and calcium concentrations and also increased milk calcium concentration. The addition of Trp to the culture medium increased intracellular CaM and the activity of CaMKII. Activation of Ca2+/calmodulin-dependent kinase by higher concentration of intracellular Ca2+ directly activates AMPK in HeLa cells, and this kinase is one of the two known upstream kinases of AMPK that is sensitive to the concentration of intracellular Ca2+ (Hurley et al. 2005; McFadden and Corl 2009).

To summarize, there have been few studies on the influence of Trp on breastfeeding, and the results have been conflicting. According to some animal research, Trp administration may improve mammary gland growth and development as well as milk production. Because Trp is an essential component of proteins required to produce breast tissue, its consumption may influence breast function. However, in human investigations, the effect of Trp on breastfeeding has not been firmly established. Some preliminary research suggests that Trp supplementation may enhance Trp levels in milk, although the effect on milk production is unknown. Other studies have revealed that Trp does not affect milk production. It should be emphasized that lactation is a complex physiological process that is influenced by a variety of elements such as hormones, neuromodulation, and breast tissue growth, among others. In some circumstances, Trp may affect lactation, although additional research is needed to better understand the precise effects and processes of Trp on lactation (Table 2.1).

Table 2.1 The effect of supplementation with Trp or its metabolites at different times on sows and piglets

	Addition period	Additive amount	Effect on the mother body	Effect on offspring	References
Tryptophan	Late pregnancy and lactation	0.42–0.56% Trp	Increase melatonin level in sows	Improve the survival rate and growth performance of piglets	Munn et al. 2021)
	Gestation period	Each sow was supplemented with 36 mg melatonin every day	Improve antioxidant status, inflammatory response, and mitochondrial dysfunction in sow placenta	Improve fetal growth performance	Peng et al. (2021)
		1.2–4.2 g Trp	The milk yield of sows increased first and then decreased	–	Paulicks et al. (2006)
		Trp supplement	Enlarge the sow's uterus and ovaries	Increase in litter size	Sein and Chepelev (1990)
	Lactation period	0.15%, 0.18%, 2.4%, 3.0% Trp	The milk yield of sows increased with the increase of Trp supplemental level	–	Paulicks et al. (2006)
		0.046%, 0.066%, 0.091%, 0.121%, 0.161% Trp	Increase sow milk production, reach a stable level at 0.091%	–	Lewis and Speer (1974)

Table 2.1 (continued)

	Addition period	Additive amount	Effect on the mother body	Effect on offspring	References
5-HT	Post-lactation period	5-HT	Improve the calcium state in milk production and improve the lactating ability of dairy cows	–	Weaver et al. (2016)
	The transition period from pregnancy to lactation	5-HTP	Increase the concentration of serum serotonin, PTHrP, and calcium in postpartum serum and increase the concentration of milk calcium	–	Laporta et al. (2013)

5-HTP 5-hydroxytryptophan, *AANAT* arylalkylamine N-acetyltransferase, *BAX* bcl-2-associated X protein, *Bd-2* β-defensins 2, *CaM* calmodulin, *CaMKII* calmodulin-dependent protein kinase II, *CaSR* calcium-sensing receptor, *DNMT1a* DNA methyltransferase 1a, *DNMT3a* DNA methyltransferase 3a, *FAS* fatty acid synthase, *GCs* granulosa cells, *GDF9* growth differentiation factor 9, *GH* growth hormone, *GnRH* gonadotropin-releasing hormone, *GRH* growth hormone-releasing hormone, *HAS12* hyaluronan synthases 12, *HIOMT* hydroxyindole-O-methyltransferase, *5-HT* 5-hydroxytryptamine, *IL-10* interleukin-10, *IL-6* interleukin-6, *LH* luteinizing hormone, *LPS* lipopolysaccharide, *LS* lactose synthase, *MARF1* meiosis regulator and mRNA stability factor 1, *MT* melatonin, *MT1* melatonin 1, *MT2* melatonin 2, *NAD* nicotinamide adenine dinucleotide, *PMECs* porcine mammary epithelial cells, *PRL* prolactin, *PTHrP* parathyroid hormone related-protein, *PTX3* pentraxin 3, *ROS* reactive oxygen species, *SOD* superoxide dismutase, *SSCs* spermatogonia stem cells, *THI* Trp hydroxylase inhibitor, *TNF* tumor necrosis factor

2.6 Tryptophan in Mother-to-Offspring Nutritional Transmission

Trp, the least abundant essential amino acid, is required for protein synthesis and is therefore important for the growth and development of the placenta, mammary gland, and fetus (Silvano et al. 2021). Normally, very little dietary Trp (<1%) is used for protein synthesis, and the amount of protein synthesized matches exactly that degraded (Badawy 2014). The bulk of dietary Trp is metabolized via four pathways, including the serotonin pathway, tryptamine pathway, and indole pathway (Bender 1983). Trp requirements in pregnancy are several-fold: (1) the need for increased protein synthesis by the mother and fetal growth and development, (2) serotonin (5-HT) for signaling pathways, (3) KA for neuronal protection, (4) QA for NAD+ synthesis, and (5) other KYNs for suppressing fetal rejection. These goals could not be achieved if maternal plasma Trp is depleted. In this chapter, we summarize the most important information concerning the effect of Trp on nutritional transmission and immunity acquisition from mother to offspring, focusing on the mechanisms and consequences of the degradation of Trp in the placenta during pregnancy, as well as discuss the Trp-related pathophysiological mechanisms of pregnancy complications and offspring's health and development.

2.6.1 From Placenta to Fetus

The placenta is a disc-shaped, hemochorial, endocrine organ. The chorionic villi project into the intervillous space, in contact with the maternal blood (Maltepe and Fisher 2015). Fetal blood vessels branch in the placenta and dive into the chorionic villi, where gas, nutrients, and waste exchanges occur. Free Trp, the form immediately available for tissue uptake, is elevated in pregnancy (Morita et al. 1992), suggesting that Trp availability to the fetus is enhanced. The postulated decrease in maternal total Trp availability during pregnancy (Kudo 2013; Schröcksnadel et al. 1996; Maes et al. 2001) would be disadvantageous to the fetus, whose Trp levels are much higher than those in maternal plasma. In rats, maternal Trp availability is enhanced throughout pregnancy (Badawy 1988). As stated above, a decrease in maternal plasma total Trp and an increase in free Trp have been reported. The increase in free Trp may be due to decreased serum albumin and a non-esterified fatty acid (NEFA) elevation.

A fetus requires immune privilege and surveillance in the presence of paternally derived antigens. The upregulation or downregulation of maternal immunity also depends on cells of the placenta that are sensitive to the concentration of Trp and to the concentration of its metabolites in the cellular microenvironment (Silvano et al. 2021). Mechanisms of tolerance are activated in pregnancy to allow embryo implantation and fetal development (Hsu and Nanan 2014). The semi-allogeneic fetus expresses maternal and paternal antigens so that the immune system of the mother can defend both herself and the fetus from foreign pathogens, but it also has to tolerate the semi-allogeneic fetus (Silvano et al. 2021). There are many mechanisms that allow the maintenance of active immunological tolerance in pregnancy, such as the placental release of anti-inflammatory and pro-tolerogenic hormones, cytokines, and immunomodulatory molecules (Robertson et al. 2018). Trp catabolism is involved in immune regulation within the placenta, where IDO1 is highly expressed since it regulates mechanisms of tolerance in pregnancy (Hsu and Nanan 2014). The KP has an antimicrobial action, contributing to the protection of the fetus against infection (Adams et al. 2004), and is involved in decidualization via AHR signaling (Wang et al. 2020a). It has been observed that the rate of Trp degradation increases during pregnancy (Schröcksnadel et al. 1996), while after birth, immune activation ceases and Trp concentration returns to physiological concentration. Under physiological conditions, total Trp concentrations decrease throughout pregnancy in maternal blood, while KYN concentrations remain constant (Schröcksnadel et al. 2006). The first step in the oxidative metabolism of Trp along the KP is catalyzed independently by three different enzymes: IDO1, IDO2, and Trp 2,3-dioxygenase (TDO) (Sedlmayr et al. 2014). In vitro data showed that estrogen may upregulate the expression of IDO in the chorionic villi and in the decidua of women in early pregnancy, suggesting that it may contribute to the prevention of allogeneic fetal rejection (Li et al. 2020). During pregnancy, IDO1 defends embryonic implantation and fetal development from rejection by reducing Trp levels and suppressing T-cell activity or inducing energy of reactive T-cells (Spinelli et al. 2019; Yeung et al. 2015). Therefore, KP metabolites are mainly localized in placental endothelial cells, where they may regulate vasodilation and innate immune responses. IDO1 and IDO2 have also been localized in cytotrophoblasts and in decidual immune cells, where they control both innate and adaptative immune responses, providing a protective response against pathogens. In addition to immune tolerance, KP metabolites have pro- and antioxidant effects and are involved in physiological processes that play a key role in the regulation of vascular tone in the mother and placenta and neuroprotection. The KP, and equally important the transport of its metabolites across the placenta, affects the placental function and pregnancy outcome (Badawy 2014; Broekhuizen et al. 2020). KYN is further degraded into QA, which could facilitate the synthesis of NAD^+ (P^+) from niacin (Nishizuka and Hayaishi 1963; Williams et al. 1950). Although QA is excitotoxic, it is nevertheless the obligatory precursor of the all-important cofactor NAD^+, necessary for a wide array of redox reactions and DNA repair, and its phosphorylated form $NADP^+$ is involved in many biosynthetic redox reactions.

The monoamine serotonin (5-HT) functions as a neurotransmitter in the central, peripheral, and ENS of animals and humans. During the first half of pregnancy, 5-HT can be transferred from the mother's placenta, via the fetal periphery, to the fetal brain (Bonnin et al. 2011; Côté et al. 2007; Muller et al. 2017).

In humans, 5-HT-ergic neurons are first evident in the hindbrain at gestational week 5 (Sundström et al. 1993), after which they project to the forebrain, reaching the cerebral cortex around gestational week 10 and the most rostral telencephalon around gestational week 13 (Verney et al. 2002). At gestational week 15, during the second trimester of human pregnancy, the typical organization of clustered 5-HT cell bodies into the raphe nuclei is observed (Takahashi et al. 1986). Similarly, in rodents, 5-HT-positive cells are observed in the raphe nuclei around embryonic day (E) 10–12, which corresponds to the second trimester of pregnancy in humans. One day after their appearance, these cells are able to synthesize 5-HT (Gaspar et al. 2003). 5-HT-ergic axons reach the forebrain in mice around E16.5 and in rats around E17.5 (Bonnin et al. 2011; Zhou et al. 2000). Interestingly, 5-HT-ergic signaling molecules are already present in the forebrain before 5-HT-ergic axons reach it. Human data show the expression of 5-HT in the brain around gestational week 8, i.e., the first trimester of pregnancy (Booij et al. 2015). Similarly, in mice and rats, brain 5-HT is detected around E8–E9 in the forebrain (Buznikov et al. 2001; Gaspar et al. 2003; Choi et al. 1997; Fiorica-Howells et al. 2000). This suggests that an exogenous source may provide the fetal forebrain with 5-HT before the actual 5-HT-ergic projections reach the forebrain. Experimentally induced maternal depletion of 5-HT levels during pregnancy was found to reduce serotonin receptor expression in the developing rat forebrain (Chen et al. 2012) and induce dilated cardiomyopathy in the offspring, with the severity being negatively correlated to the maternal plasma 5-HT concentration during pregnancy (Fligny et al. 2008). Less enriched Trp diets (i.e., ±10 g/kg) also resulted in peripheral hyperserotonemia and increased TPH activity in the gastrointestinal tract of dams and their pups (Castrogiovanni et al. 2014; Huether et al. 1992). However, opposed effects in the offspring's brain (in contrast to the gastrointestinal tract) were reported due to low Trp-enriched diets (i.e., ±10 g/kg) (Flores-Cruz and Escobar 2012; Musumeci et al. 2014). However, the impact of excessive maternal serotonin concentrations on fetal development is less clear. A diet enriched with Trp (i.e., 50 g/kg) resulted in increased maternal, placental, and fetal Trp levels. As a result, both placental development and fetal growth were reduced, and pup death rate increased (Tsuji et al. 2013). Feeding of a Trp-enriched diet to pregnant rats resulted in hyperserotonemia (Castrogiovanni et al. 2014; Musumeci et al. 2014), low growth hormone concentration (Musumeci et al. 2014), changes in the expression of 5-HT-regulating genes, and a delayed maturation of the central serotonergic system (Huether et al. 1992) in the developing offspring.

Melatonin, a pineal hormone, is associated with the function of several, if not all, physiological systems (CNS, activity/rest and wake/sleep cycles, energy metabolism, thermoregulation, and immune, reproductive, endocrine, cardiovascular, respiratory, and excretory systems) in order to fulfil its functional integrative role as a circadian, seasonal, and transgenerational timing messenger. Good oocyte quality is vital to female fertility, but, with the aging process, it drops as a consequence of the reduced number of follicles and the increased oxidative stress in the ovaries (Lim and Luderer 2011). Likewise, pineal melatonin synthesis diminishes with age, and its direct and indirect antioxidant actions seem to be significant in the control of oxidative stress and improvement of oocyte quality (Reiter et al. 2014). Maternal melatonin easily transfers to the fetal circulation (Okatani et al. 1998). It is well known that newborn infants lack a circadian rhythm of melatonin secretion until roughly 3 months of age (Jaldo-Alba et al. 1993). Increased maternal melatonin circulating levels have been reported during pregnancy in sheep (McMillen et al. 1995), rats (Deguchi 1975), and humans (Wierrani et al. 1997). It was reported that maternal serum melatonin levels showed a diurnal rhythm, which is an important signal for the fetus to entrain the light-dark rhythm of the newborns after delivery. In parallel, nighttime serum melatonin levels in pregnant women increased after 24 weeks of gestation until term, being related to the feto-placental unit and the number of conceptus (higher in twins than in singleton pregnancies), with the highest amounts being detected during the last trimester, reaching its peak right before parturition time and returning to basal levels immediately after birth (Nakamura et al. 2001). Daytime serum melatonin levels, on the other hand, were low, and while the levels showed an increasing tendency toward the end of pregnancy, no statistically significant changes occurred (Nakamura et al. 2001).

Sano et al. has shown that active L-Trp could transfer from mother to fetus, and the fetus could de novo synthesize 5-HT. Plasma 5-HT was decreased in undernourished mothers. In intrauterine growth restriction (IUGR) fetuses, maternal-fetal L-Trp transfer remained unaffected, but conversion to QA was impaired, implying that NAD production also decreased. To ensure the transport of Trp from the mother to the fetus, a diet high in carbohydrates and low in proteins has been suggested (Yokogoshi and Wurtman 1986). Carbohydrates evoke insulin secretion. Insulin, in turn, reduces most large amino acids, except for Trp (Wurtman and Fernstrom 1975). In humans and rodents, the transfer of placental Trp from maternal blood into the fetal circulation is predominantly mediated by the so-called system A and system L transporters (Kudo and Boyd 2001; Winterhager and Gellhaus 2017). Excessive amounts of nonessential amino acids in the maternal blood thus might increase the maternal-fetal transfer of essential amino acids, including Trp. Multiple animal studies investigated the effects of gestational low-protein diets on an offspring's brain 5-HT system. Notably, some studies used isocaloric protein diets in which the low-protein diets were adjusted with a higher content of carbohydrates, thereby creating low-protein high-carbohydrate diets. Such low-protein high-carbohydrate diets reduced maternal plasma Trp levels and increased maternal plasma 5-HT levels and the offspring's brain 5-HT levels after birth (Honório de Melo Martimiano et al. 2017; Martin-Gronert et al. 2016). It remains unclear whether the exposure to such diets decreases or increases placental and fetal 5-HT levels (Honório de Melo Martimiano et al. 2017; Martin-Gronert et al. 2016). When using a low-protein high-carbohydrate diet supplemented with Trp, neither the maternal nor fetal 5-HT system is altered (Sano et al. 2016). Importantly, when exposed to a prenatal carbohydrate-restricted diet consisting of similar protein levels, and thus Trp levels, as the control diet, fetal whole-brain concentrations of Trp, 5-HT, and 5-HIAA were reduced (Koski et al. 1993). Similarly, ILA, a metabolite produced by probiotic B. infantis fermenting human milk Trp, has also shown anti-inflammatory effects on intestinal epithelial cells by activating AHR (Meng et al. 2020). This anti-inflammatory feature of ILA was found only in the fetal mouse intestines, not in the mature ones.

2.6.2 From Maternal Microbes to Offspring

The microbiota plays a crucial role in individuals' early and long-term health by activating and developing the immune system, contributing to the development of the CNS, and aiding in the digestion and metabolism of food (Yao et al. 2021). When in utero, offspring are protected by maternal immunity and simultaneously develop their own immunity. The initial development may be partly stimulated by maternal microbiota (Lu et al. 2021), which has indicated that maternal delivery of AHR ligands, specifically the tested ligand indole-3-carbinol, to the fetus showed the capability to prevent the development of necrotizing enterocolitis by reducing TLR4 signaling in the offspring's gut. Indole and its derivatives are natural AHR ligands and can be transformed from Trp by gut microbiota (Bosi et al. 2020).

The gut microbiota of breast-fed and formula-fed infants differ significantly, as do the risks for allergies, gut dysfunction, and upper respiratory tract infections.

Saraf et al. (2017) confirmed that a formula diet alters the colon microbiota and appears to shift Trp metabolism from serotonin to tryptamine, which may lead to greater histamine levels and the risk of allergies in infants. In their experiment, correlation analyses indicate that *Akkermansia*, *Butyricimonas*, *Clostridium*, *Holdemania*, and *Bilophila* were positively correlated with Trp metabolism in the soy-fed group, while only the *Bilophila* genus correlated with Trp metabolism in the milk group. However, unassigned genera from phyla Tenericutes, Proteobacteria, Bacteroides, and Lentisphaerae, in addition to the RFN20 genus from the Firmicutes phylum, were positively correlated with Trp metabolism in the sow-fed group but not in the formula-fed groups. In addition, *Butyricimonas* was negatively correlated with Trp metabolism in the sow-fed group, while unassigned genera from Proteobacteria and Bacteroides phyla were negatively correlated with Trp metabolism in the soy group. Furthermore, serotonin precursors Trp and 5HTP were unaffected in distal colon contents but increased in serum and urine of soy and milk groups. Moreover, 5HIAA, a catabolic product of serotonin level, was not altered in distal colon contents and urine and undetectable in serum, ruling out the possibility of low serotonin due to increased catabolism of serotonin into 5HIAA in formula-fed piglets. The results from Laursen et al. (2021) revealed that breastmilk-promoted Bifidobacterium species produce aromatic lactic acids in the gut of infants and suggest that these microbial metabolites may impact immune function in early life.

2.6.3 Tryptophan Metabolism in Maternal and Offspring's Health

Among the essential amino acids, L-Trp is the least abundant, with plasma concentrations of ~40–80 μM in humans (Silvano et al. 2021). Trp metabolism may also play an important role in abnormal pregnancies and pregnancy complications such as gestational diabetes and hypertension, as well as preeclamptic placentas. Broekhuizen et al. (2020) found that L-Trp induces IDO1- and NO-dependent relaxation in placental arteries, which is determined by L-Trp uptake rather than IDO1 expression. This suggests that increased L-Trp uptake might compensate for reduced IDO1 expression in preeclamptic placentas. Prenatal stress-caused reduction of maternal gut microbiota, including *Parasutterella excrementihominis* and Bifidobacterium, dysregulates the metabolic pathways of Trp, a well-studied precursor for multiple metabolites crucial for proper immune and neurodevelopment. This reduction leads to reduced transportation of Trp and its derivative serotonin to the fetus across the placenta, thereby inducing aberrant fetal neurodevelopment (Wang et al. 2020b). Additionally, maternal administration of KYN, another Trp derivative, rapidly increases levels of KYN in fetal plasma and brain, leading to offspring with ASD-like behavioral abnormalities (Murakami et al. 2021).

2.7 Tryptophan Metabolism in the Interorgan Interaction

2.7.1 The Impact of Tryptophan Metabolism on the Gut-Brain Axis

The gut-brain axis is a bidirectional communication network between the intestine and the CNS, which plays an important role in host metabolism and immune function (Sarkar et al. 2016). Microbial regulation of Trp metabolism can produce serotonin, a key neurotransmitter participating in the regulation of enteric function and central neurotransmission. Trp-derived metabolites are also metabolized through the KYN and indole pathways (Gao et al. 2020a). These metabolites constitute the interface of the gut-brain axis via AHR signaling (Roth et al. 2021). Dysregulation of Trp metabolites may contribute to the pathogenesis of several neurologic and psychiatric disorders.

2.7.1.1 Serotonin and the Gut-Brain Axis

Serotonin is produced in enterochromaffin cells via TPH1, and indigenous spore-forming microbes colonized in the human and mouse gut stimulate serotonin production in colonic enterochromaffin cells. TPH1 is required for the degradation of Trp into 55-HT, which is further decarboxylated to serotonin by aromatic amino acid decarboxylase (Ma et al. 2018; Yano et al. 2015).

Microbial-derived serotonin plays a crucial role in modulating neuronal differentiation, migration, and synapse formation during CNS development (O'Mahony et al. 2015). The lack of serotonin in the CNS, as seen in TPH2-deficient mice, results in growth retardation and disturbs neuronal communication in the brain, making it prone to neurodevelopmental disorders (Ugrumov 2010b). The central serotonergic system develops from the raphe nuclei in the brainstem, one of the early origins of the CNS (Briscoe et al. 1999). The uptake of serotonin is higher in the developing brain compared to the adult brain, as evidenced by the evident gene expression of the serotonin transporter from embryonic day 13 and reaching maximum uptake at 5 weeks of age (O'Mahony et al. 2015). Early colonization of the gut microbiome is necessary for the development of the central serotonergic systems (Clarke et al. 2013).

In GF mice, anxiety-like behaviors and higher concentrations of serotonin in the hippocampus were associated with elevated levels of Trp precursors in the blood circulation compared to conventional mice (Clarke et al. 2013). Gut microbiota-determined Trp metabolism significantly impacts peripheral Trp levels, thus altering central serotonin metabolism (Gao et al. 2019). Apart from directly modulating Trp metabolism, gut microbiota could indirectly modulate central serotonin synthesis through microbial SCFAs or the outer membrane of Gram-negative bacteria LPS. Butyrate could be transported into the blood and exert neuroprotective effects in stressed mice by enhancing brain serotonin levels and improving the blood-brain barrier (Sun et al. 2016). LPS, as an inflammatory stimulus, may activate the KP and hamper serotonin synthesis (Gao et al. 2020a; O'Mahony et al. 2015). The ENS is composed of vagal and sacral neural crest cells that migrate along the intestinal wall and intrinsically innervate the bowel (Lake and Heuckeroth 2013). During ENS development, serotonergic neurons guide the development and survival of neurons (O'Mahony et al. 2015). Colonization of GF mice with microbiome from conventional mice affected the neuroanatomy of the ENS and promoted intestinal transit associated with elevated mucosal and neuronal serotonin production and improved proliferation of intestinal neuronal progenitors (De Vadder et al. 2018). These neuroprotective effects were correlated with the activation of the ENS 5-HT4 receptor, establishing a potential link between gut microbiota and ENS development through serotonin-dependent activation of 5-HT4 receptor (De Vadder et al. 2018; Liu et al. 2009). The 5-HT4 receptor is required for ENS development and maintenance, as its deficiency reduces the abundance of ENS neurons after birth (Liu et al. 2009). The 5-HT2B receptor could also improve the differentiation of enteric neurons, while its antagonist blocked the beneficial effects of activated 5-HT2B expression on ENS function (Fiorica-Howells et al. 2000). The phenotypic composition of the ENS is affected by the enteric microenvironment and the activity of enteric serotonergic neurons (Lavdas et al. 1997). Cells of the adult ENS undergo increased vulnerability to neuronal degeneration and age-related neurons loss, which may be associated with a high incidence of intestinal dysfunction in older individuals (Saffrey 2013, 2014). The activation of 5-HT4 receptors improved the apoptotic rate of enteric neurons and inflammation-induced excessive autophagy in axon terminals (Liu et al. 2009). 5-HT4 receptors also promoted extraganglionic adult stem cells to differentiate into new neurons, which could substitute for injured ones (O'Mahony et al. 2015). The immune system is involved in signaling pathway of the gut-brain axis, which can be modulated by serotonin (Qi et al. 2022; Spiller 2007). In mouse models with colitis, higher 5-HT levels with a reduced enterochromaffin cell number were associated with the amelioration of intestinal inflammation (Khan et al. 2006). The low severity of colitis was observed in TPH1-knockout mice (Ghia et al. 2009). Serotonin has also been reported to modulate immune cell function, as suggested to promote actin remodeling and stimulate ERK phosphorylation of macrophages (Nakamura et al. 2008). The 5-HT receptors are also involved in modulating excitatory actions on vagal afferent fibers, suggesting serotonin signals to the brain and affects intestinal functions such as motility and pain (Mawe and Hoffman 2013; Sommer and Backhed 2013). The main stress axis, the hypothalamic-pituitary-adrenal (HPA) axis, is involved in hormonal communication (O'Mahony et al. 2015). Early life stress can result in anxiety and depressive-like behaviors, as well as intestinal disorders. Considering the impact of stress on the gut-brain axis, improved development of the HPA axis is critical for the normal function of the gut-brain axis (O'Mahony et al. 2015). Activation of HPA axis is associated with an altered serotonergic system (Ohta et al. 2014). The weaning stress at the early stage impacted 5-HT through the gut-brain axis, with increased 5-HT in offspring following rectal distension in the colon and spinal cord (Ohta et al. 2014; Ren et al. 2007). A higher number of activated 5-HT-positive neurons was observed in stressed rats on the dorsal raphe nucleus, associated with reducing the modulation of pain (Ren et al. 2007).

2.7.1.2 Kynurenine and the Gut-Brain Axis

The KP is the major Trp metabolic pathway, accounting for 90% of degraded Trp in both immune and epithelial cells (Clarke et al. 2013; Vecsei et al. 2013). Not only does the KP regulate serotonin availability, but it is also implicated in cognitive and behavioral-related neurological diseases (Cervenka et al. 2017). KYN can be metabolized into two distinct pathways, eventually producing KA and QA, respectively. The former product could exert neuroprotective effects against QA-induced excessively elevated neurotoxicity, which causes cognitive impairments (Kennedy et al. 2017). Thus, apart from regulating Trp availability for serotonin biosynthesis, activating the KP also plays an important role in modulating brain function through its downstream neurotoxic/neuroprotective products (Gao et al. 2020a). KYN accumulation has been suggested to cause social deficits in mice, such as memory and behavioral injury (Pocivavsek et al. 2014). This may depend on its effects on the α7-nicotinic acetylcholine receptor (Szalardy et al. 2012). However, lower plasma levels of KYN and higher Trp were observed in gut microbiota-depleted mice that had cognitive deficits (Desbonnet et al. 2015). When receiving gut microbes from depressed patients, mice are prone to anxiety-associated behaviors, which aligns with higher plasma ratios of KYN to Trp levels (Kelly et al. 2016). The intervention of microbial composition is highly associated with an altered KYN profile in the brain, evidenced by elevated KYN, KA, and QA in the brains of mice infected with the parasite Toxoplasma gondii (Notarangelo et al. 2014). Dietary supplementation of Bifidobacterium infants in rats can protect against depressive behavior by increasing circulating Trp and KA levels (Desbonnet et al. 2008). In addition, supplementation of anthocyanins increased Trp and KA by promoting colonization of *Ruminococcus* and decreasing the abundance of *Oscillibacter*, which ameliorated high fat-induced neuroinflammation and behavioral deficits in rats (Gao et al. 2020a). Gut microbiota can also impact the brain's KP by directly modulating the activity of key enzymes involved. Recolonization with gut microbiota in GF mice could restore IDO activity compared with normal mice (Clarke et al. 2013). The gut microbiota-induced elevated immune cytokines activate the central KP by stimulating IDO activity (Mu et al. 2016). The KYN produced from the periphery can directly pass the blood-brain barrier (Cervenka et al. 2017). Microbially produced SCFAs can inhibit IDO activity and modulate the KP depending on STAT1/HDAC pathway (Martin-Gallausiaux et al. 2018). Probiotics *Lactobacillus* strains can directly suppress intestinal IDO activity and reduce circulating levels of KYN, further inhibiting the progression of depressive behaviors in stressed mice (Marin et al. 2017). Moreover, the KP is highly associated with inflammatory mediators, and enzymes in the pathway are involved in immunoregulation. For example, activation of TLRs may increase KP metabolism through IFN-γ-dependent IDO1 induction (Campbell et al. 2014). In patients with IBS, the altered microbial profile and increased KYN metabolism were associated with low-grade inflammatory responses (Clarke et al. 2012). The AHR itself could directly modulate IDO and TDO expression (Bessede et al. 2014). Endogenous KYN levels were increased in AHR-deficient mice, while KYN regulated AHR activation in the brain challenged with experimental stroke (Cuartero et al. 2014; Garcia-Lara et al. 2015). Astrocyte activity and CNS inflammation were modulated by KYN metabolism through AHR signaling, and administration of an AHR agonist ameliorated intestinal injury in a preclinical model of colitis (Lamas et al. 2016; Rothhammer et al. 2016).

2.7.1.3 Microbial Tryptophan Metabolites and the Gut-Brain Axis

The tryptophanase from gut microbiota can catabolize Trp into indoles, which serve as important signaling molecules responsible for maintaining intestinal homeostasis (Kiely et al. 2018). Gut microbial-derived Trp metabolites can signal to distant organs, including the brain, potentially modulating communication between gut microbiota and the CNS (Agus et al. 2018).

Indole is metabolized through the action of the microbial enzyme TnaA (Gao et al. 2020a). Once passed into circulation, indole accumulation has been studied for its impact on brain function and mood disorders. When rats received an acute rectal injection of indole, motor activity significantly decreased, and c-Fos protein expression increased in the dorsal vagal complex (Jaglin et al. 2018). Colonization of GF rats with indole-generating bacteria led to exacerbated anxiety in open-field tests (Jaglin et al. 2018).

Indole derivatives, including tryptamine, IAA, and IPA, have been reported to impact intestinal barrier function and host immune responses (Gao et al. 2020a). For example, IPA has been identified as a beneficial metabolite contributing to microbial immunomodulatory effects (Jennis et al. 2018). These indole derivatives can affect neuronal differentiation and proliferation by binding with AHR. AHR is not only expressed along the gut but also highly expressed in cells in the CNS, such as neurons and microglial cells (Juricek and Coumoul 2018). They can also cross the blood-brain barrier and ameliorate neuronal inflammation (Rothhammer et al. 2016). Their functions, involving reducing CNS autoimmunity, can also be achieved by activating AHR and inhibiting the NF-κB pathway (Marsland 2016).

2.7.1.4 Aryl Hydrocarbon Receptor in Tryptophan Metabolites to Brain Signaling

AHR is a ligand-mediated transcription factor that senses endogenous and exogenous stimuli to eliminate and detoxify xenobiotics (Li and Wang 2010; Mackowiak and Wang 2016). When binding with ligands such as Trp metabolites, AHR is translocated from the cytoplasm into the nucleus. It then forms a heterodimer with AHR nuclear translocator (ARNT). This dimer binds to the promoter of target genes and activates transcription, thereby modulating AHR metabolism (Wincent et al. 2012).

Trp metabolites, including serotonin, KYN, and indole derivatives, serve as ligands for AHR in the microbiota-brain axis, mediated through vagal, immune, and metabolic mechanisms, with critical effects on behavior and CNS function (Ma et al. 2020b).

Serotonin and its catabolites, 5-HIAA, serve as endogenous agonists of AHR, and disrupted serotonin metabolism in the gut is associated with intestinal disorders, such as IBS (Jin et al. 2016; Manzella et al. 2018; Zhao et al. 2017). In addition, KYN and its metabolites, KA, act as AHR agonists, playing an important role in modulating cancer and immune responses (Liu et al. 2018). The excess production of XA and overactivated TDO facilitate the migration of tumor cells, which can be reversed by inhibiting AHR (Novikov et al. 2016). Another AHR ligand, 3-hydroxyanthranilic acid, can ameliorate oxidative stress-induced cell apoptosis, dependent on AHR signaling (Joshi et al. 2015). Moreover, indole and indole-associated chemicals are important AHR agonists (Kawai et al. 2017). The aminotransferases increased AHR activity through the metabolization of IPA into several AHR activators, including IAA, skatole, and IAld. Tryptamine, the product of Trp degradation through decarboxylase, has also been reported as a bacteria-derived AHR ligand (Jin et al. 2017). Microbial-derived indoxyl sulfate (IS) can directly activate AHR as an endogenous ligand (Schroeder et al. 2010).

The AHR acts as a critical node in Trp metabolism to brain signaling. Intestinal-derived Trp metabolites could regulate the microbiome-brain axis through AHR-dependent immune signaling (Ma et al. 2020b). The three major metabolic pathways (serotonin, KYN, and indoles) in the gut are regulated directly or indirectly by the gut microbiome, which promotes anti-inflammatory effects and ameliorates CNS inflammation by activating AHR (Agus et al. 2018; Nicolas and Chang 2019). It was also reported that the coevolution of KYN-TDO/IDO-AHR realizes neuroactive functions such as improving inflammation and autoimmune diseases (Cervenka et al. 2017; Kennedy et al. 2017). In the absence of AHR activation, IDO1 activity is inhibited, accompanied by increasing immune tolerance and reducing antigen-specific T-lymphocyte activity (Platten et al. 2005; Sznurkowski et al. 2011). Emerging evidence suggests that the pathogenesis of autism spectrum disorder (ASD) is associated with the ARNT-facilitated AHR-IDO1 pathway (Fujisawa et al. 2016; Lim et al. 2016). Moreover, indole derivatives activate microglia, which mediate response signals to CNS inflammation and suppress CNS autoimmunity by binding with AHR in astrocyte cells (Nicolas and Chang 2019). Microbial-derived Trp metabolites activate AHR by modulating IFN-I signaling in astrocyte cells. The activated AHR further inhibits NF-κB activation by promoting the secretion of suppressor of cytokine signaling 2, an anti-inflammatory molecule, implicating the protective effect of AHR ligands against CNS autoimmunity (Marsland 2016; Rothhammer et al. 2016). In addition to astrocytes, Trp metabolites activated AHR expressed on microglia, along with the production of TGFα and vascular endothelial growth factor B, intervening in CNS-associated diseases and the transcriptional program of astrocytes (Lee et al. 2015; Rothhammer et al. 2018). Gut microbial Trp metabolites can be involved in key regions of the brain's reward network through the AHR-associated metabolic pathway, in which the amygdala-nucleus accumbens circuit and the amygdala-anterior insula (aINS) are critical in affecting non-homeostatic food intake (Mayer 2011; Osadchiy et al. 2018). The activated AHR positively connects indoles and the amygdala-aINS circuit with food addiction scores (Osadchiy et al. 2018). Meanwhile, indoles bidirectionally regulate the production of the gut hormone glucagon-like peptide 1 (GLP-1) involving AHR (Chimerel et al. 2014). GLP-1 acts locally on vagal afferent nerve terminals and then signals to the brain circuits. Based on AHR involvement in glucose and insulin-related metabolism, activated AHR contributes to low fasting glucose levels and promotes GLP-1 secretion, suggesting the regulatory roles of Trp metabolites on food intake behavior (Natividad et al. 2018; Nicolas and Chang 2019). Trp metabolites are also involved in microbiome-brain axis by regulating the AHR-based vagal pathway (Fulling et al. 2019). The vagus nerve is important for maintaining corporeal homeostasis by affecting hunger, neurotransmitter levels, satiety, and inflammation in the brain (Ressler and Mayberg 2007). The SCFAs or microbial KYN-activated receptors expressed on gut enterochromaffin cells facilitate the release of serotonin in a calcium-dependent manner. In turn, the local serotonin regulated the activity of synapse-like contact of the vagal nerve through 5-HT3 receptors on gut mucosal unmyelinated vagal afferents, mediating satiety or brain inflammation (Osadchiy et al. 2018; Zubcevic et al. 2019).

2.7.1.5 Tryptophan Metabolism in Neurologic Disorders

Trp metabolism has important implications for neurogastroenterology in intestinal and CNS function through gut-brain axis signaling, such as IBS. IBS is comorbid with mood and anxiety problems and is accompanied by changes in bowel habits (Drossman 2016). The gut microbial community may be critical for the progression of IBS, as evidenced by the different abundances of bacterial taxa (e.g., higher ratio of Firmicutes to Bacteroidetes) in IBS patients, suggesting an associated role of intestinal microbes in the occurrence of IBS (Zhuang et al. 2017). Brain networks including emotional regulation, executive control networks, and sensorimotor have been identified as important for the development of symptom-associated anxiety and visceral pain in IBS patients (Mayer et al. 2015). Impaired Trp metabolism is associated with IBS pathogenesis (Agus et al. 2018). Plasma KYN levels, the ratio of KYN to Trp, and peripheral IDO1 activity were increased in IBS patients compared with healthy controls, illustrating the higher activity of the KP in IBS patients (Clarke et al. 2012). The KYN acts as a modulator of the glutamatergic N-methyl-D-aspartate (NMDA) receptor pathway, which is critical in the development of visceral pain in IBS. Upregulation of NMDA receptors in the CNS responds to noxious colonic distention during visceromotor responses, and this impact could be suppressed by the antagonist of NMDA receptor, 7-chlorokynurenic acid, a derivative of KA (Kolhekar and Gebhart 1996). In preclinical models of IBS, dysregulated NMDA receptor expression was associated with dysbiosis in the gut and CNS (Chen et al. 2017). In addition to the KP, disrupted 5-HT homeostasis was observed in IBS patients and may be associated with dysmotility. The mucosal 5-HT level is lower in IBS patients and is associated with reduced expression of TPH1 and serotonin reuptake transporter (Foley et al. 2011; Kerckhoffs et al. 2012). However, the changes in 5-HT levels differ regarding IBS subtypes, with a lower colonic 5-HT level in constipation-predominant IBS and higher in diarrhea-predominant IBS (Manocha and Khan 2012). In this condition, bacterial infection-induced activation of the 5-HT pathway associated with microbial stimulation of enterochromaffin cells signaling continuously causes symptoms in post-infectious IBS (Wheatcroft et al. 2005). Serotonin receptors, such as the 5-HT3 type, are altered in IBS patients, and treatment with its antagonists to IBS patients improved colonic transit, increased small intestinal absorption, and ameliorated visceral pain by mediating gut-brain axis (Camilleri and Boeckxstaens 2017). In IBS patients, altered microbial flora-induced disorder in Trp metabolism underlie visceral pain. Microbiota manipulation through probiotics is beneficial to visceral pain perception in dysbiotic or GF murine models (McKernan et al. 2010; Vermeulen et al. 2014). Trp metabolites, including 5-HT and KYN, are reported to be involved in both local and ascending neuronal pathways that transmit visceral pain. The colonic mucosal production of 5-HT is associated with the severity of discomfort in IBS patients because 5-HT locally exerts effects on visceral hyperalgesia by modulating vagal afferents and inflammatory responses in the gut (Browning 2015; Cremon et al. 2011; Shajib et al. 2017). Administration of 5-HT3 antagonist inhibited the excitatory effect that colonic mucosal supernatants of IBS patients could activate rat mesenteric sensory neurons in vitro (Cremon et al. 2011). Other studies also demonstrated that both 5-HT4 and 5-HT3 modulated visceral pain in the development of IBS, which may represent a potential therapeutic strategy (O'Mahony et al. 2015). Moreover, IBS patients may manifest depression or anxiety compared with healthy controls, which is associated with alterations in Trp metabolism. The reduced KYN and 5-HT in the intestinal mucosal biopsy significantly correlate with the psychological index of IBS patients (Bosi et al. 2020; Keszthelyi et al. 2013). The effects of KYN on stress-associated disorders depended on its modulation of NMDA receptors located in CNS (Bosi et al. 2020).

Disordered Trp metabolism is also associated with various neuropsychiatric diseases, including depression and ASD, as evidenced by the decreased central serotonin availability in patients with depressive disorder (Agus et al. 2018; Kaluzna-Czaplinska et al. 2019). Alterations in Trp metabolism, such as decreased plasma Trp level and the ratio of KYNA to QA, also correlate with learning injury and processing efficiency in patients with depressive disorder (Zhou et al. 2019). ASD is characterized by social and behavioral impairment, and it also develops intestinal functional abnormalities. A shift in Trp metabolism from 5-HT synthesis to the KP was observed in ASD patients (Kaluzna-Czaplinska et al. 2019). These alterations in Trp and serotonin levels were significantly associated with gut dysbiosis (e.g., reduced *Alistipes* and *Veillonella*, increased *Collinsella*), suggesting that altered Trp metabolism in ASD may be regulated by the gut microbiome (Luna et al. 2017).

2.7.2 *The Impact of Tryptophan Metabolism on the Gut-Liver Axis*

The disorder of crosstalk between the gut and liver contributes to both gut and liver disease. It has been revealed that microbial dysbiosis and gut barrier dysfunction cause the translocation of microbe and their derivatives, further inducing chronic inflammation and leading to liver injury and host metabolic syndrome (Li et al. 2021b; McLaughlin et al. 2017). Trp metabolites, under microbial bioconversion, such as indole and its derivatives, interact with the host and impact gut and hepatic function by circulating in the plasma (Schroeder and Backhed 2016). Trp derivatives become intercellular signal molecules through further catalysis in the liver and affect liver metabolism and immune response (Brunius et al. 2016; Hwang et al. 2013). Targeting Trp metabolism via the gut-liver axis may be a potential biotechnological intervention to improve liver function.

2.7.2.1 Tryptophan Metabolic Pathway in the Liver

The Trp metabolism in the liver mainly involves three pathways: the KP, the indole pathway, and the serotonin pathway (Teunis et al. 2022). The majority of dietary Trp is converted through the KP by the hepatic rate-limiting enzymes TDO and IDO outside of the liver. The KP is involved in the biosynthesis of nicotinic acid and NAD+ (Marszalek-Grabska et al. 2021). It is important in mediating immune responses, while inflammatory factors such as LPS upregulate IDO expression and promote KYN production (Cervenka et al. 2017).

Non-absorbed dietary Trp is transformed into indole by the gut microbiome in the distal colon. The enzymes involved in these metabolic activities are derived from various gut bacteria including genera *Clostridium*, *Bacteroides*, and *Peptostreptococcus* (Dodd et al. 2017a). The indole derivatives flux into the liver and undergo further metabolism. Initially, indole is oxidized by hepatic metabolic enzymes cytochrome P450 (CYP450) to indican and indoxyl (Banoglu et al. 2001). Indole can also be oxidized by CYP2A6 and synthesize indoxyl, oxindole, and 6-hydroxyindole (Gillam et al. 2000). The hepatic sulfotransferases conjugate indoxyl to produce indoxyl-3-sulfate, which is excreted in urine (Banoglu and King 2002). Apart from indole, IAA can be converted to cytotoxic species indole-3-acetic acid aspartate through aspartate conjugation or combined with glutamine to generate indoleacetic glutamine (Folkes et al. 1999; Gao et al. 2018). IPA can be finally converted to indoleacryloyl glycine through indole-3-acrylate (IA) conjugated with glycine (Gao et al. 2018; Li et al. 2021b). A variety of Trp metabolites such as tryptamine, indoles, and KYN are ligands for the AHR, which could regulate the immune system. Indoles have been identified as antagonists of pregnane X receptor (PXR), which is highly expressed in the liver. This intracellular receptor is involved in lipid metabolism and inflammatory suppression (Illes et al. 2020; Moreau et al. 2008).

The third Trp metabolic pathway is the serotonin pathway, which converts Trp to 5-HT via the intestinal enzyme TPH1 (Agus et al. 2018). Apart from the roles of 5-HT as a hormone and neurotransmitter, 5-HT also plays an important role in digestion and becomes a mitogen for different tissues involving the liver, especially during liver injury (Ruddell et al. 2008). Decreased serotonin levels indeed contributed to lower mRNA expression of liver fibrotic genes and pro-inflammatory cytokines in vitro (Kyritsi et al. 2020). Melatonin, the product of the 5-HT pathway, has been studied for its anti-inflammatory effects. Oral gavage of melatonin ameliorated liver inflammation induced by ochratoxin in ducks (Xia et al. 2021). This effect was also found in mice: melatonin supplementation significantly attenuated LPS-induced liver biopsies (Bellanti et al. 2017).

2.7.2.2 Microbial Tryptophan Derivatives Involved in Liver Disease

The significant role of the gut-liver axis in the development of liver disease has been well-established, and the impacts of Trp metabolites contributing to liver disease will be discussed.

The altered Trp metabolism by bacteria has been associated with alcoholic liver disease both in human and mouse models, with AHR being identified as an important mediator (Hendrikx et al. 2019). Lower intestinal levels of IAA, a known ligand for AHR, were observed in mouse models with ethanol-induced steatohepatitis. This could be rescued by daily gavage of IAA, which promoted the production of IL-22 in the intestinal lamina propria, induced antimicrobial peptides like Reg3g, and inhibited bacterial translocation to the liver (Zelante et al. 2013). Ethanol administration induced severe liver injury in intestine-specific AHR knockout mice (Qian et al. 2022). Fecal samples from patients with alcoholic liver disease showed reduced mRNA and protein expression of intestinal AHR (Qian et al. 2022). Lower intestinal levels of IAA were also determined in patients with alcoholic hepatitis, suggesting disturbed Trp metabolism's involvement in the occurrence of human alcoholic liver disease (Hendrikx et al. 2019). Moreover, indole-3-carbinol, a precursor of AHR ligand, exerted hepatoprotective effects during alcoholic liver disease. Oral gavage of indole-3-carbinol to mice with liver injury significantly improved hepatic apoptosis and oxidative stress by reducing hepatic neutrophil infiltration and ameliorating leaky gut (Choi et al. 2018b). Gut microbial interventions suggested the therapeutic effect of targeting AHR signaling for alcoholic liver disease. A pectin-enriched diet promoted the colonization of *Bacteroides*, increased intestinal indole derivative levels, and modulated the AHR pathway, contributing to resistance to liver injury (Ferrere et al. 2017). Lactobacillus rhamnosus GG, one of Trp-degrading bacteria, improved liver injury by modulating the AHR/IL-22 pathway, which decreased bacterial translocation and LPS secretion (Gu et al. 2021b). Indole has also been studied to prevent LPS-induced hepatic inflammation in vivo and in vitro by downregulating the NF-κB signaling pathway (Beaumont et al. 2018b).

Nonalcoholic fatty liver disease (NAFLD), which develops from nonalcoholic fatty liver (NAFL) to nonalcoholic steatohepatitis (NASH), has become one of the major causes of chronic liver disease (Friedman et al. 2018). The KP is activated via IDO by inflammatory molecules, which may play a crucial role in the occurrence of NAFLD. IDO-deficient mice were prone to high levels of pro-inflammatory cytokines and hepatic fibrosis, suggesting the anti-inflammatory role of IDO activity (Nagano et al. 2013). However, contradictory results shows that inhibition of IDO ameliorated methionine/choline-deficient diet-induced steatosis by downregulating pro-inflammatory genes and alpha-smooth muscle actin, suggesting a positive association between IDO activity and inflammation (Vivoli et al. 2015). In NAFLD patients, fecal Trp levels were reduced accompanied by high activity of the KP. The higher abundance of *Collinsella*, *Acinetobacter*, and *Actinomyces* in patients was significantly associated with higher levels of KYN (Sui et al. 2021). Transplantation of fecal microbiome from NAFLD patients to GF mice caused an increase in KYN levels and hepatic lipid accumulation (Sui et al. 2021).

Microbiota-derived indole pathway has been identified to affect the development of NAFLD (Li et al. 2021b). Oral gavage of indoles ameliorated LPS-induced pro-inflammatory cytokine production in residential liver macrophages and infiltration of immune cells associated with inhibition of the NF-κB pathway (Beaumont et al. 2018b). IA treatment improved high-fat diet-induced hepatosteatosis in mice by reducing hepatic macrophage number and mRNA levels involved in inflammatory responses such as TNF-α and MCP-1 (Ji et al. 2019). IA could also activate thermogenesis and decrease hepatic lipid accumulation, thus preventing diet-induced obesity in mice (Choi et al. 2012). In addition, IAA could reduce inflammatory levels in vitro through an AHR-dependent mechanism (Ritze et al. 2014).

The levels of 5-HIAA, the serotonin marker, were increased in NAFLD patients, which is associated with the severity of liver-related complications such as hepatocellular carcinoma (Wegermann et al. 2021). Serotonin aggravated hepatic fibrosis and inflammation. In a mouse model with NAFLD, the gene expression of serotonin receptors was highly upregulated in the liver, while knockout of serotonin receptor or block of gene expression rescued hepatic steatosis (Choi et al. 2018a). The in vitro study also suggested that reduced serotonin concentration contributed to lower gene expression involved in fibrosis (e.g., α-SMA and COL1a1) and pro-inflammatory response (e.g., IL-1β, IL-8) (Kyritsi et al. 2020). Melatonin, downstream metabolites of serotonin, exerted a protective role against liver fibrosis (Celinski et al. 2014).

2.7.2.3 Plausible Therapeutic Target for the Gut-Liver Axis

As intestinal dysbiosis is a critical factor affecting hepatic function, modulating the microbial community to promote Trp metabolism homeostasis could be a potential treatment for liver injury. Probiotics and symbiotics have been identified to modulate Trp generation. *Peptostreptococcus*-derived IA improved intestinal barrier function and reduced inflammation in mice with colitis (Wlodarska et al. 2017).

Administration of *Clostridium* sporogenes and Trp-supplemented diets also prevented mice from colitis by promoting IPA production and activating PXR (Venkatesh et al. 2014). Bifidobacteria sp. and Faecalibacterium prausnitzii, along with dietary fibers, have been proposed to treat NASH by increasing microbial butyrate levels and ameliorating intestinal inflammation (Ralli et al. 2021). Lower levels of intestinal inflammation impact Trp metabolism, as evidenced by reduced pro-inflammatory Trp metabolites.

Modulation of microbial Trp metabolites could also be an effective treatment for improving liver function. There are several approaches to target Trp metabolism, which involve supplementation of Trp, indole and its derivatives, indole mimicry, modulating the KP, and microbial engineering.

Supplementation of Trp has been reported to improve NAFLD in fructose-fed mice by modulating intestinal barrier and serotonergic pathway, further decreasing hepatic fat deposition (Ritze et al. 2014). However, as Trp could also be metabolized to pro-inflammatory metabolites (e.g., 5-HT), it increased steatosis in high-fat high-sugar-fed mice associated with the activation of inflammatory responses (Osawa et al. 2011). Therefore, the controversial effects of Trp supplementation suggest the importance of modulating key enzymes in the Trp metabolism pathway for treating liver disease.

Direct supplementation of indole and its derivatives is also used in the treatment of liver disease. In animal models, gavage of indole significantly reduced LPS-induced hepatic inflammation by suppressing the NF-κB pathway (Beaumont et al. 2018a). Supplementation of IPA promoted the differentiation of hepatic M2 macrophages, thus reducing the ratio of M1/M2, which was also confirmed by in vitro macrophage and hepatocyte culture (Silva et al. 2021). Oral administration of IAA could reduce alcoholic fatty liver disease by increasing IL-22 production from innate lymphoid cells and preventing hepatic bacterial translocation (Hendrikx et al. 2019). Moreover, oral administration of IPA significantly suppressed the NF-κB pathway, improved intestinal barrier function, and further attenuated liver injury in high-fat diet-induced steatohepatitis and metabolic disorders in rats (Zhao et al. 2019b). The IPA-enriched diet can also serve as a nutritional intervention for metabolic disorders with insulin resistance, as confirmed by its ability to lower blood glucose and insulin in rats (Abildgaard et al. 2018). Regarding the anti-inflammatory effects of IAA and IPA in animal models, a clinical trial developing indole derivatives in NASH patients warranted confirmation (Mole et al. 2016). Indoles may also be beneficial for liver regeneration as endogenous ligands of AHR, such as IAA, IPA, and tryptamine (Gutierrez-Vazquez and Quintana 2018). Due to the production site, microbial Trp metabolism dominantly activated intestinal AHR. Activation of AHR by indoles improves intestinal barrier function through the upregulation of genes involved in tight junction, goblet cell differentiation, and mucus secretion (Bansal et al. 2010; Wlodarska et al. 2017). The enhanced intestinal barrier prevents the translocation of gut-derived toxins, which participate in liver regeneration (Zheng and Wang 2021). Moreover, indoles can modulate the production of GLP-1 derived from intestinal L cells and impact host metabolism in an AHR-dependent way. GLP-1 has been considered to play a critical role in improving glucose balance and liver regeneration (Chimerel et al. 2014; Natividad et al. 2018). As AHR is expressed in various immune cells, microbial IAA and tryptamine inhibit inflammation in hepatocytes and macrophages, suggesting the protective role of indoles in regulating hepatic immune response (Krishnan et al. 2018).

Modulation of the KP contributes to improving anti-inflammatory ability. KYN 3-monooxygenase (KMO), one of the markers of neurological disorders, has been demonstrated to have pro-inflammatory effects, while inhibition of KMO could decrease toxic 3-HK and elevate protective KA in a mouse model with Huntington's disease (Bondulich et al. 2021). Many KMO inhibitors have been developed targeting neurological diseases, and the experience may also be helpful for NASH.

Microbial metabolic mimicry could overcome the defects of natural products, which are limited by their short life cycle and lower oral bioavailability (Kumar et al. 2020). For example, multiheterocyclic AHR agonists achieved anti-inflammatory effects at very low doses according to the analysis of diverse structures of indole and indazole and computational studies (Chen et al. 2020; Li et al. 2021b). Indole mimics could be a potential treatment for hepatic steatosis and inflammation through the activation of PEKFB3 expression (Ma et al. 2020a). Synthetic indole has been shown to have antidiabetic activities. For example, compound 13 m showed high anti-lipogenesis activity through the upregulation of cholesterol transport, which may be a potent therapy for hepatic metabolic disease (Rajan et al. 2018). Apart from mimicry, bacterial engineering could also be an effective intervention for Trp metabolism through gene clones encoding critical enzymes involved in Trp metabolism in bacteria with genetic manipulation. Supplementation of the engineered *Lactobacillus reuteri* aimed to produce IL-22, which significantly improved alcohol fatty liver and had clinical applications (Hendrikx et al. 2019). Colonization of engineered *Bacteroides thetaiotaomicron* Trp D+ ameliorated colitis-induced intestinal barrier dysfunction by promoting tryptamine levels (Bhattarai et al. 2018).

The side effects of modulating Trp metabolism should be considered. For example, a Trp-enriched diet increased the risk of portal hypertension in patients with liver cirrhosis. Oral gavage of IPA exacerbated liver injury through the disordered TGF-β1/Smad signaling pathway in mouse models with CCL4-induced liver fibrosis (Liu et al. 2021a). Overall, the disruption of Trp metabolism could be an important factor causing liver disease, and targeting these pathways has therapeutic implications in liver disease.

2.7.3 The Impact of Tryptophan Metabolism on the Gut-Kidney Axis

The gut microbiota-derived Trp metabolites such as KYN, IS, and IAA are important uremic toxins implicated in the progression of kidney disease (KD) (Roager and Licht 2018). Indolic and KP disturbances, highly impacted by immunological system activity, are observed in the most common KD (Dolivo et al. 2018). Accumulated Trp metabolites act as ligands to activate AHR signaling and affect disease progression, making them potential therapeutic targets (Gao et al. 2018).

2.7.3.1 Uremic Toxins from Tryptophan Metabolic Pathway

Microbial-derived Trp metabolites have been positively correlated with kidney function decline (Pawlak et al. 2003). Transplantation of feces from KD patients into GF mice resulted in higher serum levels of uremic toxins and oxidative stress in the kidneys. This establishes a causal role of aberrant microbiota in the development of renal disease (Wang et al. 2020c). Patients with KD are exposed to a variety of uremic toxins, which are largely (95%) derived from the KP and others from the indolic and serotonin pathway (Mo et al. 2020). Two rate-limiting enzymes, TDO and IDO, mediate the KP, leading to the production of the intermediate N-formyl KYN and further KYN. Then KYN can be converted to 3-hydroxykynurenine (3-HKYN), KA, 3-hydroxy anthranilic acid, anthranilic acid, and QA (Dolivo et al. 2018). These metabolites are involved in regulating the host immune response. In KD patients, higher activity of IDO and concentrations of KYN metabolites were observed, while Trp was reduced (Zakrocka and Zaluska 2022). IDO activity was also associated with pro-inflammatory cytokines, providing evidence that chronic inflammation led to the activation of the KP and impaired kidney function (Debnath et al. 2017; Pawlak et al. 2009). Apart from IDO, another KP enzyme, kynureninase, mediates downstream metabolites and induces inflammation. Its activity increases with the severity of KD by producing excessive 3-HKYN and secreting pro-inflammatory cytokines (Pawlak et al. 2010). During KD development, an excess of KYN metabolites markedly suppresses renal mesangial cell proliferation (Yoshimura et al. 2009). In contrast, KYN and QA promote their proliferation associated with upregulated insulin-like growth factor-1 and downregulated hepatocyte growth factor levels. Abnormalities in the KP are associated with the dysfunction of mesangial cells and may be involved in the progression of renal failure and glomerular fibrosis (Mor et al. 2020). Oxidative stress induced by 3-HKYN and QA intensifies cell injury and increases the apoptotic rate in renal tissue and glomerular fibrosis processes (Mor et al. 2020; Reyes-Ocampo et al. 2015; Yoshimura et al. 2009). Two other Trp metabolic pathways are also described: the serotonin pathway, which converts Trp to melatonin, and indolic components degraded by intestinal bacteria before absorption (Zhu et al. 2011). IAA is directly produced in the intestine or endogenously in tissue via tryptamine, while IS and indoxyl-β-D glucuronide (IDG) are oxidized and sulfated in the liver (Addi et al. 2018; Meyer and Hostetter 2012). In KD patients, IS, IAA, and IDG are commonly excreted and accumulated in urine (Duranton et al. 2012). However, a high-sulfur-amino-acid diet could abrogate the toxic effects of IS on the kidneys by inhibiting microbial tryptophanase activity in a murine model of KD (Lobel et al. 2020). These intestinal metabolites of Trp can act as AHR agonists, and AHR expression has a positive correlation with KD severity (Mo et al. 2020).

2.7.3.2 Gut Microbiota-Derived Tryptophan Metabolites as AHR Ligands in KD

Tryptophan metabolites derived from gut microbiota participate in kidney function as AHR ligands. For example, IS and IAA, indolic compounds metabolized by the gut microbiome and accumulated in KD patients, upregulated eight downstream genes of AHR (Chen et al. 2019). Activation of AHR contributed to progressive glomerular and renal tubular cell injury, which can lead to glomerulosclerosis and renal fibrosis, further exacerbating KD (Milanesi et al. 2019).

IAA, mainly produced by intestinal *Clostridium* and *Bacteroides*, exerted deleterious effects on renal cells by interacting with AHR (Zhao et al. 2019a). However, IAA could only indirectly activate the AHR pathway by inducing tissue factor expression, which required NF-κB pathway activation (Liu et al. 2021b). Moreover, uremic solute IAA has been verified to activate the AHR/p38 MAPK/NF-κB pathway and increase COX-2 activity, causing oxidative damage and inflammation in renal endothelial cells (Dou et al. 2015). In diabetic kidney disease rat models, oxidative injury, metabolic disorder, and matrix accumulation activated glomerular mesangial cells and altered tubular cell function. These impacts can be alleviated in AHR-knockout mice (Lee et al. 2016). AHR protein expression was positively correlated with plasma IAA levels in KD patients, while higher IAA was prone to inducing oxidative stress and immune dysfunction in KD patients (de Brito et al. 2019).

IS played a crucial role in the progress of KD via AHR activation. IS was generated in the liver by CYP2E1 and sulfotransferases by metabolizing bacterial Trp metabolites, which gradually accumulated in the body and aggravated renal injury (Liu et al. 2021b). IS upregulated AHR pathway-associated genes CYP1A1 and CYP1B1 expression in cells from KD mice (Xiang et al. 2020). The activation of AHR caused podocyte injury and glomerular damage along with upregulated CYP1A1 expression in mice exposed to IS (Zhao et al. 2019a). The levels of IS were significantly reduced in Tangshen Formula-treated KD rats, accompanied by improvement in renal function. Mechanically, AHR expression was downregulated, and nuclear translocation of NF-κB p65 was inhibited after Tangshen Formula treatment (Zhao et al. 2020). IS has been shown to alter the balance in the antioxidant system and cause the occurrence of oxidative stress (Lin et al. 2019). For example, IS-activated AHR promoted monocytes to produce pro-inflammatory cytokines and induced oxidative stress, which increased the mortality of KD patients (Nakano et al. 2019). For other gut microbiota-derived Trp metabolites, tryptamine could activate AHR and regulate the intestinal immune response (Islam et al. 2017). Indole-3-aldehyde combined with AHR promoted the secretion of IL-22 in lamina propria lymphocytes and reduced macrophage-derived TNF-α and interleukin-1β (Krishnan et al. 2019; Zelante et al. 2013). Moreover, indole-3-lactic acid, a *Bifdobacterium* spp.-derived AHR ligand, could activate immunoregulatory T cells and develop anti-inflammatory effects (Cervantes-Barragan et al. 2017; Sakurai et al. 2019). In the serotonin pathway, although serotonin is not an AHR ligand, it acts as a CYP1A1 substrate to promote sustained AHR signaling (Liu et al. 2021b). A therapeutic target for KD could be developed based on the regulation of AHR.

2.7.3.3 Potential Nutritional Interventions Targeting Tryptophan Metabolism in KD

Considering that Trp catabolites produced by gut microbiota may form uremic toxins, with the prominent example of IS in KD, therapeutic approaches should aim to reduce the levels of deleterious Trp catabolites and increase beneficial catabolites. The potential nutritional approaches involve modification of diet and the use of probiotics, prebiotics, and synbiotics.

Results from diet modifications are promising. Reducing protein intake significantly decreased plasma IS levels and urine IS, QA, and indoxyl glucuronide in healthy subjects (Poesen et al. 2015). However, restricted protein consumption has been shown to reduce IS levels and improve bacterial diversity, but no beneficial effect was observed on kidney function improvement (Di Iorio et al. 2019). An increase in dietary fibers could reduce plasma IS levels in hemodialyzed patients (Sirich et al. 2014). A higher proportion of protein from natural plants was associated with lower IS levels and mortality rates in patients with KD (Chen et al. 2016).

Dysregulated AHR signaling could exacerbate the progression of KD, and potential ligands should be identified to regulate the activity of AHR and improve KD. Inhibition of AHR signaling has been shown to lead to ameliorations in proteinuria, hypoalbuminuria, and hyperlipidemia in nephritic rats (Mo et al. 2020). AHR ligands are abundant in either natural products or endogenous metabolites (Liu et al. 2021b). Endogenous AHR ligands include Trp metabolites, arachidonic acid and heme metabolites, and LPS (Zhao et al. 2019a). Indole, a potent antagonist of AHR, was able to mitigate the effects of dysregulated IS production by binding AHR (Jin et al. 2014). Natural products are vital sources for drug discovery and have beneficial effects on KD by binding AHR (Yang et al. 2019). For example, flavonoids derived from natural plants such as genistein, daidzein, and quercetin could regulate KD by activating AHR and modulating downstream target gene expression (Xue et al. 2017). Barleriside A and rhoifolin isolated from Plantaginis semen have been studied to inhibit AHR expression and profibrotic protein expression, further improving renal fibrosis in vitro and in vivo (Miao et al. 2020). Baicalin and tanshinone I have also been identified to alleviate kidney injury through the regulation of AHR-dependent CYP1A1 and CYP1A2 induction (Feng et al. 2013; Wang et al. 2015b).

Supplementation of probiotics, prebiotics, and synbiotics has been investigated in both preclinical and clinical trials to promote intestinal health, which could further improve kidney function (Madella et al. 2022). For example, commonly used combinations involving Lactobacillus and inulin, or Bifidobacteria with fructooligosaccharides, could enhance the relative abundances of SCFA-producing bacteria. SCFAs have been widely considered to maintain gut integrity, prevent harmful microbial metabolites and toxins from entering circulation, reduce uremic toxins, and alleviate gut indole load. These could represent attractive therapeutic strategies to improve KD (Markowiak and Slizewska 2017; Melekoglu et al. 2021; Yang et al. 2021). Identification of beneficial microbes involved in Trp metabolism and developing microbiome-targeted regulation for renal health are under further investigation.

References

Abecia JA, Forcada F, Zúiga O (2002) The effect of melatonin on the secretion of progesterone in sheep and on the development of ovine embryos in vitro. Vet Res Commun 26(2):151–158

Abildgaard A, Elfving B, Hokland M, Wegener G, Lund S (2018) The microbial metabolite indole-3-propionic acid improves glucose metabolism in rats, but does not affect behaviour. Arch Physiol Biochem 124(4):306–312

Adams O, Besken K, Oberdörfer C, MacKenzie CR, Rüssing D, Däubener W (2004) Inhibition of human herpes simplex virus type 2 by interferon gamma and tumor necrosis factor alpha is mediated by indoleamine 2,3-dioxygenase. Microbes Infect 6(9):806–812

Addi T, Dou L, Burtey S (2018) Tryptophan-derived uremic toxins and thrombosis in chronic kidney disease. Toxins (Basel) 10(10)

Adeola O, King D (2006) Developmental changes in morphometry of the small intestine and jejunal sucrase activity during the first nine weeks of postnatal growth in pigs. J Anim Sci 84(1):112–118

Adhyaru BB, Jacobson TA (2018) Safety and efficacy of statin therapy. Nat Rev Cardiol 15(12):757–769

Agus A, Planchais J, Sokol H (2018) Gut microbiota regulation of tryptophan metabolism in health and disease. Cell Host Microbe 23(6):716–724

Anders HJ, Andersen K, Stecher B (2013) The intestinal microbiota, a leaky gut., and abnormal immunity in kidney disease Kidney Int 83:1010–1016

Anderson G, Seo M, Berk M, Carvalho F, A., & Maes, M. (2016) Gut permeability and microbiota in Parkinson's disease: role of depression, tryptophan catabolites, oxidative and nitrosative stress and melatonergic pathways. Curr Pharm Des 22(40):6142–6151

Arentsen T, Qian Y, Gkotzis S, Femenia T, Wang T, Udekwu K et al (2017) The bacterial peptidoglycan-sensing molecule Pglyrp2 modulates brain development and behavior. Mol Psychiatry 22(2):257–266

Arsenescu R, Arsenescu V, Zhong J, Nasser M, Melinte R, Dingle CR et al (2011) Role of the xenobiotic receptor in inflammatory bowel disease. Inflamm Bowel Dis 17(5):1149–1162

Arumugam M, Raes J, Pelletier E, Le Paslier D, Yamada T, Mende DR et al (2011) Enterotypes of the human gut microbiome. Nature 473(7346):174–180

Badawy AA (1988) Effects of pregnancy on tryptophan metabolism and disposition in the rat. Biochem J 255(1):369–372

Badawy AA (2014) The tryptophan utilization concept in pregnancy. Obstet Gynecol Sci 57(4):249–259

Baião NC, Lara L (2005) Oil and fat in broiler nutrition. Braz J Poul Sci 7:129–141

Bailey MT (2014) Influence of stressor-induced nervous system activation on the intestinal microbiota and the importance for immunomodulation. Adv Exp Med Biol 817:255–276

Bailey MT, Dowd SE, Galley JD, Hufnagle AR, Allen RG, Lyte M (2011) Exposure to a social stressor alters the structure of the intestinal microbiota: implications for stressor-induced immunomodulation. Brain Behav Immun 25(3):397–407

Banoglu E, King RS (2002) Sulfation of indoxyl by human and rat aryl (phenol) sulfotransferases to form indoxyl sulfate. Eur J Drug Metab Pharmacokinet 27(2):135–140

Banoglu E, Jha GG, King RS (2001) Hepatic microsomal metabolism of indole to indoxyl, a precursor of indoxyl sulfate. Eur J Drug Metab Pharmacokinet 26(4):235–240

Bansal T, Alaniz RC, Wood TK, Jayaraman A (2010) The bacterial signal indole increases epithelial-cell tight-junction resistance and attenuates indicators of inflammation. Proc Natl Acad Sci USA 107(1):228–233

Bauchart-Thevret C, Stoll B, Chacko S, Burrin DG (2009) Sulfur amino acid deficiency upregulates intestinal methionine cycle activity and suppresses epithelial growth in neonatal pigs. Am J Physiol Endocrinol Metab 296(6):E1239–E1250

Beaumont M, Neyrinck AM, Olivares M, Rodriguez J, de Rocca Serra A, Roumain M, Bindels LB, Cani PD, Evenepoel P, Muccioli GG, Demoulin JB, Delzenne NM (2018a) The gut microbiota metabolite indole alleviates liver inflammation in mice. FASEB J 32(12):fj201800544

Beaumont M, Neyrinck AM, Olivares M, Rodriguez J, Serra AD, Roumain M, Bindels LB, Cani PD, Evenepoel P, Muccioli GG, Demoulin JB, Delzenne NM (2018b) The gut microbiota metabolite indole alleviates liver inflammation in mice. Faseb J 32(12):6681–6693

Bellanti F, Villani R, Facciorusso A, Vendemiale G, Serviddio G (2017) Lipid oxidation products in the pathogenesis of non-alcoholic steatohepatitis. Free Radic Biol Med 111:173–185

Bellono NW, Bayrer JR, Leitch DB, Castro J, Zhang CC, O'Donnell TA et al (2017) Enterochromaffin cells are gut chemosensors that couple to sensory neural pathways. Cell 170(1)

Bender DA (1983) Biochemistry of tryptophan in health and disease. Mol Aspects Med 6(2):101–197

Bessede A, Gargaro M, Pallotta MT, Matino D, Servillo G, Brunacci C et al (2014) Aryl hydrocarbon receptor control of a disease tolerance defence pathway. Nature 511(7508):184–190

Bhattarai Y, Muniz Pedrogo DA, Kashyap PC (2017) Irritable bowel syndrome: a gut microbiota-related disorder? Am J Physiol Gastrointest Liver Physiol 312(1):G52–G62

Bhattarai Y, Williams BB, Battaglioli EJ, Whitaker WR, Till L, Grover M, Linden DR, Akiba Y, Kandimalla KK, Zachos NC, Kaunitz JD, Sonnenburg JL, Fischbach MA, Farrugia G, Kashyap PC (2018) Gut microbiota-produced tryptamine activates an epithelial G-protein-coupled receptor to increase colonic secretion. Cell Host Microbe 23(6):775–785 e775

Bikker P, Dirkzwager A, Fledderus J, Trevisi P, le Huerou-Luron I, Lalles J, Awati A (2006) The effect of dietary protein and fermentable carbohydrates levels on growth performance and intestinal characteristics in newly weaned piglets. J Anim Sci 84(12):3337–3345

Bindelle J, Buldgen A, Lambotte D, Wavreille J, Leterme P (2007) Effect of pig faecal donor and of pig diet composition on in vitro fermentation of sugar beet pulp. Anim Feed Sci Technol 132(3-4):212–226

Bittinger MA, Nguyen LP, Bradfield CA (2003) Aspartate aminotransferase generates proagonists of the aryl hydrocarbon receptor. Mol Pharmacol 64(3):550–556

Biviano A, Martinez del Rio C, Phillips D (1993) Ontogenesis of intestine morphology and intestinal disaccharidases in chickens (Gallus gallus) fed contrasting purified diets. J Comp Physiol B 163:508–518

Blumberg R, Powrie F (2012) Microbiota, disease, and back to health: a metastable journey. Sci Transl Med 4(137):137rv137

Bondulich MK, Fan Y, Song Y, Giorgini F, Bates GP (2021) Ablation of kynurenine 3-monooxygenase rescues plasma inflammatory cytokine levels in the R6/2 mouse model of Huntington's disease. Sci Rep 11(1):5484

Bonnin A, Goeden N, Chen K, Wilson ML, King J, Shih JC et al (2011) A transient placental source of serotonin for the fetal forebrain. Nature 472(7343):347–350

Booij L, Tremblay RE, Szyf M, Benkelfat C (2015) Genetic and early environmental influences on the serotonin system: consequences for brain development and risk for psychopathology. J Psychiatry Neurosci 40(1):5–18

Bosi A, Banfi D, Bistoletti M, Giaroni C, Baj A (2020) Tryptophan metabolites along the microbiota-gut-brain axis: an interkingdom communication system influencing the gut in health and disease. Int J Tryptophan Res 13:1178646920928984

Briscoe J, Sussel L, Serup P, Hartigan-O'Connor D, Jessell TM, Rubenstein JL, Ericson J (1999) Homeobox gene Nkx2.2 and specification of neuronal identity by graded Sonic hedgehog signalling. Nature 398(6728):622–627

Broekhuizen M, Klein T, Hitzerd E, de Rijke YB, Schoenmakers S, Sedlmayr P et al (2020) l-tryptophan-induced vasodilation is enhanced in preeclampsia: studies on its uptake and metabolism in the human placenta. Hypertension 76(1):184–194

Browning KN (2015) Role of central vagal 5-HT3 receptors in gastrointestinal physiology and pathophysiology. Front Neurosci 9. <Go to ISI>://WOS:000366712700001

Brunius C, Vidanarachchi JK, Tomankova J, Lundstrom K, Andersson K, Zamaratskaia G (2016) Skatole metabolites in urine as a biological marker of pigs with enhanced hepatic metabolism. Animal 10(10):1734–1740

Buddington RK, Elnif J, Puchal-Gardiner AA, Sangild PT (2001) Intestinal apical amino acid absorption during development of the pig. Am J Physiol-Regul Integrat Comp Physiol 280(1):R241–R247

Buznikov GA, Lambert HW, Lauder JM (2001) Serotonin and serotonin-like substances as regulators of early embryogenesis and morphogenesis. Cell Tissue Res 305(2):177–186

Camilleri M, Boeckxstaens G (2017) Dietary and pharmacological treatment of abdominal pain in IBS. Gut 66(5):966–974

Campbell BM, Charych E, Lee AW, Moller T (2014) Kynurenines in CNS disease: regulation by inflammatory cytokines. Front Neurosci 8:12

Castro N, Spengler M, Lollivier V, Wellnitz O, Bruckmaier RM (2012) Diurnal pattern of melatonin in blood and milk of dairy cows. Milchwissenschaft-milk Sci Int 66(4)

Castrogiovanni P, Musumeci G, Trovato FM, Avola R, Magro G, Imbesi R (2014) Effects of high-tryptophan diet on pre- and postnatal development in rats: a morphological study. Eur J Nutr 53(1):297–308

Celinski K, Konturek PC, Slomka M, Cichoz-Lach H, Brzozowski T, Konturek SJ, Korolczuk A (2014) Effects of treatment with melatonin and tryptophan on liver enzymes, parameters of fat metabolism and plasma levels of cytokines in patients with non-alcoholic fatty liver disease – 14 months follow up. J Physiol Pharmacol 65(1):75–82

Cervantes-Barragan L, Chai JN, Tianero MD, Di Luccia B, Ahern PP, Merriman J et al (2017) Lactobacillus reuteri induces gut intraepithelial CD4(+)CD8alphaalpha(+) T cells. Science 357(6353):806–810

Cervenka I, Agudelo LZ, Ruas JL (2017) Kynurenines: Tryptophan's metabolites in exercise, inflammation, and mental health. Science 357(6349)

Chang L (2006) Liver function interpretation. Dr. Hepatitis. 5:36–37. Retrieved from

Charney DS, Heninger GR, Reinhard JF, Hafstead SKM (1982) The effect of IV l-tryptophan on prolactin, growth hormone, and mood in healthy subjects. Psychopharmacology

Chelakkot C, Ghim J, Ryu SH (2018) Mechanisms regulating intestinal barrier integrity and its pathological implications. Exp Mol Med 50

Chen A, Kelley LD, Janušonis S (2012) Effects of prenatal stress and monoaminergic perturbations on the expression of serotonin 5-HT$_4$ and adrenergic β_2 receptors in the embryonic mouse telencephalon. Brain Res 1459:27–34

Chen X, Wei G, Jalili T, Metos J, Giri A, Cho ME, Boucher R, Greene T, Beddhu S (2016) The associations of plant protein intake with all-cause mortality in CKD. Am J Kidney Dis 67(3):423–430

Chen A, Chen Y, Tang Y, Bao C, Cui Z, Xiao M, Lin C (2017) Hippocampal AMPARs involve the central sensitization of rats with irritable bowel syndrome. Brain Behav 7(3):e00650

Chen YY, Chen DQ, Chen L, Liu JR, Vaziri ND, Guo Y, Zhao YY (2019) Microbiome-metabolome reveals the contribution of gut-kidney axis on kidney disease. J Transl Med 17(1):5

Chen J, Haller CA, Jernigan FE, Koerner SK, Wong DJ, Wang Y, Cheong JE, Kosaraju R, Kwan J, Park DD, Thomas B, Bhasin S, De La Rosa RC, Premji AM, Liu L, Park E, Moss AC, Emili A, Bhasin M et al (2020) Modulation of lymphocyte-mediated tissue repair by rational design of heterocyclic aryl hydrocarbon receptor agonists. Sci Adv 6(3):eaay8230

Cheng L, Wang L, Xu Q, Huang L, Zhou D, Li Z et al (2015a) Chitooligosaccharide supplementation improves the reproductive performance and milk composition of sows. Livest Sci 174:74–81

Cheng Y, Jin UH, Allred CD, Jayaraman A, Chapkin RS, Safe S (2015b) Aryl hydrocarbon receptor activity of tryptophan metabolites in young adult mouse colonocytes. Drug Metab Dispos 43(10):1536–1543

Chimerel C, Emery E, Summers DK, Keyser U, Gribble FM, Reimann F (2014) Bacterial metabolite indole modulates incretin secretion from intestinal enteroendocrine L cells. Cell Rep 9(4):1202–1208

Choera T, Zelante T, Romani L, Keller NP (2017) A multifaceted role of tryptophan metabolism and indoleamine 2,3-dioxygenase activity in aspergillus fumigatus-host interactions. Front Immunol 8:1996

Choi DS, Ward SJ, Messaddeq N, Launay JM, Maroteaux L (1997) 5-HT2B receptor-mediated serotonin morphogenetic functions in mouse cranial neural crest and myocardiac cells. Development 124(9):1745–1755

Choi Y, Kim Y, Park S, Lee KW, Park T (2012) Indole-3-carbinol prevents diet-induced obesity through modulation of multiple genes related to adipogenesis, thermogenesis or inflammation in the visceral adipose tissue of mice. J Nutr Biochem 23(12):1732–1739

Choi W, Namkung J, Hwang I, Kim H, Lim A, Park HJ, Lee HW, Han KH, Park S, Jeong JS, Bang G, Kim YH, Yadav VK, Karsenty G, Ju YS, Choi C, Suh JM, Park JY, Park S, Kim H (2018a) Serotonin signals through a gut-liver axis to regulate hepatic steatosis. Nat Commun 9:4824

Choi Y, Abdelmegeed MA, Song BJ (2018b) Preventive effects of indole-3-carbinol against alcohol-induced liver injury in mice via antioxidant, anti-inflammatory, and anti-apoptotic mechanisms: role of gut-liver-adipose tissue axis. J Nutr Biochem 55:12–25

Chu R, Liu C (1984) Morphological and functional comparisons of Peyer's patches in different parts of the swine small intestine. Vet Immunol Immunopathol 6(3–4):391–403

Chwalibog A, Jakobsen K, Thorbek G (1994) Loss of body fat in piglets in spite of gain in live weight. J Anim Physiol Anim Nutr 72(1-5):80–85

Clarke G, McKernan DP, Gaszner G, Quigley EM, Cryan JF, Dinan TG (2012) A distinct profile of tryptophan metabolism along the kynurenine pathway downstream of toll-like receptor activation in irritable bowel syndrome. Front Pharmacol 3:90

Clarke G, Grenham S, Scully P, Fitzgerald P, Moloney RD, Shanahan F et al (2013) The microbiome-gut-brain axis during early life regulates the hippocampal serotonergic system in a sex-dependent manner. Mol Psychiatry 18(6):666–673

Clarke G, Stilling RM, Kennedy PJ, Stanton C, Cryan JF, Dinan TG (2014) Minireview: gut microbiota: the neglected endocrine organ. Mol Endocrinol 28(8):1221–1238

Coëffier M, Marion-Letellier R, Déchelotte P (2010) Potential for amino acids supplementation during inflammatory bowel diseases. Inflamm Bowel Dis 16(3):518–524

Collino S, Montoliu I, Martin FP, Scherer M, Mari D, Salvioli S et al (2013) Metabolic signatures of extreme longevity in northern Italian centenarians reveal a complex remodeling of lipids, amino acids, and gut microbiota metabolism. PLoS One 8(3):e56564

Comai S, Bertazzo A, Brughera M, Crotti S (2020) Tryptophan in health and disease. Adv Clin Chem 95:165–218

Cook KL, Rothrock MJ Jr, Loughrin JH, Doerner KC (2007) Characterization of skatole-producing microbial populations in enriched swine lagoon slurry. FEMS Microbiol Ecol 60(2):329–340

Côté F, Fligny C, Bayard E, Launay JM, Gershon MD, Mallet J, Vodjdani G (2007) Maternal serotonin is crucial for murine embryonic development. Proc Natl Acad Sci USA 104(1):329–334

Cox AJ, Pyne DB, Saunders PU, Fricker PA (2010) Oral administration of the probiotic Lactobacillus fermentum VRI-003 and mucosal immunity in endurance athletes. Br J Sports Med 44(4):222–226

Cremon C, Carini G, Wang B, Vasina V, Cogliandro RF, De Giorgio R, Stanghellini V, Grundy D, Tonini M, De Ponti F, Corinaldesi R, Barbara G (2011) Intestinal serotonin release, sen-

sory neuron activation, and abdominal pain in irritable bowel syndrome. Am J Gastroenterol 106(7):1290–1298
Cryan JF, Clarke G, Dinan TG, Schellekens H (2018) A microbial drugstore for motility. Cell Host Microbe 23(6):691–692
Cryan JF, O'Riordan KJ, Cowan CSM, Sandhu KV, Bastiaanssen TFS, Boehme M et al (2019) The microbiota-gut-brain axis. Physiol Rev 99(4):1877–2013
Cuartero MI, Ballesteros I, de la Parra J, Harkin AL, Abautret-Daly A, Sherwin E, Fernandez-Salguero P, Corbi AL, Lizasoain I, Moro MA (2014) L-kynurenine/aryl hydrocarbon receptor pathway mediates brain damage after experimental stroke. Circulation 130(23):2040–2051
Cummings J, Stephen A (2007) Carbohydrate terminology and classification. Eur J Clin Nutr 61(1):S5–S18
Dahl GE, Connor EE (2021) Grand challenges to livestock physiology and management. Front Anim Sci 2. Retrieved from
Dai Z-L, Li X-L, Xi P-B, Zhang J, Wu G, Zhu W-Y (2012a) Metabolism of select amino acids in bacteria from the pig small intestine. Amino Acids 42:1597–1608
Dai Z-L, Li X-L, Xi P-B, Zhang J, Wu G, Zhu W-Y (2012b) Regulatory role for L-arginine in the utilization of amino acids by pig small-intestinal bacteria. Amino Acids 43:233–244
Dai Z-L, Li X-L, Xi P-B, Zhang J, Wu G, Zhu W-Y (2013) L-Glutamine regulates amino acid utilization by intestinal bacteria. Amino Acids 45:501–512
Daly K, Al-Rammahi M, Moran A, Marcello M, Ninomiya Y, Shirazi-Beechey SP (2013) Sensing of amino acids by the gut-expressed taste receptor T1R1-T1R3 stimulates CCK secretion. Am J Physiol-Gastrointest Liver Physiol 304(3):G271–G282
de Brito JS, Borges NA, dos Anjos JS, Nakao LS, Stockier-Pinto MB, Paiva BR, Cardoso-Weide LD, Cardozo LFMD, Mafra D (2019) Aryl hydrocarbon receptor and uremic toxins from the gut microbiota in chronic kidney disease patients: is there a relationship between them? Biochemistry 58(15):2054–2060
De Vadder F, Grasset E, Manneras Holm L, Karsenty G, Macpherson AJ, Olofsson LE, Backhed F (2018) Gut microbiota regulates maturation of the adult enteric nervous system via enteric serotonin networks. Proc Natl Acad Sci USA 115(25):6458–6463
Debnath S, Velagapudi C, Redus L, Thameem F, Kasinath B, Hura CE, Lorenzo C, Abboud HE, O'Connor JC (2017) Tryptophan metabolism in patients with chronic kidney disease secondary to type 2 diabetes: relationship to inflammatory markers. Int J Tryptophan Res 10:1178646917694600
Degroote J, Vergauwen H, Wang W, Van Ginneken C, De Smet S, Michiels J (2020) Changes of the glutathione redox system during the weaning transition in piglets, in relation to small intestinal morphology and barrier function. J Anim Sci Biotechnol 11:1–17
Deguchi T (1975) Ontogenesis of a biological clock for serotonin: acetyl coenzyme A N-acetyltransferase in pineal gland of rat. Proc Natl Acad Sci USA 72(7):2814–2818
Deng S-L, Chen S-R, Wang Z-P, Zhang Y, Tang J-X, Li J, Wang X-X, Cheng J-M, Jin C, Li X-Y, Zhang B-L, Yu K, Lian Z-X, Liu G-S, Liu Y-X (2016) Melatonin promotes development of haploid germ cells from early developing spermatogenic cells of Suffolk sheep under in vitro condition. J Pineal Res 60(4):435–447
Deng SL, Zhang BL, Reiter RJ, Liu YX (2020) Melatonin ameliorates inflammation and oxidative stress by suppressing the p38MAPK signaling pathway in LPS-induced sheep orchitis. Antioxidants 9(12):1277
Desbonnet L, Garrett L, Clarke G, Bienenstock J, Dinan TG (2008) The probiotic Bifidobacteria infantis: an assessment of potential antidepressant properties in the rat. J Psychiatr Res 43(2):164–174
Desbonnet L, Garrett L, Clarke G, Kiely B, Cryan JF, Dinan TG (2010) Effects of the probiotic Bifidobacterium infantis in the maternal separation model of depression. Neuroscience 170(4):1179–1188
Desbonnet L, Clarke G, Traplin A, O'Sullivan O, Crispie F, Moloney RD, Cotter PD, Dinan TG, Cryan JF (2015) Gut microbiota depletion from early adolescence in mice: implications for brain and behaviour. Brain Behav Immun 48:165–173

Di Iorio BR, Rocchetti MT, De Angelis M, Cosola C, Marzocco S, Di Micco L, di Bari I, Accetturo M, Vacca M, Gobbetti M, Di Iorio M, Bellasi A, Gesualdo L (2019) Nutritional therapy modulates intestinal microbiota and reduces serum levels of total and free indoxyl sulfate and p-cresyl sulfate in chronic kidney disease (medika study). J Clin Med 8(9)

Dodd D, Spitzer MH, Van Treuren W, Merrill BD, Hryckowian AJ, Higginbottom SK et al (2017a) A gut bacterial pathway metabolizes aromatic amino acids into nine circulating metabolites. Nature 551(7682):648–652

Dodd D, Spitzer MH, Van Treuren W, Merrill BD, Hryckowian AJ, Higginbottom SK, Le A, Cowan TM, Nolan GP, Fischbach MA, Sonnenburg JL (2017b) A gut bacterial pathway metabolizes aromatic amino acids into nine circulating metabolites. Nature 551(7682):648

Dolivo DM, Larson SA, Dominko T (2018) Tryptophan metabolites kynurenine and serotonin regulate fibroblast activation and fibrosis. Cell Mol Life Sci 75(20):3663–3681

Dona AC, Pages G, Gilbert RG, Kuchel PW (2010) Digestion of starch: in vivo and in vitro kinetic models used to characterise oligosaccharide or glucose release. Carbohydr Polym 80(3):599–617

Dong F, Hao F, Murray IA, Smith PB, Koo I, Tindall AM et al (2020) Intestinal microbiota-derived tryptophan metabolites are predictive of Ah receptor activity. Gut Microbes 12(1):1–24

Doron I, Leonardi I, Li XV, Fiers WD, Semon A, Bialt-DeCelie M et al (2021) Human gut mycobiota tune immunity via CARD9-dependent induction of anti-fungal IgG antibodies. Cell 184(4):1017–1031 e1014

Dou L, Sallee M, Cerini C, Poitevin S, Gondouin B, Jourde-Chiche N, Fallague K, Brunet P, Calaf R, Dussol B, Mallet B, Dignat-George F, Burtey S (2015) The cardiovascular effect of the uremic solute indole-3 acetic acid. J Am Soc Nephrol 26(4):876–887

Drossman DA (2016) Functional gastrointestinal disorders: history, pathophysiology, clinical features and Rome IV. Gastroenterology

Duan X, Chen D, Zheng P, Tian G, Wang J, Mao X et al (2016) Effects of dietary mannan oligosaccharide supplementation on performance and immune response of sows and their offspring. Anim Feed Sci Technol 218:17–25

Duranton F, Cohen G, De Smet R, Rodriguez M, Jankowski J, Vanholder R, Argiles A, Work EUT, G. (2012) Normal and pathologic concentrations of uremic toxins. J Am Soc Nephrol 23(7):1258–1270

Dürk T, Panther E, Müller T, Sorichter S, Ferrari D, Pizzirani C et al (2005) 5-Hydroxytryptamine modulates cytokine and chemokine production in LPS-primed human monocytes via stimulation of different 5-HTR subtypes. Int Immunol 17(5):599–606

El Aidy S, Kunze W, Bienenstock J, Kleerebezem M (2012) The microbiota and the gut-brain axis: insights from the temporal and spatial mucosal alterations during colonisation of the germfree mouse intestine. Benef Microbes 3(4):251–259

Elbeldi-Ferchiou A, Ben Ahmed M, Smiti-Khanfir M, Houman MH, Abdeladhim M, Belhadj Hmida N et al (2011) Resistance to exogenous TGF-beta effects in patients with systemic lupus erythematosus. J Clin Immunol 31(4):574–583

Elizei SS, Poormasjedi-Meibod MS, Wang X, Kheirandish M, Ghahary A (2017) Kynurenic acid downregulates IL-17/1L-23 axis in vitro. Mol Cell Biochem 431(1-2):55–65

Eriksson L, Valtonen M, Laitinen JT, Paananen M, Kaikkonen M (1998) Diurnal rhythm of melatonin in bovine milk: pharmacokinetics of exogenous melatonin in lactating cows and goats. Acta Vet Scand 39(3):301–310

Etienne-Mesmin L, Chassaing B, Gewirtz AT (2017) Tryptophan: a gut microbiota-derived metabolites regulating inflammation. World J Gastrointest Pharmacol Ther 8(1):7–9

Fang J, Wang H, Zhou YP, Zhang H, Zhou HT, Zhang XH (2021) Slimy partners: the mucus barrier and gut microbiome in ulcerative colitis. Exp Mol Med 53(5):772–787

Farzi A, Frohlich EE, Holzer P (2018) Gut microbiota and the neuroendocrine system. Neurotherapeutics 15(1):5–22

Fasano A, Shea-Donohue T (2005) Mechanisms of disease: the role of intestinal barrier function in the pathogenesis of gastrointestinal autoimmune diseases. Nat Clin Pract Gastroenterol Hepatol 2(9):416–422

Feng C, Xie X, Wu M, Li C, Gao M, Liu M, Qi X, Ren J (2013) Tanshinone I protects mice from aristolochic acid I-induced kidney injury by induction of CYP1A. Environ Toxicol Pharmacol 36(3):850–857

Fenghua W, Xinghong Y, Meiping L (2012) Trial analysis of the teaching process of Physiology "Digestion in the stomach". J Hubei Univ Trad Chin Med 14(4):80–80

Fengli W (2014) Composition and action of the gastric juice of monogastric animals. Farm Technol Adviser 10:46–46

Ferrere G, Wrzosek L, Cailleux F, Turpin W, Puchois V, Spatz M, Ciocan D, Rainteau D, Humbert L, Hugot C, Gaudin F, Noordine ML, Robert V, Berrebi D, Thomas M, Naveau S, Perlemuter G, Cassard AM (2017) Fecal microbiota manipulation prevents dysbiosis and alcohol-induced liver injury in mice. J Hepatol 66(4):806–815

Fiorica-Howells E, Maroteaux L, Gershon MD (2000) Serotonin and the 5-HT receptor in the development of enteric neurons. J Neurosci 20(1):294–305

Fleming SA, Monaikul S, Patsavas AJ, Waworuntu RV, Berg BM, Dilger RN (2019) Dietary polydextrose and galactooligosaccharide increase exploratory behavior, improve recognition memory, and alter neurochemistry in the young pig. Nutr Neurosci 22(7):499–512

Fligny C, Fromes Y, Bonnin P, Darmon M, Bayard E, Launay JM et al (2008) Maternal serotonin influences cardiac function in adult offspring. Faseb J 22(7):2340–2349

Flores-Cruz GM, Escobar A (2012) Reduction of serotonergic neurons in the dorsal raphe due to chronic prenatal administration of a tryptophan-free diet. Int J Dev Neurosci 30(2):63–67

Foley S, Garsed K, Singh G, Duroudier NP, Swan C, Hall IP, Zaitoun A, Bennett A, Marsden C, Holmes G, Walls A, Spiller RC (2011) Impaired uptake of serotonin by platelets from patients with irritable bowel syndrome correlates with duodenal immune activation. Gastroenterology 140(5):1434–1443 e1431

Folkes LK, Dennis MF, Stratford MR, Candeias LP, Wardman P (1999) Peroxidase-catalyzed effects of indole-3-acetic acid and analogues on lipid membranes, DNA, and mammalian cells in vitro. Biochem Pharmacol 57(4):375–382

Forsythe P, Sudo N, Dinan T, Taylor VH, Bienenstock J (2010) Mood and gut feelings. Brain Behav Immun 24(1):9–16

Foster JA, Rinaman L, Cryan JF (2017) Stress & the gut-brain axis: regulation by the microbiome. Neurobiol Stress 7:124–136

Fothergill LJ, Galiazzo G, Hunne B, Stebbing MJ, Fakhry J, Weissenborn F et al (2019) Distribution and co-expression patterns of specific cell markers of enteroendocrine cells in pig gastric epithelium. Cell Tissue Res 378:457–469

Fouhse JM, Gänzle MG, Regmi PR, van Kempen TA, Zijlstra RT (2015) High amylose starch with low in vitro digestibility stimulates hindgut fermentation and has a bifidogenic effect in weaned pigs. J Nutr 145(11):2464–2470

Franceschi C, Bonafe M, Valensin S, Olivieri F, De Luca M, Ottaviani E, De Benedictis G (2000) Inflamm-aging. An evolutionary perspective on immunosenescence. Ann N Y Acad Sci 908:244–254

Freewan M, Rees MD, Plaza TS, Glaros E, Lim YJ, Wang XS et al (2013) Human indoleamine 2,3-dioxygenase is a catalyst of physiological heme peroxidase reactions: implications for the inhibition of dioxygenase activity by hydrogen peroxide. J Biol Chem 288(3):1548–1567

Friedman M (2018) Analysis, nutrition, and health Benefits of tryptophan. Int J Tryptophan Res 11:1178646918802282

Friedman SL, Neuschwander-Tetri BA, Rinella M, Sanyal AJ (2018) Mechanisms of NAFLD development and therapeutic strategies. Nat Med 24(7):908–922

Frungieri MB, Artur M, Karina Z, Pignataro OP, Calandra RS, Gonzalez-CalvaR SI (2005) Direct effect of melatonin on Syrian hamster testes: melatonin subtype 1a receptors, inhibition of androgen production, and interaction with the local corticotropin-releasing hormone system. Endocrinology 3:1541–1552

Fujisawa TX, Nishitani S, Iwanaga R, Matsuzaki J, Kawasaki C, Tochigi M, Sasaki T, Kato N, Shinohara K (2016) Association of aryl hydrocarbon receptor-related gene variants with the severity of autism spectrum disorders. Front Psychiatry 7:184

Fulling C, Dinan TG, Cryan JF (2019) Gut microbe to brain signaling: what happens in vagus. Neuron 101(6):998–1002
Galenza A, Moreno-Roman P, Su YH, Acosta-Alvarez L, Debec A, Guichet A et al (2023) Basal stem cell progeny establish their apical surface in a junctional niche during turnover of an adult barrier epithelium. Nat Cell Biol 25(5)
Gao J, Xu K, Liu H, Liu G, Bai M, Peng C et al (2018) Impact of the gut microbiota on intestinal immunity mediated by tryptophan metabolism. Front Cell Infect Microbiol 8:13
Gao K, Pi Y, Mu CL, Farzi A, Liu Z, Zhu WY (2019) Increasing carbohydrate availability in the hindgut promotes hypothalamic neurotransmitter synthesis: aromatic amino acids linking the microbiota-brain axis. J Neurochem 149(5):641–659
Gao K, Mu CL, Farzi A, Zhu WY (2020a) Tryptophan metabolism: a link between the gut microbiota and brain. Adv Nutr 11(3):709–723
Gao X, Yu B, Yu J, Mao X, Huang Z, Luo Y et al (2020b) Influences of dietary starch structure on intestinal morphology, barrier functions, and epithelium apoptosis in weaned pigs. Food Funct 11(5):4446–4455
Garcia-Lara L, Perez-Severiano F, Gonzalez-Esquivel D, Elizondo G, Segovia J (2015) Absence of aryl hydrocarbon receptors increases endogenous kynurenic acid levels and protects mouse brain against excitotoxic insult and oxidative stress. J Neurosci Res 93(9):1423–1433
Garrett JR, Ekström J, Anderson LC (1999) Neural mechanisms of salivary gland secretion. Karger Medical and Scientific Publishers
Gaspar P, Cases O, Maroteaux L (2003) The developmental role of serotonin: news from mouse molecular genetics. Nat Rev Neurosci 4(12):1002–1012
Geng D, Li XF, Bi YL, Jiang Y, Wang Z, Chang G et al (2023) Research progress on microbial-gut-brain axis regulation mechanism in livestock and poultry. J Agric Biotechnol 30(02):404–415
Ghia JE, Li N, Wang HQ, Collins M, Deng YK, El-Sharkawy RT, Cote F, Mallet J, Khan WI (2009) Serotonin has a key role in pathogenesis of experimental colitis. Gastroenterology 137(5):1649–1660
Gilbert E, Wong E, Webb K Jr (2008) Board-invited review: peptide absorption and utilization: implications for animal nutrition and health. J Anim Sci 86(9):2135–2155
Gillam EM, Notley LM, Cai H, De Voss JJ, Guengerich FP (2000) Oxidation of indole by cytochrome P450 enzymes. Biochemistry 39(45):13817–13824
Gkouskou KK, Deligianni C, Tsatsanis C, Eliopoulos AG (2014) The gut microbiota in mouse models of inflammatory bowel disease. Front Cell Infect Microbiol 4:28
Golubeva AV, Joyce SA, Moloney G, Burokas A, Sherwin E, Arboleya S et al (2017) Microbiota-related changes in bile acid & tryptophan metabolism are associated with gastrointestinal dysfunction in a mouse model of autism. EBioMedicine 24:166–178
Gordon JI (2012) Honor thy gut symbionts redux. Science 336(6086):1251–1253
Gou H-Z, Zhang Y-L, Ren L-F, Li Z-J, Zhang L (2022) How do intestinal probiotics restore the intestinal barrier? Front Microbiol 13:929346
Greenhill C (2014) IBD: glucocorticoids revealed to augment intestinal epithelial barrier function. Nat Rev Gastroenterol Hepatol 11(2):75
Greig CJ, Zhang L, Cowles RA (2019) Potentiated serotonin signaling in serotonin re-uptake transporter knockout mice increases enterocyte mass and small intestinal absorptive function. Physiol Rep 7(21):e14278
Gribble FM, Reimann F (2016) Enteroendocrine cells: chemosensors in the intestinal epithelium. Annu Rev Physiol 78(78):277–299
Grivennikov SI, Wang K, Mucida D, Stewart CA, Schnabl B, Jauch D et al (2012) Adenoma-linked barrier defects and microbial products drive IL-23/IL-17-mediated tumour growth. Nature 491(7423):254–258
Gross ER, Gershon MD, Margolis KG, Gertsberg ZV, Cowles RA (2012) Neuronal serotonin regulates growth of the intestinal mucosa in mice. Gastroenterology 143(2):408-+
Gu K, Liu GM, Wu CM, Jia G, Zhao H, Chen XL et al (2021a) Tryptophan improves porcine intestinal epithelial cell restitution through the CaSR/Rac1/PLC-γ1 signaling pathway. Food Funct 12(18):8787–8799

Gu ZL, Li FY, Liu YH, Jiang MW, Zhang LH, He LQ, Wilkey DW, Merchant M, Zhang X, Deng ZB, Chen SY, Barve S, McClain CJ, Feng WK (2021b) Exosome-like nanoparticles from Lactobacillus rhamnosus GG protect against alcohol-associated liver disease through intestinal aryl hydrocarbon receptor in mice. Hepatol Commun 5(5):846–864

Guangyan Y (2003) Current status of research on salivary gland diseases in China. Chin J Oral Med 38(2):81–84

Guo Z, Zhao N, Wang B, Lv J, Li H, He Z (2021) Progress in investigating the role of tryptophan and its metabolites in the animal gut-brain axis. China Anim Husband J 57(09):15–21

Gupta NK, Thaker AI, Kanuri N, Riehl TE, Rowley CW, Stenson WF, Ciorba MA (2012) Serum analysis of tryptophan catabolism pathway: correlation with Crohn's disease activity. Inflamm Bowel Dis 18(7):1214–1220

Gurkar AU, Niedernhofer LJ (2015) Comparison of mice with accelerated aging caused by distinct mechanisms. Exp Gerontol 68:43–50

Gutierrez-Vazquez C, Quintana FJ (2018) Regulation of the immune response by the aryl hydrocarbon receptor. Immunity 48(1):19–33

Hamard A, Sève B, Le Floc'h, N. (2007) Intestinal development and growth performance of early-weaned piglets fed a low-threonine diet. Animal 1(8):1134–1142

Hampson D, Kidder D (1986) Influence of creep feeding and weaning on brush border enzyme activities in the piglet small intestine. Res Vet Sci 40(1):24–31

Hannon J, Hoyer D (2008) Molecular biology of 5-HT receptors. Behav Brain Res 195(1):198–213

Harrington L, Srikanth CV, Antony R, Rhee SJ, Mellor AL, Shi HN, Cherayil BJ (2008) Deficiency of indoleamine 2, 3-dioxygenase enhances commensal-induced antibody responses and protects against Citrobacter rodentium-induced colitis. Infect Immun 76(7):3045–3053

Hartmann C, Thüring E-M, Greune L, Michels BE, Pajonczyk D, Leußink S et al (2022) Intestinal brush border formation requires a TMIGD1-based intermicrovillar adhesion complex. Sci Signal 15(751):eabm2449

Hashimoto T, Perlot T, Rehman A, Trichereau J, Ishiguro H, Paolino M et al (2012) ACE2 links amino acid malnutrition to microbial ecology and intestinal inflammation. Nature 487(7408):477–481

Hata T, Asano Y, Yoshihara K, Kimura-Todani T, Miyata N, Zhang XT et al (2017) Regulation of gut luminal serotonin by commensal microbiota in mice. PLoS One 12(7):e0180745

Hendrikx T, Duan Y, Wang Y, Oh JH, Alexander LM, Huang W, Starkel P, Ho SB, Gao B, Fiehn O, Emond P, Sokol H, van Pijkeren JP, Schnabl B (2019) Bacteria engineered to produce IL-22 in intestine induce expression of REG3G to reduce ethanol-induced liver disease in mice. Gut 68(8):1504–1515

Henning SJ (1985) Ontogeny of enzymes in the small intestine. Annu Rev Physiol 47(1):231–245

Hernandez LL, Stiening CM, Wheelock JB, Baumgard LH, Parkhurst AM, Collier RJ (2008) Evaluation of serotonin as a feedback inhibitor of lactation in the bovine. J Dairy Sci 91(5):1834–1844

Hernandez LL, Limesand SW, Collier JL, Horseman ND, Collier RJ (2009) The bovine mammary gland expresses multiple functional isoforms of serotonin receptors. J Endocrinol 203(1):123–131

Hernandez LL, Gregerson KA, Horseman ND (2012) Mammary gland serotonin regulates parathyroid hormone-related protein and other bone-related signals. Am J Physiol Endocrinol Metabol 302(8):E1009–E1015

Högberg A, Lindberg JE (2004) Influence of cereal non-starch polysaccharides and enzyme supplementation on digestion site and gut environment in weaned piglets. Anim Feed Sci Technol 116(1-2):113–128

Hold GL (2016) Gastrointestinal microbiota and colon cancer. Dig Dis 34(3):244–250

Hongyang W, Zhenjun B (1998) Limiting factors affecting crude fibre digestion and utilisation. Anim Husband Abroad Feeds 5:13–16

Honório de Melo Martimiano P, de Sa Braga Oliveira A, Ferchaud-Roucher V, Croyal M, Aguesse A, Grit I et al (2017) Maternal protein restriction during gestation and lactation in the rat

results in increased brain levels of kynurenine and kynurenic acid in their adult offspring. J Neurochem 140(1):68–81
Horowitz A, Chanez-Paredes SD, Haest X, Turner JR (2023) Paracellular permeability and tight junction regulation in gut health and disease. Nat Rev Gastroenterol Hepatol 20(7):417–432
Hou YQ, Yin YL, Wu GY (2015) Dietary essentiality of "nutritionally non-essential amino acids" for animals and humans. Exp Biol Med 240(8):997–1007
Hsu P, Nanan R (2014) Foetal immune programming: hormones, cytokines, microbes and regulatory T cells. J Reprod Immunol 104-105:2–7
Hua Z, Bing Y, Dongting H, Daiwen C (2022) Advances in carbohydrate nutritional balance and intestinal health in piglets. J Anim Nutr 34(10):6160–6166. Retrieved from
Hubbard TD, Murray IA, Perdew GH (2015a) Indole and tryptophan metabolism: endogenous and dietary routes to Ah receptor activation. Drug Metab Dispos 43(10):1522–1535
Hubbard TD, Murray IA, Bisson WH, Lahoti TS, Gowda K, Amin SG et al (2015b) Adaptation of the human aryl hydrocarbon receptor to sense microbiota-derived indoles. Sci Rep 5(1):12689
Huether G, Thömke F, Adler L (1992) Administration of tryptophan-enriched diets to pregnant rats retards the development of the serotonergic system in their offspring. Brain Res Dev Brain Res 68(2):175–181
Hurley RL, Anderson KA, Franzone JM, Kemp BE, Means AR, Witters LA (2005) The Ca2+/calmodulin-dependent protein kinase kinases are AMP-activated protein kinase kinases*. J Biolo Chem 280(32):29060–29066
Hussain T, Tan B, Ren W, Rahu N, Dad R, Kalhoro DH, Yin Y (2017) Polyamines: therapeutic perspectives in oxidative stress and inflammatory diseases. Amino Acids 49:1457–1468
Huycke TR, Miller BM, Gill HK, Nerurkar NL, Sprinzak D, Mahadevan L, Tabin CJ (2019) Genetic and mechanical regulation of intestinal smooth muscle development. Cell 179(1):90
Hwang SJ, Hwang YJ, Yun MO, Kim JH, Oh GS, Park JH (2013) Indoxyl 3-sulfate stimulates Th17 differentiation enhancing phosphorylation of c-Src and STAT3 to worsen experimental autoimmune encephalomyelitis. Toxicol Lett 220(2):109–117
Ihekweazu FD, Versalovic J (2018) Development of the pediatric gut microbiome: impact on health and disease. Am J Med Sci 356(5):413–423
Illes P, Krasulova K, Vyhlidalova B, Poulikova K, Marcalikova A, Pecinkova P, Sirotova N, Vrzal R, Mani S, Dvorak Z (2020) Indole microbial intestinal metabolites expand the repertoire of ligands and agonists of the human pregnane X receptor. Toxicol Lett 334:87–93
Illnerová H, Buresová M, Presl J (1993) Melatonin rhythm in human milk. J Clin Endocrinol Metabol 77(3):838–841
Ishizuka B, Fusama S, Hirai K, Hosaka T, Hamada N, Amemiya A, Itoh MT (2000) Melatonin secretion from organ-cultured pineal glands of rats: modulation by gonadectomy and gonadotropin-releasing hormone agonist administration. Eur J Endocrinol 142(4):387–392
Islam J, Sato S, Watanabe K, Watanabe T, Ardiansyah HK et al (2017) Dietary tryptophan alleviates dextran sodium sulfate-induced colitis through aryl hydrocarbon receptor in mice. J Nutr Biochem 42:43–50
Israelyan N, Margolis KG (2018) Serotonin as a link between the gut-brain-microbiome axis in autism spectrum disorders. Pharmacol Res 132:1–6
Jaglin M, Rhimi M, Philippe C, Pons N, Bruneau A, Goustard B, Dauge V, Maguin E, Naudon L, Rabot S (2018) Indole, a signaling molecule produced by the gut microbiota, negatively impacts emotional behaviors in rats. Front Neurosci 12:216
Jaldo-Alba F, Muñóz-Hoyos A, Molina-Carballo A, Molina-Font JA, Acuña-Castroviejo D (1993) Light deprivation increases plasma levels of melatonin during the first 72 h of life in human infants. Acta Endocrinol (Copenh) 129(5):442–445
Jansen M, Bang Jensen B, McLoughlin M, Rodger H, Taksdal T, Sindre H et al (2017) The epidemiology of pancreas disease in salmonid aquaculture: a summary of the current state of knowledge. J Fish Dis 40(1):141–155
Jenkins TA, Nguyen JC, Polglaze KE, Bertrand PP (2016) Influence of tryptophan and serotonin on mood and cognition with a possible role of the gut-brain axis. Nutrients 8(1)

Jennis M, Cavanaugh CR, Leo GC, Mabus JR, Lenhard J, Hornby PJ (2018) Microbiota-derived tryptophan indoles increase after gastric bypass surgery and reduce intestinal permeability in vitro and in vivo. Neurogastroenterol Motil 30(2)

Ji Y, Gao Y, Chen H, Yin Y, Zhang W (2019) Indole-3-acetic acid alleviates nonalcoholic fatty liver disease in mice via attenuation of hepatic lipogenesis, and oxidative and inflammatory stress. Nutrients 11(9)

Ji Y, Hou YQ, Blachier F, Wu ZL (2023) Editorial: amino acids in intestinal growth and health. Front Nutr 10

Jiang Q, Adebowale TO, Tian J, Yin Y, Yao K (2020) Effects of dietary alpha-ketoglutarate on bacteria profiles in the faeces of lactating sows and their suckling piglets. Arch Anim Nutr 74(1):39–56

Jianying Y, Maoyong F, Zhouwei X, Lei L (2013) A study of serum IL-1β, IL-6 and IL-10 in the prognosis of acute pancreatitis. Anhui Med 17(7):1185–1187

Jin UH, Lee SO, Sridharan G, Lee K, Davidson LA, Jayaraman A, Chapkin RS, Alaniz R, Safe S (2014) Microbiome-derived tryptophan metabolites and their aryl hydrocarbon receptor-dependent agonist and antagonist activities. Mol Pharmacol 85(5):777–788

Jin DC, Cao HL, Xu MQ, Wang SN, Wang YM, Yan F, Wang BM (2016) Regulation of the serotonin transporter in the pathogenesis of irritable bowel syndrome. World J Gastroenterol 22(36):8137–8148

Jin UH, Cheng Y, Park H, Davidson LA, Callaway ES, Chapkin RS, Jayaraman A, Asante A, Allred C, Weaver EA, Safe S (2017) Short chain fatty acids enhance aryl hydrocarbon (Ah) responsiveness in mouse colonocytes and Caco-2 human colon cancer cells. Sci Rep 7(1):10163

Johnston JD, Klosen P, Barrett P, Hazlerigg DG (2006) Regulation of MT melatonin receptor expression in the foetal rat pituitary. J Neuroendocrinol 18(1):50–56

Joshi AD, Carter DE, Harper TA Jr, Elferink CJ (2015) Aryl hydrocarbon receptor-dependent stanniocalcin 2 induction by cinnabarinic acid provides cytoprotection against endoplasmic reticulum and oxidative stress. J Pharmacol Exp Ther 353(1):201–212

Juricek L, Coumoul X (2018) The aryl hydrocarbon receptor and the nervous system. Int J Mol Sci 19(9)

Kaczor, M., & Skalski, M. (2016). Treatment of behavioral sleep problems in children

Kaluzna-Czaplinska J, Gatarek P, Chirumbolo S, Chartrand MS, Bjorklund G (2019) How important is tryptophan in human health? Critical Rev Food Sci Nutr 59(1):72–88

Kaneko JJ, Harvey JW, Bruss ML (2008) Clinical biochemistry of domestic animals. Academic Press

Kang KY, Lee SH, Jung SM, Park SH, Jung BH, Ju JH (2015) Downregulation of tryptophan-related metabolomic profile in rheumatoid arthritis synovial fluid. J Rheumatol 42(11):2003–2011

Kasuya E, Kushibiki S, Yayou KI, Ito S, Saito T, Hodate K, Sutoh M (2008) Effects of serotonin injected into the third ventricle on prolactin and growth hormone secretion in Holstein steers. Anim Sci J 79(3):362

Kasuya E, Yayou KI, Hashizume T, Kitagawa S, Sutoh M (2010) A possible role of central serotonin in L-tryptophan-induced GH secretion in cattle. Anim Sci J 81(3):345–351

Kawai S, Iijima H, Shinzaki S, Hiyama S, Yamaguchi T, Araki M, Iwatani S, Shiraishi E, Mukai A, Inoue T, Hayashi Y, Tsujii M, Motooka D, Nakamura S, Iida T, Takehara T (2017) Indigo Naturalis ameliorates murine dextran sodium sulfate-induced colitis via aryl hydrocarbon receptor activation. J Gastroenterol 52(8):904–919

Keating DJ, Spencer NJ (2019) What is the role of endogenous gut serotonin in the control of gastrointestinal motility? Pharmacol Res 140:50–55

Kelly JR, Borre Y, O'Brien C, Patterson E, El Aidy S, Deane J, Kennedy PJ, Beers S, Scott K, Moloney G, Hoban AE, Scott L, Fitzgerald P, Ross P, Stanton C, Clarke G, Cryan JF, Dinan TG (2016) Transferring the blues: depression-associated gut microbiota induces neurobehavioural changes in the rat. J Psychiatr Res 82:109–118

Kendig DM, Grider JR (2015) Serotonin and colonic motility. Neurogastroenterol Motil 27(7):899–905

Kennedy PJ, Cryan JF, Dinan TG, Clarke G (2017) Kynurenine pathway metabolism and the microbiota-gut-brain axis. Neuropharmacology 112(Pt B):399–412

Kerckhoffs AP, ter Linde JJ, Akkermans LM, Samsom M (2012) SERT and TPH-1 mRNA expression are reduced in irritable bowel syndrome patients regardless of visceral sensitivity state in large intestine. Am J Physiol Gastrointest Liver Physiol 302(9):G1053–G1060

Keszthelyi D, Troost FJ, Jonkers DM, Kruimel JW, Leue C, Masclee AA (2013) Decreased levels of kynurenic acid in the intestinal mucosa of IBS patients: relation to serotonin and psychological state. J Psychosom Res 74(6):501–504

Khan WI, Motomura Y, Wang H, El-Sharkawy RT, Verdu EF, Verma-Gandhu M, Rollins BJ, Collins SM (2006) Critical role of MCP-1 in the pathogenesis of experimental colitis in the context of immune and enterochromaffin cells. Am J Physiol Gastrointest Liver Physiol 291(5):G803–G811

Kiely CJ, Pavli P, O'Brien CL (2018) The role of inflammation in temporal shifts in the inflammatory bowel disease mucosal microbiome. Gut Microbes 9(6):477–485

Kim W-T, Shinde P, Chae BJ (2008) Effect of lecithin with or without chitooligosaccharide on the growth performance, nutrient digestibility, blood metabolites and pork quality of finishing pigs. Can J Anim Sci 88(2):283–292

Kim CJ, Kovacs-Nolan JA, Yang C, Archbold T, Fan MZ, Mine Y (2010) l-Tryptophan exhibits therapeutic function in a porcine model of dextran sodium sulfate (DSS)-induced colitis. J Nutr Biochem 21(6):468–475

Kimura I, Ozawa K, Inoue D, Imamura T, Kimura K, Maeda T et al (2013) The gut microbiota suppresses insulin-mediated fat accumulation via the short-chain fatty acid receptor GPR43. Nat Commun 4:1829

Kleiman SC, Bulik-Sullivan EC, Glenny EM, Zerwas SC, Huh EY, Tsilimigras MCB et al (2019) Correction: the gut-brain axis in healthy females: lack of significant association between microbial composition and diversity with psychiatric measures. PLoS One 14(8):e0221724

Knudsen KB, Hansen I (1991) Gastrointestinal implications in pigs of wheat and oat fractions: 1. Digestibility and bulking properties of polysaccharides and other major constituents. Br J Nutr 65(2):217–232

Knudsen KEB, Hedemann MS, Lærke HN (2012) The role of carbohydrates in intestinal health of pigs. Anim Feed Sci Technol 173(1-2):41–53

Kolhekar R, Gebhart GF (1996) Modulation of spinal visceral nociceptive transmission by NMDA receptor activation in the rat. J Neurophysiol 75(6):2344–2353

Kong X, Wang X, Yin Y, Li X, Gao H, Bazer FW, Wu G (2014) Putrescine stimulates the mTOR signaling pathway and protein synthesis in porcine trophectoderm cells. Biol Reprod 91(5):106–110

Korecka A, Dona A, Lahiri S et al (2016) Bidirectional communication between the Aryl hydrocarbon Receptor (AhR) and the microbiome tunes host metabolism NPJ. Biofilms Microbiomes 2:16014

Koski KG, Lanoue L, Young SN (1993) Restriction of maternal dietary carbohydrate decreases fetal brain indoles and glycogen in rats. J Nutr 123(1):42–51

Kostic AD, Xavier RJ, Gevers D (2014) The microbiome in inflammatory bowel disease: current status and the future ahead. Gastroenterology 146(6):1489–1499

Kovacs CS, Fuleihan GE-H (2006) Calcium and bone disorders during pregnancy and lactation. Endocrinol Metabol Clin 35(1):21–51

Krishnan S, Ding Y, Saedi N, Choi M, Sridharan GV, Sherr DH, Yarmush ML, Alaniz RC, Jayaraman A, Lee K (2018) Gut microbiota-derived tryptophan metabolites modulate inflammatory response in hepatocytes and macrophages. Cell Rep 23(4):1099–1111

Krishnan S, Ding Y, Saeidi N, Choi M, Sridharan GV, Sherr DH, Yarmush ML, Alaniz RC, Jayaraman A, Lee K (2019) Gut microbiota-derived tryptophan metabolites modulate inflammatory response in hepatocytes and macrophages. Cell Rep 28(12):3285

Kudo Y (2013) The role of placental indoleamine 2,3-dioxygenase in human pregnancy. Obstet Gynecol Sci 56(4):209–216

Kudo Y, Boyd CA (2001) Characterisation of L-tryptophan transporters in human placenta: a comparison of brush border and basal membrane vesicles. J Physiol 531(Pt 2):405–416

Kumar D, Sharma S, Kalra S, Singh G, Monga V, Kumar B (2020) Medicinal perspective of indole derivatives: recent developments and structure-activity relationship studies. Curr Drug Targets 21(9):864–891

Kwon YH, Wang H, Denou E, Ghia JE, Rossi L, Fontes ME et al (2019) Modulation of gut microbiota composition by serotonin signaling influences intestinal immune response and susceptibility to colitis. Cell Mol Gastroenterol Hepatol 7(4):709–728

Kyritsi K, Chen L, O'Brien A, Francis H, Hein TW, Venter J, Wu N, Ceci L, Zhou T, Zawieja D, Gashev AA, Meng F, Invernizzi P, Fabris L, Wu C, Skill NJ, Saxena R, Liangpunsakul S, Alpini G, Glaser SS (2020) Modulation of the tryptophan hydroxylase 1/monoamine oxidase-A/5-hydroxytryptamine/5-hydroxytryptamine receptor 2A/2B/2C axis regulates biliary proliferation and liver fibrosis during cholestasis. Hepatology 71(3):990–1008

Lake JI, Heuckeroth RO (2013) Enteric nervous system development: migration, differentiation, and disease. Am J Physiol Gastrointest Liver Physiol 305(1):G1–G24

Lamas B, Richard ML, Leducq V, Pham HP, Michel ML, Da Costa G, Bridonneau C, Jegou S, Hoffmann TW, Natividad JM, Brot L, Taleb S, Couturier-Maillard A, Nion-Larmurier I, Merabtene F, Seksik P, Bourrier A, Cosnes J, Ryffel B et al (2016) CARD9 impacts colitis by altering gut microbiota metabolism of tryptophan into aryl hydrocarbon receptor ligands. Nat Med 22(6):598

Lamas B, Richard ML, Sokol H (2017) Caspase recruitment domain 9, microbiota, and tryptophan metabolism: dangerous liaisons in inflammatory bowel diseases. Curr Opin Clin Nutr Metab Care 20(4):243–247

Laporta J, Peters TL, Weaver SR, Merriman KE, Hernandez LL (2013) Feeding 5-hydroxy-l-tryptophan during the transition from pregnancy to lactation increases calcium mobilization from bone in rats. Domestic animal endocrinology 44(4):176–184

Laporta J, Keil KP, Vezina CM, Hernandez LL (2014a) Peripheral serotonin regulates maternal calcium trafficking in mammary epithelial cells during lactation in mice. PLoS One 9(10):e110190. Retrieved 2014, from

Laporta J, Keil KP, Weaver SR, Cronick CM, Prichard AP, Crenshaw TD, Heyne GW, Vezina CM, Lipinski RJ, Hernandez LL (2014b) Serotonin regulates calcium homeostasis in lactation by epigenetic activation of hedgehog signaling. Molecular endocrinology (Baltimore, Md.) 28(11):1866–1874

Laursen MF, Sakanaka M, von Burg N, Mörbe U, Andersen D, Moll JM et al (2021) Bifidobacterium species associated with breastfeeding produce aromatic lactic acids in the infant gut. Nat Microbiol 6(11):1367–1382

Lavdas AA, Blue ME, Lincoln J, Parnavelas JG (1997) Serotonin promotes the differentiation of glutamate neurons in organotypic slice cultures of the developing cerebral cortex. J Neurosci 17(20):7872–7880

Lawrence RC, Felson DT, Helmick CG, Arnold LM, Choi H, Deyo RA et al (2008) Estimates of the prevalence of arthritis and other rheumatic conditions in the United States. Part II. Arthritis Rheum 58(1):26–35

Lee JS, Cella M, McDonald KG, Garlanda C, Kennedy GD, Nukaya M et al (2012) AHR drives the development of gut ILC22 cells and postnatal lymphoid tissues via pathways dependent on and independent of Notch. Nat Immunol 13(2):144–151

Lee YH, Lin CH, Hsu PC, Sun YY, Huang YJ, Zhuo JH, Wang CY, Gan YL, Hung CC, Kuan CY, Shie FS (2015) Aryl hydrocarbon receptor mediates both pro-inflammatory and anti-inflammatory effects in lipopolysaccharide-activated microglia. Glia 63(7):1138–1154

Lee WJ, Liu SH, Chiang CK, Lin SY, Liang KW, Chen CH, Tien HR, Chen PH, Wu JP, Tsai YC, Lai DW, Chang YC, Sheu WH, Sheu ML (2016) Aryl hydrocarbon receptor deficiency attenuates oxidative stress-related mesangial cell activation and macrophage infiltration and extracellular matrix accumulation in diabetic nephropathy. Antioxid Redox Signal 24(4):217–231

Lee C, Doo E, Choi JM, Jang SH, Ryu HS, Lee JY et al (2017) The increased level of depression and anxiety in irritable bowel syndrome patients compared with healthy controls: systematic review and meta-analysis. J Neurogastroenterol Motil 23(3):349–362

Lee JY, Li S, Shin NE, Na Q, Dong J, Jia B, Jones-Beatty K, McLane MW, Ozen M, Lei J, Burd I (2019) Melatonin for prevention of placental malperfusion and fetal compromise associated with intrauterine inflammation-induced oxidative stress in a mouse model. J Pineal Res 67(3):e12591

Levin AD, van den Brink GR (2014) Selective inhibition of mucosal serotonin as treatment for IBD? Gut 63(6):866–867

Lewis AJ, Speer VC (1974) Tryptophan requirement of the lactating sow. J Anim Sci 38(4):778–784

Li HS, Wang HB (2010) Activation of xenobiotic receptors: driving into the nucleus. Exp Opin Drug Metabol Toxicol 6(4):409–426

Li YZ, Baidoo SK, Johnston LJ, Anderson JE (2011) Effects of tryptophan supplementation on aggression among group-housed gestating sows. J Anim Sci 89(6):1899–1907

Li S, Zhao S, Luo J, Zhou H, Xiao Z, Weng Y et al (2020) Estrogen induces indoleamine 2,3-dioxygenase expression via suppressors of cytokine signaling 3 in the chorionic villi and decidua of women in early pregnancy. Am J Reprod Immunol 83(1):e13197

Li JJ, Zhang L, Wu T, Li YF, Zhou XJ, Ruan Z (2021a) Indole-3-propionic acid improved the intestinal barrier by enhancing epithelial barrier and mucus barrier. J Agric Food Chem 69(5):1487–1495

Li XJ, Zhang BB, Hu YY, Zhao Y (2021b) New insights into gut-bacteria-derived indole and its derivatives in intestinal and liver diseases. Front Pharmacol 12:769501. https://doi.org/10.3389/fphar.2021.769501

Liang HW, Dai ZL, Liu N, Ji Y, Chen JQ, Zhang YC et al (2018) Dietary L-tryptophan modulates the structural and functional composition of the intestinal microbiome in weaned piglets. Front Microbiol:9

Liang HW, Dai ZL, Kou J, Sun KJ, Chen JQ, Yang Y et al (2019) Dietary L-tryptophan supplementation enhances the intestinal mucosal barrier function in weaned piglets: implication of tryptophan-metabolizing microbiota. Int J Mol Sci 20(1)

Lim J, Luderer U (2011) Oxidative damage increases and antioxidant gene expression decreases with aging in the mouse ovary. Biol Reprod 84(4):775–782

Lim Y, Williams M, Lentle R, Janssen P, Mansel B, Keen S, Chambers P (2013) An exploration of the microrheological environment around the distal ileal villi and proximal colonic mucosa of the possum (Trichosurus vulpecula). J Roy Soc Interface 10(81):20121008

Lim CK, Essa MM, de Paula Martins R, Lovejoy DB, Bilgin AA, Waly MI, Al-Farsi YM, Al-Sharbati M, Al-Shaffae MA, Guillemin GJ (2016) Altered kynurenine pathway metabolism in autism: implication for immune-induced glutamatergic activity. Autism Res 9(6):621–631

Lin YT, Wu PH, Tsai YC, Hsu YL, Wang HY, Kuo MC, Kuo PL, Hwang SJ (2019) Indoxyl sulfate induces apoptosis through oxidative stress and mitogen-activated protein kinase signaling pathway inhibition in human astrocytes. J Clin Med 8(2):191. https://doi.org/10.3390/jcm8020191

Liu MT, Kuan YH, Wang JW, Hen R, Gershon MD (2009) 5-HT receptor-mediated neuroprotection and neurogenesis in the enteric nervous system of adult mice. J Neurosci 29(31):9683–9699

Liu X, Shin N, Koblish HK, Yang G, Wang Q, Wang K et al (2010) Selective inhibition of IDO1 effectively regulates mediators of antitumor immunity. Blood 115(17):3520–3530

Liu Y, Liang X, Dong W, Fang Y, Lv J, Zhang T, Fiskesund R, Xie J, Liu J, Yin X, Jin X, Chen D, Tang K, Ma J, Zhang H, Yu J, Yan J, Liang H, Mo S et al (2018) Tumor-repopulating cells induce PD-1 expression in CD8(+) T cells by transferring kynurenine and AhR activation. Cancer Cell 33(3):480–494 e487

Liu JB, Zhang Y, Li Y, Yan HL, Zhang HF (2019) L-tryptophan enhances intestinal integrity in diquat-challenged piglets associated with improvement of redox status and mitochondrial function. Animals 9(5):266

Liu F, Sun C, Chen Y, Du F, Yang Y, Wu G (2021a) Indole-3-propionic acid-aggravated CCl(4)-induced liver fibrosis via the TGF-beta1/Smads signaling pathway. J Clin Transl Hepatol 9(6):917–930

Liu JR, Miao H, Deng DQ, Vaziri ND, Li P, Zhao YY (2021b) Gut microbiota-derived tryptophan metabolism mediates renal fibrosis by aryl hydrocarbon receptor signaling activation. Cell Mol Life Sci 78(3):909–922

Liu N, Sun SQ, Wang PJ, Sun YA, Hu QJ, Wang XY (2021c) The mechanism of secretion and metabolism of gut-derived 5-hydroxytryptamine. Int J Mol Sci 22(15)

Liu Z, Mu C, Zhu W (2022) Progress in the interaction between gut microbes and serotonin. J Microbiol 62(04):1334–1344

Llopis M, Antolin M, Carol M, Borruel N, Casellas F, Martinez C et al (2009) Lactobacillus casei downregulates commensals' inflammatory signals in Crohn's disease mucosa. Inflamm Bowel Dis 15(2):275–283

Lloyd DA, Gabe SM (2008) Intestinal morphology, intestinal regeneration and the promise of tissue engineering. Intestinal failure: diagnosis, management and transplantation. Wiley-Blackwell, p 13

Lobel L, Cao YG, Fenn K, Glickman JN, Garrett WS (2020) Diet posttranslationally modifies the mouse gut microbial proteome to modulate renal function. Science 369(6510):1518–1524

Long YS, Peng H, Yingjun Z, Zhe Y (2022) Plant extracts for healthy farming through inflammation control. Feed Ind 43(02):1–7. Retrieved from. https://doi.org/10.13302/j.cnki.fi.2022.02.001

Lowe MM, Mold JE, Kanwar B, Huang Y, Louie A, Pollastri MP et al (2014) Identification of cinnabarinic acid as a novel endogenous aryl hydrocarbon receptor ligand that drives IL-22 production. PLoS One 9(2):e87877

Lu P, Yamaguchi Y, Fulton WB, Wang S, Zhou Q, Jia H et al (2021) Maternal aryl hydrocarbon receptor activation protects newborns against necrotizing enterocolitis. Nat Commun 12(1):1042

Lukes J, Stensvold CR, Jirku-Pomajbikova K, Wegener Parfrey L (2015) Are human intestinal eukaryotes beneficial or commensals? PLoS Pathog 11(8):e1005039

Luna RA, Oezguen N, Balderas M, Venkatachalam A, Runge JK, Versalovic J, Veenstra-VanderWeele J, Anderson GM, Savidge T, Williams KC (2017) Distinct microbiome-neuroimmune signatures correlate with functional abdominal pain in children with autism spectrum disorder. Cell Mol Gastroenterol Hepatol 3(2):218–230

Lund ML, Egerod KL, Engelstoft MS, Dmytriyeva O, Theodorsson E, Patel BA, Schwartz TW (2018) Enterochromaffin 5-HT cells – a major target for GLP-1 and gut microbial metabolites. Mol Metabol 11:70–83

Ma H, Zhang W, Zhu X, Song W, Liu J, Jia Z (2010) Effects of rumen-protected tryptophan on growth performance, fibre characteristics, nutrient utilization and plasma essential amino acids in Cashmere goats during the cashmere slow growth period. Livest Sci 131(2–3):227–233

Ma N, Guo P, Zhang J, He T, Kim SW, Zhang G, Ma X (2018) Nutrients mediate intestinal bacteria-mucosal immune crosstalk. Front Immunol 9:5

Ma L, Li H, Hu J, Zheng J, Zhou J, Botchlett R, Matthews D, Zeng T, Chen L, Xiao X, Athrey G, Threadgill DW, Li Q, Glaser S, Francis H, Meng F, Li Q, Alpini G, Wu C (2020a) Indole alleviates diet-induced hepatic steatosis and inflammation in a manner involving myeloid cell 6-phosphofructo-2-kinase/fructose-2,6-biphosphatase 3. Hepatology 72(4):1191–1203

Ma N, He T, Johnston LJ, Ma X (2020b) Host-microbiome interactions: the aryl hydrocarbon receptor as a critical node in tryptophan metabolites to brain signaling. Gut Microbes 11(5):1203–1219

Mackowiak B, Wang H (2016) Mechanisms of xenobiotic receptor activation: direct vs. indirect. Biochim Biophys Acta 1859(9):1130–1140

Madella AM, Van Bergenhenegouwen J, Garssen J, Masereeuw R, Overbeek SA (2022) Microbial-derived tryptophan catabolites, kidney disease and gut inflammation. Toxins (Basel) 14(9)

Maes M, Ombelet W, Verkerk R, Bosmans E, Scharpé S (2001) Effects of pregnancy and delivery on the availability of plasma tryptophan to the brain: relationships to delivery-induced immune activation and early post-partum anxiety and depression. Psychol Med 31(5):847–858

Maltepe E, Fisher SJ (2015) Placenta: the forgotten organ. Annu Rev Cell Dev Biol 31:523–552
Manocha M, Khan WI (2012) Serotonin and GI disorders: an update on clinical and experimental studies. Clin Transl Gastroenterol 3(4):e13
Manzella C, Singhal M, Alrefai WA, Saksena S, Dudeja PK, Gill RK (2018) Serotonin is an endogenous regulator of intestinal CYP1A1 via AhR. Sci Rep 8:6103. https://doi.org/10.1038/s41598-018-24213-5
Margolis KG, Li Z, Stevanovic K, Saurman V, Israelyan N, Anderson GM et al (2016) Serotonin transporter variant drives preventable gastrointestinal abnormalities in development and function. J Clin Invest 126(6):2221–2235
Marin IA, Goertz JE, Ren TT, Rich SS, Onengut-Gumuscu S, Farber E, Wu M, Overall CC, Kipnis J, Gaultier A (2017) Microbiota alteration is associated with the development of stress-induced despair behavior. Sci Rep 7:43859. https://doi.org/10.1038/srep43859
Marion J, Biernat M, Thomas F, Savary G, Le Breton Y, Zabielski R et al (2002) Small intestine growth and morphometry in piglets weaned at 7 days of age. Effects of level of energy intake. Reprod Nutr Dev 42(4):339–354
Markowiak P, Slizewska K (2017) Effects of probiotics, prebiotics, and synbiotics on human health. Nutrients 9(9)
Marsland BJ (2016) Regulating inflammation with microbial metabolites. Nat Med 22(6):581–583
Marszalek-Grabska M, Walczak K, Gawel K, Wicha-Komsta K, Wnorowska S, Wnorowski A, Turski WA (2021) Kynurenine emerges from the shadows – current knowledge on its fate and function. Pharmacol Ther 225:107845
Martin AM, Lumsden AL, Young RL, Jessup CF, Spencer NJ, Keating DJ (2017) The nutrient-sensing repertoires of mouse enterochromaffin cells differ between duodenum and colon. Neurogastroenterol Motility 29(6)
Martin-Gallausiaux C, Larraufie P, Jarry A, Beguet-Crespel F, Marinelli L, Ledue F, Reimann F, Blottiere HM, Lapaque N (2018) Butyrate produced by commensal bacteria down-regulates indolamine 2,3-dioxygenase 1 (IDO-1) expression via a dual mechanism in human intestinal epithelial cells. Front Immunol 9:2838
Martin-Gronert MS, Stocker CJ, Wargent ET, Cripps RL, Garfield AS, Jovanovic Z et al (2016) 5-HT2A and 5-HT2C receptors as hypothalamic targets of developmental programming in male rats. Dis Model Mech 9(4):401–412
Matsuda M, Imaoka T, Vomachka AJ, Gudelsky GA, Hou Z, Mistry M, Bailey JP, Nieport KM, Walther DJ, Bader M, Horseman ND (2004) Serotonin regulates mammary gland development via an autocrine-paracrine loop. Dev Cell 6(2):193–203
Matsuyoshi H, Kuniyasu H, Okumura M, Misawa H, Katsui R, Zhang GX et al (2010) A 5-HT-receptor activation-induced neural plasticity enhances reconstructs of enteric nerve circuit insult. Neurogastroenterol Motility 22(7)
Mawe GM, Hoffman JM (2013) Serotonin signalling in the gut – functions, dysfunctions and therapeutic targets. Nat Rev Gastroenterol Hepatol 10(8):473–486
Mayer EA (2011) Gut feelings: the emerging biology of gut-brain communication. Nat Rev Neurosci 12(8):453–466
Mayer EA, Labus JS, Tillisch K, Cole SW, Baldi P (2015) Towards a systems view of IBS. Nat Rev Gastroenterol Hepatol 12(10):592–605
McFadden JW, Corl BA (2009) Activation of AMP-activated protein kinase (AMPK) inhibits fatty acid synthesis in bovine mammary epithelial cells. Biochem Biophys Res Commun 390(3):388–393
McGettrick AF, Corcoran SE, Barry PJ, McFarland J, Cres C, Curtis AM et al (2016) Trypanosoma brucei metabolite indolepyruvate decreases HIF-1alpha and glycolysis in macrophages as a mechanism of innate immune evasion. Proc Natl Acad Sci USA 113(48):E7778–E7787
McKernan DP, Fitzgerald P, Dinan TG, Cryan JF (2010) The probiotic Bifidobacterium infantis 35624 displays visceral antinociceptive effects in the rat. Neurogastroenterol Motil 22(9):1029–1035, e1268
McLaughlin T, Ackerman SE, Shen L, Engleman E (2017) Role of innate and adaptive immunity in obesity-associated metabolic disease. J Clin Invest 127(1):5–13

McMillen IC, Houghton DC, Young IR (1995) Melatonin and the development of circadian and seasonal rhythmicity. J Reprod Fertil Suppl 49:137–146

Meimandipour A, Soleimani A, Houshmand M, Kasim A, Hair-Bejo M, Shuhaimi M, Yazid A (2011) Effects of rough handling on short chain fatty acid production and gastrointestinal pH in broilers and modulatory role of Lactobacilli. Afr J Biotechnol 10(74):17030–17037

Melekoglu E, Cetinkaya MA, Kepekci-Tekkeli SE, Kul O, Samur G (2021) Effects of prebiotic oligofructose-enriched inulin on gut-derived uremic toxins and disease progression in rats with adenine-induced chronic kidney disease. PLoS One 16(10):e0258145

Meng D, Sommella E, Salviati E, Campiglia P, Ganguli K, Djebali K et al (2020) Indole-3-lactic acid, a metabolite of tryptophan, secreted by Bifidobacterium longum subspecies infantis is anti-inflammatory in the immature intestine. Pediatr Res 88(2):209–217

Metghalchi S, Ponnuswamy P, Simon T, Haddad Y, Laurans L, Clément M et al (2015) Indoleamine 2,3-dioxygenase fine-tunes immune homeostasis in atherosclerosis and colitis through repression of interleukin-10 production. Cell Metabol 22(3):460–471

Meyer TW, Hostetter TH (2012) Uremic solutes from colon microbes. Kidney Int 81(10):949–954

Meyer A, Hess B, Paisley S, Du M, Caton J (2014) Small intestinal growth measures are correlated with feed efficiency in market weight cattle, despite minimal effects of maternal nutrition during early to midgestation. J Anim Sci 92(9):3855–3867

Miao J, Adewole D, Liu S, Xi P, Yang C, Yin Y (2019) Tryptophan supplementation increases reproduction performance, milk yield, and milk composition in lactating sows and growth performance of their piglets. J Agric Food Chem 67(18):5096–5104

Miao H, Cao G, Wu XQ, Chen YY, Chen DQ, Chen L, Vaziri ND, Feng YL, Su W, Gao Y, Zhuang SG, Yu XY, Zhang L, Guo Y, Zhao YY (2020) Identification of endogenous 1-aminopyrene as a novel mediator of progressive chronic kidney disease via aryl hydrocarbon receptor activation. Br J Pharmacol 177(15):3415–3435

Milanesi S, Garibaldi S, Saio M, Ghigliotti G, Picciotto D, Ameri P, Garibotto G, Barisione C, Verzola D (2019) Indoxyl sulfate induces renal fibroblast activation through a targetable heat shock protein 90-dependent pathway. Oxid Med Cell Long 2019:050183. https://doi.org/10.1155/2019/2050183

Milani C, Duranti S, Bottacini F, Casey E, Turroni F, Mahony J et al (2017) The first microbial colonizers of the human gut: composition, activities, and health implications of the infant gut microbiota. Microbiol Mol Biol Rev 81(4):10

Mo Y, Lu Z, Wang L, Ji C, Zou C, Liu X (2020) The aryl hydrocarbon receptor in chronic kidney disease: friend or foe? Front Cell Dev Biol 8:589752

Moeser AJ, Klok CV, Ryan KA, Wooten JG, Little D, Cook VL, Blikslager AT (2007) Stress signaling pathways activated by weaning mediate intestinal dysfunction in the pig. Am J Physiol Gastrointest Liver Physiol 292(1):G173–G181

Mohammad-Zadeh LF, Moses L, Gwaltney-Brant SM (2008) Serotonin: a review. J Vet Pharmacol Ther 31(3):187–199

Mole DJ, Webster SP, Uings I, Zheng X, Binnie M, Wilson K, Hutchinson JP, Mirguet O, Walker A, Beaufils B, Ancellin N, Trottet L, Beneton V, Mowat CG, Wilkinson M, Rowland P, Haslam C, McBride A, Homer NZ et al (2016) Kynurenine-3-monooxygenase inhibition prevents multiple organ failure in rodent models of acute pancreatitis. Nat Med 22(2):202–209

Moloney GM, O'Leary OF, Salvo-Romero E, Desbonnet L, Shanahan F, Dinan TG et al (2017) Microbial regulation of hippocampal miRNA expression: Implications for transcription of kynurenine pathway enzymes. Behav Brain Res 334:50–54

Mor A, Kalaska B, Pawlak D (2020) Kynurenine pathway in chronic kidney disease: what's old, what's new, and what's next? Int J Tryptophan Res 13:1178646920954882

Moreau A, Vilarem MJ, Maurel P, Pascussi JM (2008) Xenoreceptors CAR and PXR activation and consequences on lipid metabolism, glucose homeostasis, and inflammatory response. Mol Pharm 5(1):35–41

Morita I, Kawamoto M, Yoshida H (1992) Difference in the concentration of tryptophan metabolites between maternal and umbilical foetal blood. J Chromatogr 576(2):334–339

Mu C, Yang Y, Zhu W (2016) Gut microbiota: the brain peacekeeper. Front Microbiol 7:345
Muller CL, Anacker AM, Rogers TD, Goeden N, Keller EH, Forsberg CG et al (2017) Impact of maternal serotonin transporter genotype on placental serotonin, fetal forebrain serotonin, and neurodevelopment. Neuropsychopharmacology 42(2):427–436
Munn AL, Weaver AC, van Wettere WHEJ (2021) Supplementary tryptophan fed to sows prior to and after farrowing to improve piglet growth and survival. Anim Open Access J MDPI 11(9):2540. Retrieved 2021/08
Murakami Y, Imamura Y, Kasahara Y, Yoshida C, Momono Y, Fang K et al (2021) The effects of maternal interleukin-17A on social behavior, cognitive function, and depression-like behavior in mice with altered kynurenine metabolites. Int J Tryptophan Res 14:11786469211026639
Musumeci G, Loreto C, Trovato FM, Giunta S, Imbesi R, Castrogiovanni P (2014) Serotonin (5HT) expression in rat pups treated with high-tryptophan diet during fetal and early postnatal development. Acta Histochem 116(2):335–343
Nagano J, Shimizu M, Hara T, Shirakami Y, Kochi T, Nakamura N, Ohtaki H, Ito H, Tanaka T, Tsurumi H, Saito K, Seishima M, Moriwaki H (2013) Effects of indoleamine 2,3-dioxygenase deficiency on high-fat diet-induced hepatic inflammation. PLoS One 8(9):e73404
Nagata S, Asahara T, Wang C, Suyama Y, Chonan O, Takano K et al (2016) The effectiveness of lactobacillus beverages in controlling infections among the residents of an aged care facility: a randomized placebo-controlled double-blind trial. Ann Nutr Metab 68(1):51–59
Nakamura Y, Tamura H, Kashida S, Takayama H, Yamagata Y, Karube A et al (2001) Changes of serum melatonin level and its relationship to feto-placental unit during pregnancy. J Pineal Res 30(1):29–33
Nakamura K, Sato T, Ohashi A, Tsurui H, Hasegawa H (2008) Role of a serotonin precursor in development of gut microvilli. Am J Pathol 172(2):333–344
Nakano T, Katsuki S, Chen MX, Decano JL, Halu A, Lee LH, Pestana DVS, Kum AST, Kuromoto RK, Golden WS, Boff MS, Guimaraes GC, Higashi H, Kauffman KJ, Maejima T, Suzuki T, Iwata H, Barabasi AL, Aster JC et al (2019) Uremic toxin indoxyl sulfate promotes pro-inflammatory macrophage activation via the interplay of OATP2B1 and Dll4-Notch signaling potential mechanism for accelerated atherogenesis in chronic kidney disease. Circulation 139(1):78–96
Natividad JM, Agus A, Planchais J, Lamas B, Jarry AC, Martin R, Michel ML, Chong-Nguyen C, Roussel R, Straube M, Jegou S, McQuitty C, Le Gall M, da Costa G, Lecornet E, Michaudel C, Modoux M, Glodt J, Bridonneau C et al (2018) Impaired aryl hydrocarbon receptor ligand production by the gut microbiota is a key factor in metabolic syndrome. Cell Metab 28(5):737-+
Nicolas GR, Chang PV (2019) Deciphering the chemical lexicon of host-gut microbiota interactions. Trends Pharmacol Sci 40(6):430–445
Nishizuka Y, Hayaishi O (1963) studies on the biosynthesis of nicotinamide adenine dinucleotide. I. Enzymic synthesis of niacin ribonucleotides from 3-hydroxyanthranilic acid in mammalian tissues. J Biol Chem 238:3369–3377
Nohr MK, Pedersen MH, Gille A, Egerod KL, Engelstoft MS, Husted AS et al (2013) GPR41/FFAR3 and GPR43/FFAR2 as cosensors for short-chain fatty acids in enteroendocrine cells vs FFAR3 in enteric neurons and FFAR2 in enteric leukocytes. Endocrinology 154(10):3552–3564
Notarangelo FM, Wilson EH, Horning KJ, Thomas MA, Harris TH, Fang Q, Hunter CA, Schwarcz R (2014) Evaluation of kynurenine pathway metabolism in Toxoplasma gondii-infected mice: implications for schizophrenia. Schizophr Res 152(1):261–267
Novikov O, Wang Z, Stanford EA, Parks AJ, Ramirez-Cardenas A, Landesman E, Laklouk I, Sarita-Reyes C, Gusenleitner D, Li A, Monti S, Manteiga S, Lee K, Sherr DH (2016) An aryl hydrocarbon receptor-mediated amplification loop that enforces cell migration in ER-/PR-/Her2(-) human breast cancer cells. Mol Pharmacol 90(5):674–688
Nývltová E, Šut'ák R, Žárský V, Harant K, Hrdý I, Tachezy J (2017) Lateral gene transfer of p-cresol-and indole-producing enzymes from environmental bacteria to M astigamoeba balamuthi. Environ Microbiol 19(3):1091–1102

O'Mahony SM, Clarke G, Borre YE, Dinan TG, Cryan JF (2015) Serotonin, tryptophan metabolism and the brain-gut-microbiome axis. Behav Brain Res 277:32–48

Ohta K, Miki T, Warita K, Suzuki S, Kusaka T, Yakura T, Liu JQ, Tamai M, Takeuchi Y (2014) Prolonged maternal separation disturbs the serotonergic system during early brain development. Int J Dev Neurosci 33:15–21

Okatani Y, Okamoto K, Hayashi K, Wakatsuki A, Tamura S, Sagara Y (1998) Maternal-fetal transfer of melatonin in pregnant women near term. J Pineal Res 25(3):129–134

Opitz CA, Litzenburger UM, Sahm F, Ott M, Tritschler I, Trump S et al (2011) An endogenous tumour-promoting ligand of the human aryl hydrocarbon receptor. Nature 478(7368):197–203

Osadchiy V, Labus JS, Gupta A, Jacobs J, Ashe-McNalley C, Hsiao EY, Mayer EA (2018) Correlation of tryptophan metabolites with connectivity of extended central reward network in healthy subjects. PLoS One 13(8):e0201772

Osawa Y, Kanamori H, Seki E, Hoshi M, Ohtaki H, Yasuda Y, Ito H, Suetsugu A, Nagaki M, Moriwaki H, Saito K, Seishima M (2011) L-tryptophan-mediated enhancement of susceptibility to nonalcoholic fatty liver disease is dependent on the mammalian target of rapamycin. J Biol Chem 286(40):34800–34808

Paerregaard SI, Wulff L, Schussek S, Niss K, Morbe U, Jendholm J et al (2023) The small and large intestine contain related mesenchymal subsets that derive from embryonic precursors. Nat Commun 14(1)

Paulicks BR, Pampuch FG, Roth-Maier DA (2006) Studies on the tryptophan requirement of lactating sows. Part 1: Estimation of the tryptophan requirement by performance. Blackwell Publishing Ltd 90(11–12):474–481

Pawlak D, Tankiewicz A, Matys T, Buczko W (2003) Peripheral distribution of kynurenine metabolites and activity of kynurenine pathway enzymes in renal failure. J Physiol Pharmacol 54(2):175–189

Pawlak K, Kowalewska A, Mysliwiec M, Pawlak D (2009) Kynurenine and its metabolites – kynurenic acid and anthranilic acid are associated with soluble endothelial adhesion molecules and oxidative status in patients with chronic kidney disease. Am J Med Sci 338(4):293–300

Pawlak K, Kowalewska A, Mysliwiec M, Pawlak D (2010) 3-hydroxyanthranilic acid is independently associated with monocyte chemoattractant protein-1 (CCL2) and macrophage inflammatory protein-1 beta (CCL4) in patients with chronic kidney disease. Clin Biochem 43(13–14):1101–1106

Peng L, He Z, Chen W, Holzman IR, Lin J (2007) Effects of butyrate on intestinal barrier function in a Caco-2 cell monolayer model of intestinal barrier. Pediatr Res 61(1):37–41

Peng X, Cai X, Li J, Huang Y, Liu H, He J, Fang Z, Feng B, Tang J, Lin Y, Jiang X, Hu L, Xu S, Zhuo Y, Che L, Wu D (2021) Effects of melatonin supplementation during pregnancy on reproductive performance, maternal-placental-fetal redox status, and placental mitochondrial function in a sow model. Antioxidants (Basel, Switzerland) 10(12):1867

Picinato MC, Hirata AE, Cipolla-Neto J, Curi R (2008) Activation of insulin and IGF-1 signaling pathways by melatonin through MT1 receptor in isolated rat pancreatic islets. J Pineal Res

Platten M, Ho PP, Youssef S, Fontoura P, Garren H, Hur EM, Gupta R, Lee LY, Kidd BA, Robinson WH, Sobel RA, Selley ML, Steinman L (2005) Treatment of autoimmune neuroinflammation with a synthetic tryptophan metabolite. Science 310(5749):850–855

Png CW, Linden SK, Gilshenan KS, Zoetendal EG, McSweeney CS, Sly LI et al (2010) Mucolytic bacteria with increased prevalence in IBD mucosa augment in vitro utilization of mucin by other bacteria. Am J Gastroenterol 105(11):2420–2428

Pocivavsek A, Thomas MA, Elmer GI, Bruno JP, Schwarcz R (2014) Continuous kynurenine administration during the prenatal period, but not during adolescence, causes learning and memory deficits in adult rats. Psychopharmacology (Berl) 231(14):2799–2809

Poesen R, Mutsaers HA, Windey K, van den Broek PH, Verweij V, Augustijns P, Kuypers D, Jansen J, Evenepoel P, Verbeke K, Meijers B, Masereeuw R (2015) The influence of dietary protein intake on mammalian tryptophan and phenolic metabolites. PLoS One 10(10):e0140820

Pusceddu MM, Barboza M, Keogh CE, Schneider M, Stokes P, Sladek JA et al (2019) Nod-like receptors are critical for gut-brain axis signalling in mice. J Physiol 597(24):5777–5797

Pytliak M, Vargova V, Mechirova V, Felsoci M (2011) Serotonin receptors – from molecular biology to clinical applications. Physiol Res 60(1):15–25

Peng HQ, He Y, OuY LJ et al (2021) Progress in studying short-chain fatty acids in psychiatric disorders. Int J Psychiatry 48(04):577–579

Qi H, Li Y, Yun H, Zhang T, Huang Y, Zhou J et al (2019) Lactobacillus maintains healthy gut mucosa by producing L-Ornithine. Commun Biol 2:171

Qi M, Tan B, Wang J, Liao S, Deng Y, Ji P, Song T, Zha A, Yin Y (2022) The microbiota-gut-brain axis: a novel nutritional therapeutic target for growth retardation. Crit Rev Food Sci Nutr 62(18):4867–4892

Qian M, Liu J, Zhao D, Cai P, Pan C, Jia W, Gao Y, Zhang Y, Zhang N, Zhang Y, Zhang Q, Wu D, Shan C, Zhang M, Schnabl B, Yang S, Shen X, Wang L (2022) Aryl hydrocarbon receptor deficiency in intestinal epithelial cells aggravates alcohol-related liver disease. Cell Mol Gastroenterol Hepatol 13(1):233–256

Qiu J, Heller JJ, Guo X, Zong-ming EC, Fish K, Fu Y-X, Zhou L (2012) The aryl hydrocarbon receptor regulates gut immunity through modulation of innate lymphoid cells. Immunity 36(1):92–104

Qu Z, Ling PR, Tahan SR, Sierra P, Onderdonk AB, Bistrian BR (1996) Protein and lipd refeeding changes protein metabolism and colonic but not small intestinal morphology in protein-depleted rats. J Nutr 126(4):906–912

Wang JR, Mo S, Mao HM et al (2022) Progress in investigating the effects of tryptophan metabolites of gut microorganisms on host gut health. China Anim Husband J 58(03):7–12. https://doi.org/10.19556/j.0258-7033.20210303-02

Rajan S, Puri S, Kumar D, Babu MH, Shankar K, Varshney S, Srivastava A, Gupta A, Reddy MS, Gaikwad AN (2018) Novel indole and triazole based hybrid molecules exhibit potent anti-adipogenic and antidyslipidemic activity by activating Wnt3a/beta-catenin pathway. Eur J Med Chem 143:1345–1360

Rajilic-Stojanovic M, de Vos WM (2014) The first 1000 cultured species of the human gastrointestinal microbiota. FEMS Microbiol Rev 38(5):996–1047

Ralli T, Neupane YR, Saifi Z, Kohli K (2021) Gut microbiota as an emerging therapeutic avenue for the treatment of non-alcoholic fatty liver disease. Curr Pharm Des 27(46):4677–4685

Raman M, Ambalam P, Kondepudi KK, Pithva S, Kothari C, Patel AT et al (2013) Potential of probiotics, prebiotics and synbiotics for management of colorectal cancer. Gut Microbes 4(3):181–192

Rampelli S, Candela M, Turroni S, Biagi E, Collino S, Franceschi C et al (2013) Functional metagenomic profiling of intestinal microbiome in extreme ageing. Aging (Albany NY) 5(12):902–912

Rao JN, Platoshyn O, Li L, Guo X, Golovina VA, Yuan JX-J, Wang J-Y (2002) Activation of K+ channels and increased migration of differentiated intestinal epithelial cells after wounding. Am J Physiol Cell Physiol 282(4):C885–C898

Raybould HE, Glatzle J, Robin C, Meyer JH, Phan T, Wong H, Sternini C (2003) Expression of 5-HT receptors by extrinsic duodenal afferents contribute to intestinal inhibition of gastric emptying. Am J Physiol-Gastroint Liver Physiol 284(3):G367–G372

Reeds PJ, Burrin DG, Stoll B, Jahoor F (2000) Intestinal glutamate metabolism. J Nutr 130(4):978S–982S

Reigstad CS, Salmonson CE, Rainey JF 3rd, Szurszewski JH, Linden DR, Sonnenburg JL et al (2015) Gut microbes promote colonic serotonin production through an effect of short-chain fatty acids on enterochromaffin cells. FASEB J 29(4):1395–1403

Reiter RJ, Tan DX, Tamura H, Cruz MH, Fuentes-Broto L (2014) Clinical relevance of melatonin in ovarian and placental physiology: a review. Gynecol Endocrinol 30(2):83–89

Ren TH, Wu J, Yew D, Ziea E, Lao L, Leung WK, Berman B, Hu PJ, Sung JJ (2007) Effects of neonatal maternal separation on neurochemical and sensory response to colonic distension in a rat model of irritable bowel syndrome. Am J Physiol Gastrointest Liver Physiol 292(3):G849–G856

Ren W, Liu G, Chen S, Yin J, Wang J, Tan B et al (2017) Melatonin signaling in T cells: functions and applications. J Pineal Res 62(3)

Ressler KJ, Mayberg HS (2007) Targeting abnormal neural circuits in mood and anxiety disorders: from the laboratory to the clinic. Nat Neurosci 10(9):1116–1124

Reyes-Ocampo J, Ramirez-Ortega D, Cervantes GI, Pineda B, Balderas PM, Gonzalez-Esquivel D, Sanchez-Chapul L, Lugo-Huitron R, Silva-Adaya D, Rios C, Jimenez-Anguiano A, Perez-de la Cruz V (2015) Mitochondrial dysfunction related to cell damage induced by 3-hydroxykynurenine and 3-hydroxyanthranilic acid: non-dependent-effect of early reactive oxygen species production. Neurotoxicology 50:81–91

Reynolds LA, Finlay BB, Maizels RM (2015) Cohabitation in the intestine: interactions among helminth parasites, bacterial microbiota, and host immunity. J Immunol 195(9):4059–4066

Rhee S, Walker W, Cherayil B (2012) Salmonella infection and the age-specific response to enteric expression of IFN-γ and its target genes developmentally regulated intestinal. J Immunol 175:1127–1136

Ritze Y, Bardos G, Hubert A, Bohle M, Bischoff SC (2014) Effect of tryptophan supplementation on diet-induced non-alcoholic fatty liver disease in mice. Br J Nutr 112(1):1–7

Roager HM, Licht TR (2018) Microbial tryptophan catabolites in health and disease. Nat Commun 9(1):3294

Robertson SA, Care AS, Moldenhauer LM (2018) Regulatory T cells in embryo implantation and the immune response to pregnancy. J Clin Invest 128(10):4224–4235

Romani L, Zelante T, De Luca A, Iannitti RG, Moretti S, Bartoli A et al (2014) Microbiota control of a tryptophan-AhR pathway in disease tolerance to fungi. Eur J Immunol 44(11):3192–3200

Rooks MG, Garrett WS (2016) Gut microbiota, metabolites and host immunity. Nat Rev Immunol 16(6):341–352

Rossi G, Pengo G, Caldin M, Palumbo Piccionello A, Steiner JM, Cohen ND et al (2014) Comparison of microbiological, histological, and immunomodulatory parameters in response to treatment with either combination therapy with prednisone and metronidazole or probiotic VSL#3 strains in dogs with idiopathic inflammatory bowel disease. PLoS One 9(4):e94699

Roth W, Zadeh K, Vekariya R, Ge Y, Mohamadzadeh M (2021) Tryptophan metabolism and gut-brain homeostasis. Int J Mol Sci 22(6)

Rothhammer V, Mascanfroni ID, Bunse L, Takenaka MC, Kenison JE, Mayo L, Chao CC, Patel B, Yan R, Blain M, Alvarez JI, Kebir H, Anandasabapathy N, Izquierdo G, Jung S, Obholzer N, Pochet N, Clish CB, Prinz M et al (2016) Type I interferons and microbial metabolites of tryptophan modulate astrocyte activity and central nervous system inflammation via the aryl hydrocarbon receptor. Nat Med 22(6):586–597

Rothhammer V, Borucki DM, Tjon EC, Takenaka MC, Chao CC, Ardura-Fabregat A, de Lima KA, Gutierrez-Vazquez C, Hewson P, Staszewski O, Blain M, Healy L, Neziraj T, Borio M, Wheeler M, Dragin LL, Laplaud DA, Antel J, Alvarez JI et al (2018) Microglial control of astrocytes in response to microbial metabolites. Nature 557(7707):724–728

Ru LC, Jiao PB, Zhoujin T (2022) Progress in the study of short-chain fatty acids associated with intestinal flora. World Chin Digest J 30(13):562–570

Ruddell RG, Mann DA, Ramm GA (2008) The function of serotonin within the liver. J Hepatol 48(4):666–675

Ruddick JP, Evans AK, Nutt DJ, Lightman SL, Rook GA, Lowry CA (2006) Tryptophan metabolism in the central nervous system: medical implications. Expert Rev Mol Med 8(20):1–27

Saffrey MJ (2013) Cellular changes in the enteric nervous system during ageing. Dev Biol 382(1):344–355

Saffrey MJ (2014) Aging of the mammalian gastrointestinal tract: a complex organ system. Age (Dordr) 36(3):9603

Sagrillo-Fagundes L, Assunção Salustiano EM, Ruano R, Markus RP, Vaillancourt C (2018) Melatonin modulates autophagy and inflammation protecting human placental trophoblast from hypoxia/reoxygenation. J Pineal Res 65(4):e12520
Sakai NS, Taylor SA, Chouhan MD (2018) Obesity, metabolic disease and the pancreas –quantitative imaging of pancreatic fat. Br J Radiol 91(1089):20180267
Sakurai T, Odamaki T, Xiao JZ (2019) Production of indole-3-lactic acid by bifidobacterium strains isolated from human infants. Microorganisms 7(9)
Salazar N, Arboleya S, Valdes L, Stanton C, Ross P, Ruiz L et al (2014) The human intestinal microbiome at extreme ages of life. Dietary intervention as a way to counteract alterations. Front Genet 5:406
Salazar N, Valdes-Varela L, Gonzalez S, Gueimonde M, de Los Reyes-Gavilan CG (2017) Nutrition and the gut microbiome in the elderly. Gut Microbes 8(2):82–97
Samanta S (2022) Physiological and pharmacological perspectives of melatonin. Arch Physiol Biochem 128(5):1346–1367
Sano M, Ferchaud-Roucher V, Kaeffer B, Poupeau G, Castellano B, Darmaun D (2016) Maternal and fetal tryptophan metabolism in gestating rats: effects of intrauterine growth restriction. Amino Acids 48(1):281–290
Santhanam S, Alvarado DM, Ciorba MA (2016) Therapeutic targeting of inflammation and tryptophan metabolism in colon and gastrointestinal cancer. Transl Res 167(1):67–79
Saraf MK, Piccolo BD, Bowlin AK, Mercer KE, LeRoith T, Chintapalli SV et al (2017) Formula diet driven microbiota shifts tryptophan metabolism from serotonin to tryptamine in neonatal porcine colon. Microbiome 5(1):77
Sarkar A, Lehto SM, Harty S, Dinan TG, Cryan JF, Burnet PWJ (2016) Psychobiotics and the manipulation of bacteria-gut-brain signals. Trends Neurosci 39(11):763–781
Sartor RB (2004) Therapeutic manipulation of the enteric microflora in inflammatory bowel diseases: antibiotics, probiotics, and prebiotics. Gastroenterology 126(6):1620–1633
Scaldaferri F, Pizzoferrato M, Gerardi V, Lopetuso L, Gasbarrini A (2012) The gut barrier: new acquisitions and therapeutic approaches. J Clin Gastroenterol 46(9):S12–S17
Schiering C, Wincent E, Metidji A, Iseppon A, Li Y, Potocnik AJ et al (2017) Feedback control of AHR signalling regulates intestinal immunity. Nature 542(7640):242–245
Schröcksnadel H, Baier-Bitterlich G, Dapunt O, Wachter H, Fuchs D (1996) Decreased plasma tryptophan in pregnancy. Obstet Gynecol 88(1):47–50
Schröcksnadel K, Wirleitner B, Winkler C, Fuchs D (2006) Monitoring tryptophan metabolism in chronic immune activation. Clin Chim Acta 364(1-2):82–90
Schroeder BO, Backhed F (2016) Signals from the gut microbiota to distant organs in physiology and disease. Nat Med 22(10):1079–1089
Schroeder JC, DiNatale BC, Murray IA, Flaveny CA, Liu Q, Laurenzana EM, Lin JM, Strom SC, Omiecinski CJ, Amin S, Perdew GH (2010) The uremic toxin 3-indoxyl sulfate is a potent endogenous agonist for the human aryl hydrocarbon receptor. Biochemistry 49(2):393–400
Sedlmayr P, Blaschitz A, Stocker R (2014) The role of placental tryptophan catabolism. Front Immunol 5:230
Sein OB, Chepelev NA (1990) Methods of stimulating reproductive activity in pigs
Sensoy I (2021) A review on the food digestion in the digestive tract and the used in vitro models. Curr Res Food Sci 4:308–319
Session V (1991) Digestive physiology of pigs
Shajib MS, Baranov A, Khan WI (2017) Diverse effects of gut-derived serotonin in intestinal inflammation. ACS Chem Neurosci 8(5):920–931
Shanley DP, Aw D, Manley NR, Palmer DB (2009) An evolutionary perspective on the mechanisms of immunosenescence. Trends Immunol 30(7):374–381
Shi JM, Tian XZ, Zhou GB, Wang L, Gao C, Zhu SE, Zeng SM, Tian JH, Liu GS (2009) Melatonin exists in porcine follicular fluid and improves in vitro maturation and parthenogenetic development of porcine oocytes. J Pineal Res 4:47

Shiyuan Y, Wentao Z, Yuanchun W (2011) The effect of X-rays on the histological structure and antioxidant enzyme activity of the duodenum in young mice. J Northwest Normal Univ Nat Sci Ed 47(3):80–86

Sigurdsson GV, Schmidt S, Mellström D, Ohlsson C, Saalman R, Lorentzon M (2023) Young adult male patients with childhood-onset IBD have increased risks of compromised cortical and trabecular bone microstructures. Inflamm Bowel Dis 29(7):1065–1072

Silva MBDE, Tustumi F, Neto AAD, Dantas ACB, Santo MA, Cecconello I (2021) Gastric bypass compared with sleeve gastrectomy for nonalcoholic fatty liver disease: a systematic review and meta-analysis. Obesity Surgery 31(6):2762–2772

Silvano A, Seravalli V, Strambi N, Cecchi M, Tartarotti E, Parenti A, Di Tommaso M (2021) Tryptophan metabolism and immune regulation in the human placenta. J Reprod Immunol 147:103361

Singh J, Dartois A, Kaur L (2010) Starch digestibility in food matrix: a review. Trends Food Sci Technol 21(4):168–180

Singhal M, Turturice BA, Manzella CR, Ranjan R, Metwally AA, Theorell J et al (2019) Serotonin transporter deficiency is associated with dysbiosis and changes in metabolic function of the mouse intestinal microbiome. Sci Rep 9(1):2138

Sinn DH, Kang D, Chang Y, Ryu S, Gu S, Kim H et al (2017) Non-alcoholic fatty liver disease and progression of coronary artery calcium score: a retrospective cohort study. Gut 66(2):323–329

Sirich TL, Plummer NS, Gardner CD, Hostetter TH, Meyer TW (2014) Effect of increasing dietary fiber on plasma levels of colon-derived solutes in hemodialysis patients. Clin J Am Soc Nephrol 9(9):1603–1610

Sittipo P, Choi J, Lee S, Lee YK (2022) The function of gut microbiota in immune-related neurological disorders: a review. J Neuroinflammation 19(1):154

Sjogren K, Engdahl C, Henning P, Lerner UH, Tremaroli V, Lagerquist MK et al (2012) The gut microbiota regulates bone mass in mice. J Bone Miner Res 27(6):1357–1367

Smith F, Clark JE, Overman BL, Tozel CC, Huang JH, Rivier JE et al (2010) Early weaning stress impairs development of mucosal barrier function in the porcine intestine. Am J Physiol-Gastrointest Liver Physiol 298(3):G352–G363

Sommer F, Backhed F (2013) The gut microbiota – masters of host development and physiology. Nat Rev Microbiol 11(4):227–238

Soret R, Chevalier J, De Coppet P, Poupeau G, Derkinderen P, Segain JP, Neunlist M (2010) Short-chain fatty acids regulate the enteric neurons and control gastrointestinal motility in rats. Gastroenterology 138(5):1772–1782

Spiller R (2007) Serotonin, inflammation, and IBS: fitting the jigsaw together? J Pediatr Gastroenterol Nutr 45(Suppl 2):S115–S119

Spinelli P, Latchney SE, Reed JM, Fields A, Baier BS, Lu X et al (2019) Identification of the novel Ido1 imprinted locus and its potential epigenetic role in pregnancy loss. Hum Mol Genet 28(4):662–674

Stoll ML, Kumar R, Morrow CD, Lefkowitz EJ, Cui X, Genin A et al (2014) Altered microbiota associated with abnormal humoral immune responses to commensal organisms in enthesitis-related arthritis. Arthritis Res Ther 16(6):486. https://doi.org/10.1186/s13075-014-0486-0

Stracke J, Otten W, Tuchscherer A, Witthahn M, Metges CC, Puppe B, Düpjan S (2017) Dietary tryptophan supplementation and affective state in pigs. J Vet Behav 20:82–90

Strandwitz P (2018) Neurotransmitter modulation by the gut microbiota. Brain Res 1693(Pt B):128–133

Stoll ML, Kumar R, Lefkowitz EJ et al (2016) Fecal metabolomics in pediatric spondyloarthritis implicate decreased metabolic diversity and altered tryptophan metabolism as pathogenic factors. Genes Immun 17:400–405

Stull MA, Pai V, Vomachka AJ, Marshall AM, Jacob GA, Horseman ND (2007) Mammary gland homeostasis employs serotonergic regulation of epithelial tight junctions. Proc Natl Acad Sci USA 104(42):16708–16713

Sudo N, Chida Y, Aiba Y, Sonoda J, Oyama N, Yu XN et al (2004) Postnatal microbial colonization programs the hypothalamic-pituitary-adrenal system for stress response in mice. J Physiol 558(Pt 1):263–275

Sui G, Jia L, Quan D, Zhao N, Yang G (2021) Activation of the gut microbiota-kynurenine-liver axis contributes to the development of nonalcoholic hepatic steatosis in nondiabetic adults. Aging (Albany NY) 13(17):21309–21324

Sun J, Wang F, Hong G, Pang M, Xu H, Li H, Tian F, Fang R, Yao Y, Liu J (2016) Antidepressant-like effects of sodium butyrate and its possible mechanisms of action in mice exposed to chronic unpredictable mild stress. Neurosci Lett 618:159–166

Sundström E, Kölare S, Souverbie F, Samuelsson EB, Pschera H, Lunell NO, Seiger A (1993) Neurochemical differentiation of human bulbospinal monoaminergic neurons during the first trimester. Brain Res Dev Brain Res 75(1):1–12

Sutter CH, Bodreddigari S, Campion C, Wible RS, Sutter TR (2011) 2, 3, 7, 8-Tetrachlorodibenzo-p-dioxin increases the expression of genes in the human epidermal differentiation complex and accelerates epidermal barrier formation. Toxicological Sciences 124(1):128–137

Szalardy L, Zadori D, Toldi J, Fulop F, Klivenyi P, Vecsei L (2012) Manipulating kynurenic acid levels in the brain – on the edge between neuroprotection and cognitive dysfunction. Curr Top Med Chem 12(16):1797–1806. <Go to ISI>://WOS:000311140100007

Sznurkowski JJ, Zawrocki A, Emerich J, Sznurkowska K, Biernat W (2011) Expression of indoleamine 2,3-dioxygenase predicts shorter survival in patients with vulvar squamous cell carcinoma (vSCC) not influencing on the recruitment of FOXP3-expressing regulatory T cells in cancer nests. Gynecol Oncol 122(2):307–312

Tabecka-Lonczynska A, Mytych J, Solek P, Kulpa M, Koziorowski M (2017) New insight on the role of melatonin receptors in reproductive processes of seasonal breeders on the example of mature male European bison (Bison bonasus, Linnaeus 1758). J Photochem Photobiol B Biol 173:84–91

Tackett JJ, Gandotra N, Bamdad MC, Muise ED, Cowles RA (2017) Enhanced serotonin signaling stimulates ordered intestinal mucosal growth. J Surg Res 208:198–203

Takahashi H, Nakashima S, Ohama E, Takeda S, Ikuta F (1986) Distribution of serotonin-containing cell bodies in the brainstem of the human fetus determined with immunohistochemistry using antiserotonin serum. Brain Dev 8(4):355–365

Takamura T, Harama D, Matsuoka S, Shimokawa N, Nakamura Y, Okumura K et al (2010) Activation of the aryl hydrocarbon receptor pathway may ameliorate dextran sodium sulfate-induced colitis in mice. Immunol Cell Biol 88(6):685–689

Tan B, Li XG, Kong X, Huang R, Ruan Z, Yao K et al (2009) Dietary L-arginine supplementation enhances the immune status in early-weaned piglets. Amino Acids 37:323–331

Tan B, Yin Y, Kong X, Li P, Li X, Gao H et al (2010) L-Arginine stimulates proliferation and prevents endotoxin-induced death of intestinal cells. Amino Acids 38:1227–1235

Tang X, Xiong K (2022) Intrauterine growth retardation affects intestinal health of suckling piglets via altering intestinal antioxidant capacity, glucose uptake, tight junction, and immune responses. Oxid Med Cell Long 2022

Tao J-L, Zhang X, Zhou J-Q, Li C-Y, Yang M-H, Liu Z-J, Zhang L-L, Deng S-L, Zhang L, Shen M, Liu G-S, Liu H-L (2021) Melatonin alleviates hypoxia-induced apoptosis of granulosa cells by reducing ROS and activating MTNR1B-PKA-Caspase8/9 pathway. Antioxidants (Basel, Switzerland) 10(2):184. Retrieved 2021/01//

Terry N, Margolis KG (2017) Serotonergic mechanisms regulating the GI tract: experimental evidence and therapeutic relevance. Handb Exp Pharmacol 239:319–342

Tester RF, Karkalas J, Qi X (2004) Starch – composition, fine structure and architecture. J Cereal Sci 39(2):151–165

Teunis C, Nieuwdorp M, Hanssen N (2022) Interactions between tryptophan metabolism, the gut microbiome and the immune system as potential drivers of non-alcoholic fatty liver disease (NAFLD) and metabolic diseases. Metabolites 12(6):514

Thorburn AN, Macia L, Mackay CR (2014) Diet, metabolites, and "western-lifestyle" inflammatory diseases. Immunity 40(6):833–842

Tian X, Wang F, He C, Zhang L, Tan D, Reiter RJ, Xu J, Ji P, Liu G (2014) Beneficial effects of melatonin on bovine oocytes maturation: a mechanistic approach. J Pineal Res 57(3):239–247
Tracing the history of pig farming (2022) Hogs Today (6), 51. Retrieved from
Tsuji A, Nakata C, Sano M, Fukuwatari T, Shibata K (2013) L-tryptophan metabolism in pregnant mice fed a high L-tryptophan diet and the effect on maternal, placental, and fetal growth. Int J Tryptophan Res 6:21–33
Tuo BG, Isenberg JI (2003) Effect of 5-hydroxytryptamine on duodenal mucosal bicarbonate secretion in mice. Gastroenterology 125(3):805–814
Ugrumov MV (2010a) Developing brain as an endocrine organ: a paradoxical reality. Neurochem Res 35(6):837–850
Ugrumov MV (2010b) Developing brain as an endocrine organ: a paradoxical reality. Neurochemical Research 35(6):837–850
Underhill DM, Iliev ID (2014) The mycobiota: interactions between commensal fungi and the host immune system. Nat Rev Immunol 14(6):405–416
Uribe A, Alam M, Johansson O, Midtvedt T, Theodorsson E (1994) Microflora modulates endocrine cells in the gastrointestinal mucosa of the rat. Gastroenterology 107(5):1259–1269
Uronis JM, Muhlbauer M, Herfarth HH, Rubinas TC, Jones GS, Jobin C (2009) Modulation of the intestinal microbiota alters colitis-associated colorectal cancer susceptibility. PLoS One 4(6):e6026
Valladares R, Bojilova L, Potts AH, Cameron E, Gardner C, Lorca G, Gonzalez CF (2013) Lactobacillus johnsonii inhibits indoleamine 2,3-dioxygenase and alters tryptophan metabolite levels in BioBreeding rats. FASEB J 27(4):1711–1720
van Beek AA, Hugenholtz F, Meijer B, Sovran B, Perdijk O, Vermeij WP et al (2017) Frontline Science: Tryptophan restriction arrests B cell development and enhances microbial diversity in WT and prematurely aging Ercc1(-/Delta7) mice. J Leukoc Biol 101(4):811–821
van Tilburg Bernardes E, Pettersen VK, Gutierrez MW, Laforest-Lapointe I, Jendzjowsky NG, Cavin JB et al (2020) Intestinal fungi are causally implicated in microbiome assembly and immune development in mice. Nat Commun 11(1):2577
Van ENJ, Loest CA, Ferreira AV, Waggoner JW, Mathis CP (2008) Limiting amino acids for growing lambs fed a diet low in ruminally undegradable protein1. J Anim Sci 10:10
VanHouten JN, Wysolmerski JJ (2013) The calcium-sensing receptor in the breast. Best Pract Res Clin Endocrinol Metabol 27(3):403–414
Varel V (1987) Activity of fiber-degrading microorganisms in the pig large intestine. J Anim Sci 65(2):488–496
Vargas-Caraveo A, Sayd A, Maus SR, Caso JR, Madrigal JLM, Garcia-Bueno B, Leza JC (2017) Lipopolysaccharide enters the rat brain by a lipoprotein-mediated transport mechanism in physiological conditions. Sci Rep 7(1):13113
Vecsei L, Szalardy L, Fulop F, Toldi J (2013) Kynurenines in the CNS: recent advances and new questions. Nat Rev Drug Discov 12(1):64–82
Venkatesh M, Mukherjee S, Wang H, Li H, Sun K, Benechet AP, Qiu Z, Maher L, Redinbo MR, Phillips RS, Fleet JC, Kortagere S, Mukherjee P, Fasano A, Le Ven J, Nicholson JK, Dumas ME, Khanna KM, Mani S (2014) Symbiotic bacterial metabolites regulate gastrointestinal barrier function via the xenobiotic sensor PXR and Toll-like receptor 4. Immunity 41(2):296–310
Vereecke L, Elewaut D (2017) Spondyloarthropathies: Ruminococcus on the horizon in arthritic disease. Nat Rev Rheumatol 13(10):574–576
Vermeulen W, De Man JG, Pelckmans PA, De Winter BY (2014) Neuroanatomy of lower gastrointestinal pain disorders. World J Gastroenterol 20(4):1005–1020
Verney C, Lebrand C, Gaspar P (2002) Changing distribution of monoaminergic markers in the developing human cerebral cortex with special emphasis on the serotonin transporter. Anat Rec 267(2):87–93
Vidal-Lletjós S, Beaumont M, Tomé D, Benamouzig R, Blachier F, Lan A (2017) Dietary protein and amino acid supplementation in inflammatory bowel disease course: what impact on the colonic mucosa? Nutrients 9(3)

Vivoli E, Cappon A, Cozzi A, Navari N, Gargano M, Fallarino F, Marra F (2015) A novel role for the kynurenine pathway in experimental steatohepatitis. Digest Liver Dis 47:E21–E21
Wai CC, Wing CS et al (2018) Clinical analysis of pulmonary function changes in patients with non-alcoholic fatty liver disease. J Clin Hepatobil Dis 34(04):825–828. Retrieved from
Wan J, Yang K, Xu Q, Chen D, Yu B, Luo Y, He J (2016) Dietary chitosan oligosaccharide supplementation improves foetal survival and reproductive performance in multiparous sows. RSC Adv 6(74):70715–70722
Wang D (2014) Introduction to the digestive physiological characteristics of domestic animals. Charm China (5). Retrieved from
Wang H (2018) Characteristics of feed digestion and nutrient absorption in pigs. China Livestock Poul Breed Ind 14(12):123. Retrieved from
Wang L, Erlandsen H, Haavik J, Knappskog PM, Stevens RC (2002) Three-dimensional structure of human tryptophan hydroxylase and its implications for the biosynthesis of the neurotransmitters serotonin and melatonin. Biochemistry 41(42):12569–12574
Wang JH, Simonavicius N, Wu XS, Swaminath G, Reagan J, Tian H, Ling L (2006) Kynurenic acid as a ligand for orphan G protein-coupled receptor GPR35. J Biol Chem 281(31):22021–22028
Wang H, Ji Y, Wu G, Sun K, Sun Y, Li W, Wang B, He B, Zhang Q, Dai Z, Wu Z (2015a) l-Tryptophan Activates Mammalian Target of Rapamycin and Enhances Expression of Tight Junction Proteins in Intestinal Porcine Epithelial Cells. J Nutr 145(6):1156–1162
Wang K, Feng CC, Li CG, Yao J, Xie XF, Gong LK, Luan Y, Xing GZ, Zhu X, Qi XM, Ren J (2015b) Baicalin protects mice from aristolochic acid I-induced kidney injury by induction of CYP1A through the aromatic hydrocarbon receptor. Int J Mol Sci 16(7):16454–16468
Wang F, Knutson K, Alcaino C, Linden DR, Gibbons SJ, Kashyap P et al (2017) Mechanosensitive ion channel Piezo2 is important for enterochromaffin cell response to mechanical forces. J Physiol London 595(1):79–91
Wang M, Yang C, Wang Q, Li J, Li Y, Ding X et al (2020a) The growth performance, intestinal digestive and absorptive capabilities in piglets with different lengths of small intestines. Animal 14(6):1196–1203
Wang PC, Chen ST, Hong ZK, Li SY, Yang ZS, Quan S, Yang ZM (2020b) Tryptophan and kynurenine stimulate human decidualization via activating Aryl hydrocarbon receptor: Short title: Kynurenine action on human decidualization. Reprod Toxicol 96:282–292
Wang X, Yang S, Li S, Zhao L, Hao Y, Qin J, Zhang L, Zhang, C., Bian W, Zuo L, Gao, X., Zhu, B., Lei XG, Gu Z, Cui W., Xu X., Li Z, Zhu, B., Li Y et al (2020c) Aberrant gut microbiota alters host metabolome and impacts renal failure in humans and rodents. Gut 69(12):2131–2142
Weaver SR, Laporta J, Moore SAE, Hernandez LL (2016) Serotonin and calcium homeostasis during the transition period. Domest Anim Endocrinol 56(Suppl):S147–S154
Wegermann K, Howe C, Henao R, Wang Y, Guy CD, Abdelmalek MF, Diehl AM, Moylan CA (2021) Serum bile acid, vitamin E, and serotonin metabolites are associated with future liver-related events in nonalcoholic fatty liver disease. Hepatol Commun 5(4):608–617
Weiwei S, Shanglian F (1989) Effect of acupuncture on plasma gastrointestinal hormones and gastric secretory functions. Chin J Gastroenterol 9(3):148–150
Wheatcroft J, Wakelin D, Smith A, Mahoney CR, Mawe G, Spiller R (2005) Enterochromaffin cell hyperplasia and decreased serotonin transporter in a mouse model of postinfectious bowel dysfunction. Neurogastroenterol Motil 17(6):863–870
Whitehead TR, Price NP, Drake HL, Cotta MA (2008) Catabolic pathway for the production of skatole and indoleacetic acid by the acetogen Clostridium drakei, Clostridium scatologenes, and swine manure. Appl Environ Microbiol 74(6):1950–1953
Wierrani F, Grin W, Hlawka B, Kroiss A, Grünberger W (1997) Elevated serum melatonin levels during human late pregnancy and labour. J Obstet Gynaecol 17(5):449–451
Wikoff WR, Anfora AT, Liu J, Schultz PG, Lesley SA, Peters EC, Siuzdak G (2009) Metabolomics analysis reveals large effects of gut microflora on mammalian blood metabolites. Proc Natl Acad Sci 106(10):3698–3703
Wilck N, Matus MG, Kearney SM, Olesen SW, Forslund K, Bartolomaeus H et al (2017) Salt-responsive gut commensal modulates T(H)17 axis and disease. Nature 551(7682):585–589

Williams JN Jr, Feigelson P, Elvehjem CA (1950) Relation of tryptophan and niacin to pyridine nucleotides of tissue. J Biol Chem 187(2):597–604

Williams BA, Verstegen MW, Tamminga S (2001) Fermentation in the large intestine of single-stomached animals and its relationship to animal health. Nutr Res Rev 14(2):207–228

Williams BB, Van Benschoten AH, Cimermancic P, Donia MS, Zimmermann M, Taketani M et al (2014) Discovery and characterization of gut microbiota decarboxylases that can produce the neurotransmitter tryptamine. Cell Host Microbe 16(4):495–503

Wincent E, Bengtsson J, Bardbori AM, Alsberg T, Luecke S, Rannug U, Rannug A (2012) Inhibition of cytochrome P4501-dependent clearance of the endogenous agonist FICZ as a mechanism for activation of the aryl hydrocarbon receptor. Proc Natl Acad Sci USA 109(12):4479–4484

Winterhager E, Gellhaus A (2017) Transplacental nutrient transport mechanisms of intrauterine growth restriction in rodent models and humans. Front Physiol 8:951

Witchley JN, Penumetcha P, Abon NV, Woolford CA, Mitchell AP, Noble SM (2019) Candida albicans morphogenesis programs control the balance between gut commensalism and invasive infection. Cell Host Microbe 25(3):432–443 e436

Wlodarska M, Luo C, Kolde R, d'Hennezel E, Annand JW, Heim CE, Krastel P, Schmitt EK, Omar AS, Creasey EA, Garner AL, Mohammadi S, O'Connell DJ, Abubucker S, Arthur TD, Franzosa EA, Huttenhower C, Murphy LO, Haiser HJ et al (2017) Indoleacrylic acid produced by commensal peptostreptococcus species suppresses inflammation. Cell Host Microbe 22(1):25–37 e26

Wonga. (2004) Experiments on the digestive-absorptive and exocrine functions of the small intestine and pancreas applied in the clinic. Tibetan Med 25(3):33–34

Wu G, Knabe DA, Yan W, Flynn NE (1995) Glutamine and glucose metabolism in enterocytes of the neonatal pig. Am J Physiol-Regulat Integrat Comp Physiol 268(2):R334–R342

Wu X, Ruan Z, Gao Y, Yin Y, Zhou X, Wang L et al (2010) Dietary supplementation with L-arginine or N-carbamylglutamate enhances intestinal growth and heat shock protein-70 expression in weanling pigs fed a corn-and soybean meal-based diet. Amino Acids 39:831–839

Wurtman RJ, Fernstrom JD (1975) Control of brain monoamine synthesis by diet and plasma amino acids. Am J Clin Nutr 28(6):638–647

Xavier RJ, Podolsky DK (2007) Unravelling the pathogenesis of inflammatory bowel disease. Nature 448(7152):427–434

Xia D, Yang L, Li Y, Chen J, Zhang X, Wang H, Zhai S, Jiang X, Meca G, Wang S, Huang L, Zhu S, Fu Y, Ma W, Zhu Y, Ye H, Wang W (2021) Melatonin alleviates Ochratoxin A-induced liver inflammation involved intestinal microbiota homeostasis and microbiota-independent manner. J Hazard Mater 413:125239

Xiang FF, Cao XS, Shen B, Chen XH, Guo M, Ding XQ, Zou JZ (2020) Transcriptome profiling reveals indoxyl sulfate should be culpable of impaired T cell function in chronic kidney disease. Front Med 7

Xiusheng L (2020) Liver function and healthy pig farming. Pig Farm Today(04):83–86. Retrieved from

Xue ZH, Li D, Yu WC, Zhang Q, Hou XN, He YL, Kou XH (2017) Mechanisms and therapeutic prospects of polyphenols as modulators of the aryl hydrocarbon receptor. Food Funct 8(4):1414–1437

Yang W, Guangyan Y, Lei Z (2005) The effect of capsaicin on the secretory function of salivary and lacrimal glands in healthy subjects. J Mod Dentist 19(3):296–298

Yang H, Fu D, Kong X, Wang W, Yang X, Nyachoti C, Yin Y (2013a) Dietary supplementation with N-carbamylglutamate increases the expression of intestinal amino acid transporters in weaned Huanjiang mini-pig piglets. J Animal Sci 91(6):2740–2748

Yang H, Xiong X, Yin Y (2013b) Development and renewal of intestinal villi in pigs. In: Nutritional and physiological functions of amino acids in pigs. Springer, pp 29–47

Yang YX, Dai ZL, Zhu WY (2014) Important impacts of intestinal bacteria on utilization of dietary amino acids in pigs. Amino Acids 46(11):2489–2501

Yang T, Feng YL, Chen L, Vaziri ND, Zhao YY (2019) Dietary natural flavonoids treating cancer by targeting aryl hydrocarbon receptor. Crit Rev Toxicol 49(5):445–460
Yang X, Yu D, Xue L, Li H, Du J (2020) Probiotics modulate the microbiota-gut-brain axis and improve memory deficits in aged SAMP8 mice. Acta Pharm Sin B 10(3):475–487
Yang CY, Chen TW, Lu WL, Liang SS, Huang HD, Tseng CP, Tarng DC (2021) Synbiotics alleviate the gut indole load and dysbiosis in chronic kidney disease. Cells 10(1)
Yano JM, Yu K, Donaldson GP, Shastri GG, Ann P, Ma L et al (2015) Indigenous bacteria from the gut microbiota regulate host serotonin biosynthesis. Cell 161(2):264–276
Yao K, Guan S, Li T, Huang R, Wu G, Ruan Z, Yin Y (2011) Dietary L-arginine supplementation enhances intestinal development and expression of vascular endothelial growth factor in weanling piglets. Br J Nutr 105(5):703–709
Yao Y, Cai X, Ye Y, Wang F, Chen F, Zheng C (2021) The role of microbiota in infant health: from early life to adulthood. Front Immunol 12:708472
Yarlagadda A, Alfson E, Clayton AH (2009) The blood brain barrier and the role of cytokines in neuropsychiatry. Psychiatry (Edgmont) 6(11):18
Yelin I, Flett KB, Merakou C, Mehrotra P, Stam J, Snesrud E et al (2019) Genomic and epidemiological evidence of bacterial transmission from probiotic capsule to blood in ICU patients. Nat Med 25(11):1728–1732
Yeung AW, Terentis AC, King NJ, Thomas SR (2015) Role of indoleamine 2,3-dioxygenase in health and disease. Clin Sci (Lond) 129(7):601–672
Yin Y, Deng Z, Huang H, Zhong H, Hou Z, Gong J, Liu Q (2004) Nutritional and health functions of carbohydrate for pigs. J Anim Feed Sci 13(4):523–538
Yokogoshi H, Wurtman RJ (1986) Meal composition and plasma amino acid ratios: effect of various proteins or carbohydrates, and of various protein concentrations. Metabolism 35(9):837–842
Yoshimura H, Sakai T, Kuwahara Y, Ito M, Tsuritani K, Hirasawa Y, Nagamatsu T (2009) Effects of kynurenine metabolites on mesangial cell proliferation and gene expression. Exp Mol Pathol 87(1):70–75
Yoshio M, Yuzuru K, Yasuhiro K, Katsuyoshi T, Tatsuhide I, Hiroo I (1986) Involvement of growth hormone (GH)-releasing factor in GH secretion induced by serotoninergic mechanisms in conscious rats. Endocrinology 3:1089
Young K, Nelson R (2001) Mediation of seasonal testicular regression by apoptosis. Reproduction 122(5):677–685
Yuan G, Xiaochun W, Shuqin Z, Yujun Z, Hailong M (2019) Melatonin receptor depletion suppressed hCG-induced testosterone expression in mouse Leydig cells. Cell Mol Biol Lett
Yuwen L, Zhang FL, Chen QH, Lin SJ, Zhao YL, Li ZY (2013) The role of aromatic L-amino acid decarboxylase in bacillamide C biosynthesis by Bacillus atrophaeus C89. Sci Rep 3:1753
Zakrocka I, Zaluska W (2022) Kynurenine pathway in kidney diseases. Pharmacol Rep 74(1):27–39
Zelante T, Iannitti RG, Cunha C, De Luca A, Giovannini G, Pieraccini G, Zecchi R, D'Angelo C, Massi-Benedetti C, Fallarino F, Carvalho A, Puccetti P, Romani L (2013) Tryptophan catabolites from microbiota engage aryl hydrocarbon receptor and balance mucosal reactivity via interleukin-22. Immunity 39(2):372–385
Zelante T, Iannitti RG, Fallarino F, Gargaro M, De Luca A, Moretti S et al (2014) Tryptophan feeding of the IDO1-AhR axis in host-microbial symbiosis. Front Immunol 5:640
Zeng L, Tan B, Yin Y, Kong X, Fen Z, Fang J, Lu X (2012) Gene expression profiles in porcine intestinal epithelial cells treated with arginine using a microarray technique. J. Food. Agric Environ 10:834–839
Zhang Y, Xu R (2006) Anatomy and histology of the gastrointestinal tract. The neonatal pig: gastrointestinal physiology and nutrition
Zhang J, Yin Y, Shu XG, Li T, Li F, Tan B et al (2013) Oral administration of MSG increases expression of glutamate receptors and transporters in the gastrointestinal tract of young piglets. Amino Acids 45:1169–1177

Zhang J, Zhu S, Ma N, Johnston LJ, Wu C, Ma X (2021) Metabolites of microbiota response to tryptophan and intestinal mucosal immunity: a therapeutic target to control intestinal inflammation. Med Res Rev 41(2):1061–1088

Zhao L, Xiao HT, Mu HX, Huang T, Lin ZS, Zhong LLD, Zeng GZ, Fan BM, Lin CY, Bian ZX (2017) Magnolol, a natural polyphenol, attenuates dextran sulfate sodium-induced colitis in mice. Molecules 22(7)

Zhao H, Chen L, Yang T, Feng YL, Vaziri ND, Liu BL, Liu QQ, Guo Y, Zhao YY (2019a) Aryl hydrocarbon receptor activation mediates kidney disease and renal cell carcinoma. J Transl Med 17(1):302

Zhao ZH, Xin FZ, Xue YQ, Hu ZM, Han YM, Ma FG, Zhou D, Liu XL, Cui AY, Liu ZS, Liu YX, Gao J, Pan Q, Li Y, Fan JG (2019b) Indole-3-propionic acid inhibits gut dysbiosis and endotoxin leakage to attenuate steatohepatitis in rats. Exp Mol Med 51

Zhao T, Zhang H, Yin X, Zhao H, Ma L, Yan M, Peng L, Wang Q, Dong X, Li P (2020) Tangshen formula modulates gut Microbiota and reduces gut-derived toxins in diabetic nephropathy rats. Biomed Pharmacother 129:110325

Zheng Z, Wang B (2021) The gut-liver axis in health and disease: the role of gut microbiota-derived signals in liver injury and regeneration. Front Immunol 12:775526

Zheng X, Hu MM, Zang XJ, Fan QL, Liu YL, Che Y et al (2019) Kynurenic acid/GPR35 axis restricts NLRP3 inflammasome activation and exacerbates colitis in mice with social stress. Brain Behav Immun 79:244–255

Zhi-yuan Q, Cheng-wei W (2021) Factors affecting cecal digestion of rabbits and its nutritional regulation measures. Feed Res 44(18)

Zhou FC, Sari Y, Zhang JK (2000) Expression of serotonin transporter protein in developing rat brain. Brain Res Dev Brain Res 119(1):33–45

Zhou L, Zhang M, Wang Y, Dorfman RG, Liu H, Yu T et al (2018) Faecalibacterium prausnitzii produces butyrate to maintain Th17/Treg balance and to ameliorate colorectal colitis by inhibiting histone deacetylase 1. Inflamm Bowel Dis 24(9):1926–1940

Zhou YL, Zheng W, Liu WJ, Wang CY, Zhan YN, Li HQ, Chen LJ, Ning YP (2019) Cross-sectional relationship between kynurenine pathway metabolites and cognitive function in major depressive disorder. Psychoneuroendocrinology 101:72–79

Zhu YZ, Ma WF, Chen XC et al (2020) Tryptophan catabolism and its application in pig diets. Journal of Animal Nutrition 32(03):1019–1024

Zhu W, Stevens AP, Dettmer K, Gottfried E, Hoves S, Kreutz M, Holler E, Canelas AB, Kema I, Oefner PJ (2011) Quantitative profiling of tryptophan metabolites in serum, urine, and cell culture supernatants by liquid chromatography-tandem mass spectrometry. Anal Bioanal Chem 401(10):3249–3261

Zhu W, Yu K, Mu C, Yang Y (2014) Gut microbiota and host nutrition metabolism in pigs. Chin J Anim Nutr 26(10):3046–3051

Zhuang X, Xiong L, Li L, Li M, Chen M (2017) Alterations of gut microbiota in patients with irritable bowel syndrome: a systematic review and meta-analysis. J Gastroenterol Hepatol 32(1):28–38

Zubcevic J, Richards EM, Yang T, Kim S, Sumners C, Pepine CJ, Raizada MK (2019) Impaired autonomic nervous system-microbiome circuit in hypertension a premise for hypertension therapy. Circ Res 125(1):104–116

Chapter 3
Application of Tryptophan in Swine Production

Chunyan Guo, Ling Zhang, Lijuan Chen, Zhao Zhuo, Chousheng Liu, Shunshun Jin, Jie Yin, Huansheng Yang, Liuqin He, and Yulong Yin

Abstract Synthetic amino acid supplementation has been shown to be an effective method to reduce dietary protein levels and lower feed costs in livestock. Tryptophan (Trp) has been commonly considered the fourth limiting amino acid after lysine, methionine+cysteine, and threonine in corn-soybean meal–based diets for pigs. The biological roles of Trp in the maintenance of gut health and nutritional balance have been well documented both in humans and animals. This chapter will discuss the application of dietary Trp on swine production at different stages, with a focus on

C. Guo (✉)
Beijing Animal Nutrition Research Center, Beijing, China
e-mail: chgu@novonesis.com

L. Zhang (✉) · L. Chen · Z. Zhuo
College of Animal Science and Technology, Anhui Agricultural University, Hefei, China
e-mail: ZL@ahau.edu.cn; zhangchenlijuan@ahau.edu.cn; zhuozhao90@ahau.edu.cn

C. Liu
National Animal Husbandry Service, Beijing, China
e-mail: lchsh-nahs@agri.gov.cn

S. Jin
University of Manitoba, Winnipeg, Manitoba, Canada
e-mail: shunshuj@myumanitoba.ca

J. Yin (✉)
Animal Nutritional Genome and Germplasm Innovation Research Center, College of Animal Science and Technology, Hunan Agricultural University, Changsha, Hunan, China
e-mail: yinjie@hunau.edu.cn

H. Yang (✉) · L. He (✉)
College of Life Sciences, Hunan Normal University, Changsha, Hunan, China
e-mail: yhs@hunnu.edu.cn; heliuqin@hunnu.edu.cn

Y. Yin (✉)
Tianjin Institute of Industrial Biotechnology, Chinese Academy of Sciences, National Center of Technology Innovation for Synthetic Biology, Tianjin, China

Institute of Subtropical Agriculture, Chinese Academy of Science, Changsha, China
e-mail: yinyulong@isa.ac.cn

Y. Yin et al. (eds.), *Tryptophan in Animal Nutrition and Human Health*,
https://doi.org/10.1007/978-981-97-4719-1_3

the growth performance in postweaning piglets, nutrient requirements in growing pigs, and gestating and lactating sows.

Abbreviations

5-HT	Serotonin
HPA	Hypothalamic-pituitary-adrenal
IDO	Indoleamine 2,3-dioxygenase
IgG	Immunoglobulin G
LNAA	Large neutral amino acids
Lys	Lysine
PepT1	Peptide transporters 1
PepT2	Peptide transporters 2
PRRS	Respiratory syndrome
QA	Quinolinic acid
ROS	Reactive oxygen species
SID	Standardized ileal digestible
Trp	Tryptophan

3.1 Tryptophan Nutrition in Weaned Piglets

3.1.1 Growth Promotion

The nutrient requirements of piglets are largely affected by their developmental stages, and piglets need sufficient protein and amino acids to ensure adequate muscle and bone development (He et al. 2013; Htoo et al. 2007). Trp, a neutral amino acid, usually acts as the fourth limiting amino acid in pig feed and has a significant impact on the growth performance of piglets. Trp is the only amino acid in animals that binds to serum albumin through a non-covalent bond, and it has essential functions in promoting growth performance. Trp deficiency reduced the absorption and utilization of protein in livestock and poultry, impaired their immune function, and increased their susceptibility to diseases (Le Floc'H et al. 2011; Marion et al. 2002). Dietary Trp supplementation significantly reduced the time and frequency of fighting behaviors in piglets, increased their lying time, and alleviated weaning stress (Wu 2009). Post-weaning piglets receiving the Trp-rich diet exhibited higher feed intake and better growth performance upon enterotoxigenic *Escherichia coli* K88 challenge, indicating that Trp may improve the appetite and promote the growth performance of weaned piglets upon bacterial infection (Koopmans et al. 2005). Piglets fed with Trp exhibited an increase in feed intake and the daily average gain. Additionally, the expression and secretion of gastrin were significantly increased in the piglets fed with Trp, indicating that dietary Trp supplementation may enhance the appetite of weaned piglets and promote food intake and weight gain by regulating the expression and secretion of gastrin (Trevisi et al. 2009). The current research on Trp nutrition in pig production is mainly focused on its application in low-protein-based diets. When the dietary protein level decreases in the feed, the Trp level also decreases accordingly. Therefore, it is important to supplement industrially synthesized Trp to adjust the ratio of essential amino acids in the diet to meet the nutrient requirements and achieve similar growth performance between low-protein and normal diets in pig production (Huawei Zhang et al. 2007). It is noted that weaned piglets are more sensitive to Trp nutrition when compared to growing pigs. Excessive addition of Trp may have a negative impact on piglet feed intake and reduce the absorption and utilization efficiency of other amino acids, thereby affecting the balance of protein synthesis (Eder et al. 2001).

3.1.2 Feed Intake

Trp affects porcine appetite mainly by acting on the hypothalamic feeding center through serotonin, a neurotransmitter synthesized in the brain and gastrointestinal tract. Additionally, the synthesis of various hormones related to feed intake such as melatonin secreted by the pineal gland, insulin secreted by the pancreas, and growth hormone–releasing peptides secreted by the stomach is also regulated by Trp metabolism. The increased feed intake can enhance the growth performance and feed conversion rate of piglets and avoid weight loss in lactating sows.

Lin et al. (1999) found that there is a strong positive correlation between the average daily feed intake and the average daily weight gain of weaned piglets and the level of Trp in the feed. Moreover, they further showed that in early-weaned piglets, the average daily weight gain and feed intake are also positively correlated with Trp concentration, while the feed-to-gain ratio exhibits a negative correlation with dietary Trp level in the range from 0.16% to 0.28%. It has also been shown that the feed intake of the low Trp group (0.13%) is about 40% lower than that of the high Trp group (0.205%). Trp affects the digestive function by influencing the secretion of gastric and pancreatic juices. Generally, dietary Trp increases the secretion of gastric and pancreatic juices, thereby promoting the digestive and absorptive function of animals. With the increased addition of Trp levels in the feed, the concentration of blood urea nitrogen and triglycerides is significantly decreased in piglet serum.

The lack of Trp leads to a decreased feed intake in piglets, due to its regulatory role in appetite. Re-supplementation of Trp to Trp-deficiency diet could greatly increase the feed intake of nursery pigs and growing-finishing pigs. Currently, the ratio of Trp to lysine in creep feed and weaning feed for piglets produced by swine feed companies in China ranges from 17% to 22%. Some commercial feed companies may adjust this ratio from 20% to 22% based on the health status of weaned piglets and market conditions. There is also a considerable amount of literature reporting that dietary Trp supplementation can significantly improve the growth performance of weaned piglets. According to the average data from references, the addition of Trp could increase the feed intake by 6% and weight gain by 9% and improve feed conversion efficiency by 3% of weaned piglets, respectively. In addition, the high levels of large neutral amino acids (LNAA) including valine, isoleucine, leucine, phenylalanine, and tyrosine in high-protein diets compete with Trp transporters during the process of intestinal absorption, reducing the amount of Trp available for serotonin synthesis (Shen et al. 2012b). Lin et al. (1999) showed that when compared to the control diet in which the Trp/LNAA ratio is 3.3%, the feed intake decreased by 55% when the Trp/LNAA ratio was dropped to 2.5%. However, when the Trp/LNAA ratio reached 2.9%, the feed intake increased to 92%. It is of great importance to optimize the Trp/LNAA ratio during porcine feed formulation.

3.1.3 Stress Alleviation

After weaning, piglets are separated from their mothers and moved into the nursery where they experience sudden changes in environment, temperature, humidity, pens, psychological state, and companions. The social hierarchy also changes, leading to competitive behaviors such as tail biting and head butting among weaned piglets. Additionally, the transition from consuming liquid milk to solid feed causes a series of negative syndromes, including anorexia and diarrhea, all of which are collectively referred to as weaned stress of piglets (Moehn et al. 2012). Trp can alleviate animals' temperament, abnormal behaviors, provocative actions, and depressive emotions by regulating the concentration of 5-HT in the central nervous system (Yin 2011; Shen et al. 2012a) Trp metabolism increases the levels of 5-HT in the brain and reduces the levels of plasma cortisol which alleviates stress responses and aggressive behaviors through the hypothalamic-pituitary-adrenal (HPA) axis. As the amount of dietary Trp increases, the levels of 5-HT in various regions of the brain also increase accordingly (Shen et al. 2012b). It has been reported that the high ratio of Trp/LNAA decreases the concentrations of cortisol, epinephrine, and norepinephrine in the blood of stressed piglets, accelerating the recovery rate of stress behaviors (Youjian et al. 2005; Shen et al. 2015). Additionally, Trp addition had a sedative effect on piglets by reducing piglets' stress responses, and the lying time of piglets was drastically increased when the dietary Trp level increased proportionally (Cortamira et al. 1991). Furthermore, a high level of dietary Trp also shortened the duration of aggressive behavior among pigs (Buddington et al. 2001). Therefore, the dietary addition of Trp not only promotes protein synthesis but also reduces the energy consumption and stress responses in piglets by increasing the concentration of 5-HT in the brain, thereby promoting growth and improving the feed conversion rate of weaned piglets. In response to infection or stress, most Trp in mammals is metabolized through the kynurenine pathway into bioactive compounds that may interact with stress responses. However, acute stress has an opposite effect on the regulation of Trp level in the mammalian brain where the increased Trp level may be related to the activated sympathetic nervous system and elevated plasma catecholamine levels.

Trp can influence tail-biting behavior in piglets caused by weaning stress (Xia 2012). The addition of Trp can slow down the decreased level of cortisol in pig saliva while moving between different pens, leading to less energy wasting and increased lying time (Liuyan et al. 2019). Dietary Trp supplementation significantly reduced the duration and frequency of fighting and stress reactions as well as increased lying time in weaned piglets (He et al. 2013; Sietse Jan Koopmans et al. 2005). To reduce the fighting and tail-biting behaviors in piglets, pig farms typically adjust the Trp and lysine levels above 22%, which is higher than the NRC-recommended Trp level.

Heating is often accompanied by severe oxidative stresses. Dietary addition of 0.15% Trp alleviated the intestinal damage, oxidative stress, and mitochondrial dysfunction caused by diquat treatment in piglets. While supplementation of a high level of Trp at 0.75% significantly increased the expression level of diamine oxidase and D-lactic acid in the serum and reduced the expression of tight junction protein occludin and ZO-1 in the duodenum, damaging the intestinal epithelium integrity. Therefore, it is worth increasing the dietary Trp level to avoid high temperature-induced stress responses in the summer (Tossou et al. 2016). Additionally, the dietary addition of Trp alleviated the negative impacts on meat quality during the process of transportation and pre-slaughter, likely by regulating serotonin synthesis.

3.1.4 Improvement of the Gut Health

The intestine is the main organ and site for nutrient digestion and absorption and acts as an important barrier to protect against pathogenic microorganism infections (Luo et al. 2019). Functional amino acids have been shown to maintain gut health by enhancing immune responses, improving mucosal morphology, and inhibiting the invasion of pathogenic microorganisms (Xiao-lan et al. 2016). Dietary addition of appropriate Trp levels improved the intestinal epithelium integrity through a series of enzymatic and microbial actions (Tossou et al. 2016). Intestinal tight junction proteins exist in the intercellular space of intestinal epithelial cells and play key roles in maintaining gut morphology, preserving the integrity of the intestinal epithelium, and regulating mucosal permeability (Wenfeng et al. 2020). Research showed that increased dietary Trp levels enhance the expression of the tight junction protein occludin in the duodenum and ileum, but excessive Trp may negatively affect the integrity of the intestinal barrier (Zhao et al. 2018).

Dietary proteins are hydrolyzed into amino acids or small peptides in the gut, which in turn are absorbed and transported into the blood or lymph through the transporters localized on the intestinal epithelial cells. Free amino acids enter the cells through amino acid transporters, while the small peptide transporters (PepT1 and PepT2) in the apical membrane of the intestinal epithelial cells facilitate the transportation of small peptides into the cells. The absorptive capacity of substrates by intestinal epithelial cells mainly depends on the expression levels of the small peptide and amino acid transporter genes. Likewise, the concentration of substrates also affects the expression of transporter genes (Willemen et al. 2014). Dietary Trp supplementation has been shown to promote the entering of alkaline amino acids into the bloodstream and enhance their transportation in the small intestine, thereby promoting the digestion and absorption of proteins and amino acids (Tao et al. 2018). It has been reported that intestinal Trp activates PepT1 expression and dietary addition of Trp improves the absorption and transportation of free amino acids in the intestines (Tang et al. 2022).

3.1.5 Immune Regulation

In the gut, tryptophan is metabolized following three major pathways (Fig. 3.1): (1) the gut microbiota metabolizes tryptophan directly and the metabolites including ligands of the aryl hydrocarbon receptor (AhR), which plays a role in immune

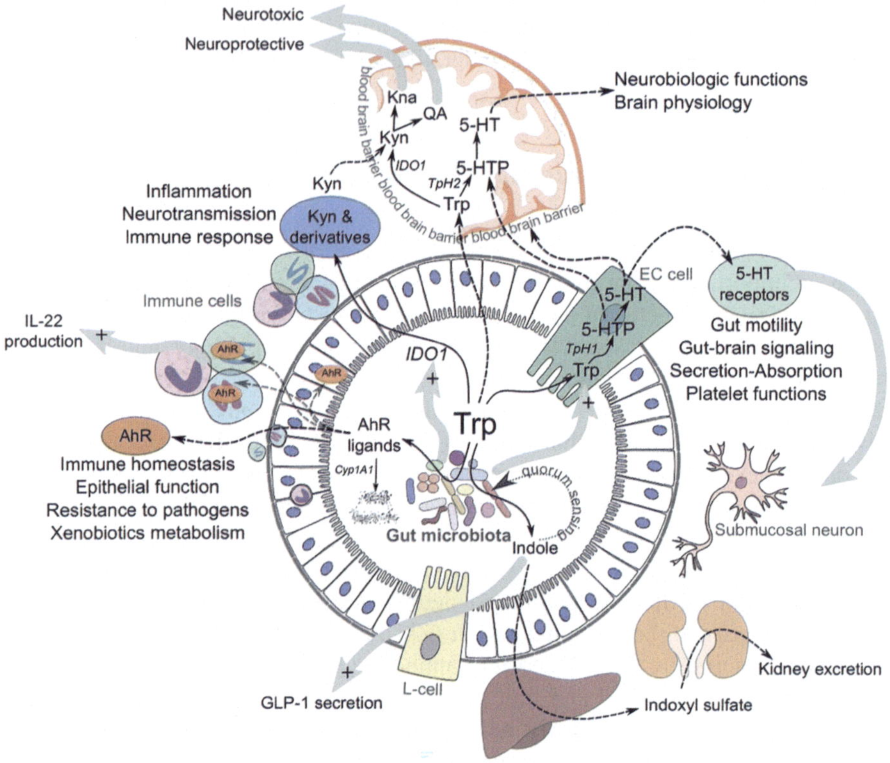

Fig. 3.1 Integrated Trp metabolism under the control of the gut microbiota in host physiology (Agus et al. 2018)

homeostasis and intestinal barrier function (Zelante et al. 2013); (2) both immune and epithelial cells via the indoleamine 2,3-dioxygenase (IDO) 1-mediated kynurenine pathway play important roles in inflammation, immune response, and neurological function (Clarke et al. 2012); and (3) enterochromaffin cells through the Trp hydroxylase 1 (TpH1) mediated the serotonin (5-hydroxytryptamine [5-HT]) production pathway, and these metabolites have functions such as stimulating intestinal motility and promoting gut-brain axis signaling (Yano et al. 2015). These metabolites are ligands for AhR, including indole and its derivatives, such as indole-3-acid-acetic (IAA), indole-3-aldehyde (IAld), indole-3-propionic acid (IPA), indole-3-acetaldehyde (IAAld), and indole acrylic acid (Alexeev et al. 2018; Hubbard et al. 2015). IAA and IPA are known to affect intestinal permeability and host immunity (Agus et al. 2018) and are promising to be used to alleviate the intestinal inflammation of weaned piglets. Moreover, Trp is necessary for T-cell proliferation and promotes the differentiation of bone marrow T-lymphocyte precursors, ultimately leading to the formation of mature T lymphocytes. Additionally, Trp may also modulate the immune responses by the involvement of indoleamine 2,3-dioxygenase (IDO), which is the only rate-limiting enzyme that catalyzes the oxidation of the indole ring in the kynurenine pathway. By degrading Trp in local tissues, IDO plays an important immune regulatory role in promoting the host immune defense, inducing the maternal immune tolerance, and anti-tumor function. IDO is expressed in tumor cells to degrade Trp which leads to Trp exhaustion in the tumor microenvironment, constraining the cytotoxic T cells at the G1 phase, inhibiting their proliferation, and resulting in immune escape of the tumor cells.

It has been suggested that piglets demand higher Trp requirements upon immune stimulation and weaning. Firstly, piglets exhibited less feed intake under environmental challenges, resulting in lower Trp levels in the blood and reduced availability of Trp for protein synthesis in the body. Secondly, the kynurenine pathway will be enhanced to degrade Trp when the body undergoes immune challenges. Therefore, dietary supplementation of Trp can alleviate the adverse effects caused by inflammation or weaning in piglets. It has been revealed that IDO could improve inflammatory symptoms and restore the intestinal immune function of weaned piglets (Jung et al. 2010). Chronic pneumonia reduces serum Trp levels and increases the IDO activity in tissues, while supplementation of Trp could significantly alleviate these adverse effects (Le Floc'h et al. 2008). The IDO activity in the jejunum of weaned piglets was significantly increased after feeding a diet containing 0.2% or 0.35% of Trp as shown by Jiangmin (2015) and Baoshi (2017). Additionally, the IDO expression in the kidney, spleen, stomach, and ileum was significantly decreased when the Trp-to-lysine ratio was increased from 0.15 to 0.18 or 0.21 (Wenfeng et al. 2020). However, when the Trp-to-lysine ratio reached 0.24, the IDO expression was significantly increased. A total of 0.5% of dietary Trp supplementation improved gut development by increasing the ratio of villus height to crypt depth (Sietse Jan Koopmans et al. 2005). In addition, Trp also regulates inflammatory responses by modulating the expression level of quinolinic acid (QA) (Yanzhi et al. 2020). The higher the expression level of QA, the faster the reduction of inflammatory responses. Increasing the dietary Trp levels induced the concentration of serum antibodies against porcine reproductive and respiratory syndrome (PRRS), which in turn enhances the humoral and cellular immune responses to protect piglets against infection (Wangjun et al. 2012).

3.1.6 Recommended Levels

Regarding the nutritional requirement of Trp for piglets less than 30 kg, commercialized Trp products need to be supplemented as the Trp level in raw materials usually cannot meet the needs of piglets. Notably, standardized ileal digestible Trp to lysine ratio (SID Trp:Lys) has been largely utilized both in scientific research and in practice to formulate pig feed rations to meet the balance of amino acids. Table 3.1 lists the recommended SID Trp:Lys ratios for piglets weighing from 5–30 kg from different institutions. Traditionally, the SID Trp:Lys ratio between 18% and 20% is recommended. According to the NRC (2012), it is recommended to use the SID Trp:Lys ratio of 16% in the early growing stage of piglets and 17% in the late growing-finishing stages. A meta-analysis was performed to estimate the SID Trp:Lys ratio that maximizes the performance of weaned piglets between 7 and 25 kg of body weight (Simongiovanni et al. 2012). The results indicated that the curvilinear plateau model appeared to be the best-adapted model to describe the response curve of a population. This model predicted that increasing the SID Trp:Lys ratio from 17% to 22% resulted in an increase in ADG by 8%.

3.2 Tryptophan Nutrition for Growing Pigs

Numerous nutritional studies have shown that increased levels of Trp in the diet of growing pig improve their growth performance. It was found that the true digestible Trp requirement of growing pigs weighing 20 to 50 kg is significantly higher than the recommended nutritional requirement from NRC (1998), based on the weight gain and nitrogen deposition (Simongiovanni et al. 2012). However, this study may overestimate the requirement of Trp because of the non-linear model used in their data analyses. Guzik et al. utilized the piecewise linear regression model to analyze

Table 3.1 Recommended SID Trp:Lys ratio for piglets (5 ~ 30 kg)

References	Recommended SID Trp:Lys ratio (%)
BSAS (UK)	18.8
CVB (Netherlands)	19
Evonik Degussa (Germany)	22
ITP (France)	19
NCPP (Denmark)	19.5
NRC (USA)	18

their data and found that the Trp requirement is only 0.15% based on weight gain. In addition, they also found that the true digestible Trp requirements of 30 and 50 kg growing pigs are 0.18% and 0.14%, respectively, based on the corn-soybean meal-pea diet (Guzik et al. 2006). These data were slightly higher than that of NRC (1998), most likely due to the increased potential of Trp requirement during the fast growth period. Tang et al. studied the balanced pattern of digestible lysine, methionine+cysteine, threonine, and Trp in growing pigs fed with a low-protein diet (M. Tang et al. 2008). They showed that the ideal digestible Lys:Trp ratio is 100:(18–21) calculated by quadratic regression analysis in the low-protein diet for growing pigs. The effects of different net energy and Trp levels in low-protein diets on pig growth have also been studied and found that there was no significant effect on the growth performance of growing-finishing pigs when reducing Trp levels in low-protein diets based on the net energy level (Yin et al. 2008). However, decreased Trp levels resulted in a decrease in the average daily weight gain in pigs during the growing-finishing stages.

Wangjun et al. investigated the effects of dietary addition of serine and Trp on the immune regulation of 30 kg of growing pigs upon a low-virulence vaccination of porcine reproductive and respiratory syndrome (PRRS) (Wangjun et al. 2012). They found that increasing the Lys:serine:Trp ratio from 100:64:18 (the NRC recommended level) to 100:80:26 significantly increased the serum level of immunoglobulin G (IgG) and inflammatory cytokines by enhancing the TLR 3/7/8 expression in the lung. Moreover, Trp and its metabolites can affect T-cell function to regulate the maturation of B cells, which in turn promotes the secretion of IgG. Therefore, the addition of serine and Trp may enhance the immune protective function of vaccination by improving both the humoral and acquired immunity.

Dietary Trp supplementation increased the expression of the tight junction protein claudin-3 which improves the intestinal barrier function, thereby improving the growth performance of growing pigs (Liu et al. 2017). The maximum serotonin concentration in the hypothalamus of growing pigs was observed within 5 days after Trp intake and dietary Trp addition before slaughter can reduce the stress response by regulating serotonin expression and decrease the PSE incidence (Adeola and Ball 1992).

Some commercial feeders did not pay much attention to the balance of other limiting amino acids besides lysine, methionine, and threonine during the growing-finishing stage. The reduction of production performance caused by Trp deficiency is particularly prominent when a large amount of DDGS or low-protein or soybean meal–free diets are used. The DDGS diet contains a high level of phosphorus but a very low level of calcium, thus requiring the addition of other mineral ingredients. Additionally, despite the abundant amount of vitamin B/E in DDGS, lysine and Trp must be additionally supplemented due to their low levels in the DDGSs.

Table 3.2 Effect of Trp nutrition in swine production

Stage	Trp level	Effects	Reference
Piglets	0.15%	Improved growth performance	Liu et al. (2019a)
	0.2% or 0.4%	Improved intestinal mucosal barrier function	Liang et al. (2018)
	0.30% or 0.45%	Enhanced the antioxidant capacity	Mao et al. (2021)
	0.5 g/kg	Increased feed intake and decreased FCR	Jansman et al. (2019)
	0.80%	Improved lipid and glucose metabolism; decrease salivary cortisol concentration	Goodarzi et al. (2021); Shen et al. (2015)
	Trp:Lys ratio (16.8%)	Improved feed-to-gain ratio	Naatjes et al. (2014)
	0.16% Trp + 0.41% Phe + 0.22% Tyr	Promoted the amino acid absorption	Duanmu et al. (2021)
	0.34% SID Trp:4.6% SID LNAA	Improved aspects of post-weaning performance	Sterndale et al. (2020)
	15 μM ICA	Enhanced intestinal epithelial proliferation	Zhang et al. (2022)
Growing pigs	SID Trp:Lys ratio (22.5%)	Increased ADG, ADFI, gain-to-feed ratio	Liu et al. (2019b)
	SID Trp:Lys (23%)	Prevented negative effects of excessive Leu in diets	Kwon et al. (2021)
	Moderate L-Lys HCl + high Ile, Val, and Trp + Val	Increased ADFI	Kerkaert et al. (2021)
	0.06% Trp in low CP diet	Increased ADG, ADFI	Sato et al. (2021)
	Trp + Thr + met	Improved growth performance and protein deposition	Valini et al. (2023)
	Trp-deficient diet	Reduced feed consumption	Eder et al. (2003)
	0.16%	Improved the growth performance and meat quality	Jiao et al. (2016); Ma et al. (2020)
Sows	0.12%	Improved reproduction performance	Miao et al. (2019)
	0.3%	Tendency of increased litter weaning weight	Mosnier et al. (2010)
	0.37%	Reduced aggressive behavior	Poletto et al. (2014)
	Trp-enriched diet	Increased exploratory activities	Lay Jr. et al. (2021)
	0.35% in gestation and 0.48% in lactation	Decreased head-to-head knocking behaviors	Li et al. (2011)
	0.36%	Decreased piglet birth weight	Xu et al. (2019)
	0.42% or 0.56%	Increased piglet survival	Munn et al. (2021)

This table was adapted from Hu et al. (2023). Reprinted permission from the KiAi Publishing Communications Ltd.

3.3 Tryptophan Nutrition for Sows

There are increasing investigations of the effect of Trp nutrition on milk production in lactating sows and reproductive performance in gestating sows as shown in Table 3.2. Traditionally, the recommended SID Trp:Lys ratio for lactating sows is approximately between 18% and 20%. However, studies have also shown that sows fed a diet with SID Trp:Lys ratio above 22% increase their feed intake and reduce their weight and backfat loss. Additionally, dietary supplementation of Trp is also capable of increasing milk production and litter weight in lactating sows. Miao et al. (2018) investigated the effects of Trp on the composition of sow milk and the secretion of hormones related to milk production. They found that the dietary addition of Trp had no significant effect on the major solid content in sow milk, except for lactose. However, the levels of growth hormones, prolactin, insulin, and other hormones related to lactation were significantly increased upon Trp supplementation. Trp metabolism may likely regulate the lactation performance by affecting the synthesis of lactose and hormones. Lactating sows fed with 0.12% Trp have been shown to increase their milk yield through the activation of 5-HT expression as well as the ADFI (Miao et al. 2019), while dietary supplementation of 0.3% Trp in lactating sows had no significant effect on their litter performance. In fact, a series of internal and external factors such as sow breeds, physiological status parity, and feeding environment largely determine the requirement of Trp for sows.

Furthermore, the regulatory effects of Trp on the reproduction performance of sows have also been reported. The requirement of Trp nutrition differs at different stages of pregnancy, with more Trp ingestion in late pregnancy than in early pregnancy. High levels of dietary Trp supplementation at either 0.42% or 0.56% during the late pregnancy stage significantly increased the survival rate of their litters (Tossou et al. 2016), while feeding sows at the late pregnancy stage with 0.39% Trp reduced the average birth weight and live farrowing rate (Tossou et al. 2016). Gestating sows fed with 0.37% Trp at an early stage did not affect the mortality and weight of their litters but reduced their aggressive behaviors. Feeding gestating sows with 0.35% Trp and lactating sows with 0.48% Trp, respectively, also reduced the duration of head-to-head knocking behaviors.

The plasma Trp levels were significantly reduced on the second and fourth day after farrowing when compared to day 58 of pregnancy. In addition, inactive litters exhibited lower plasma Trp levels when compared to active litters during the first week of lactation. These findings suggest that Trp nutrition and metabolism play important roles in the regulation of physiological behaviors and reproductive performance in sows at different stages. Therefore, it is of great importance to fully consider the Trp requirement when formulating diets for both gestating and lactating sows.

References

Adeola O, Ball RO (1992) Hypothalamic neurotransmitter concentrations and meat quality in stressed pigs offered excess dietary tryptophan and tyrosine. J Anim Sci 70:1888–1894

Agus A, Planchais J, Sokol H (2018) Gut microbiota regulation of tryptophan metabolism in health and disease. Cell Host Microbe 23(6):716–724

Alexeev EE, Lanis JM, Kao DJ, Campbell EL, Kelly CJ, Battista KD et al (2018) Microbiota-derived indole metabolites promote human and murine intestinal homeostasis through regulation of interleukin-10 receptor. Am J Pathol 188(5):1183–1194

Baoshi S (2017) Study for the signal pathway of tryptophan promoting β-defensin secretion in intestine. (Master), Southweat University, Chongqing

Buddington RK, Elnif J, Puchal-Gardiner AA, Sangild PT (2001) Intestinal apical amino acid absorption during development of the pig. Am J Phys Regul Integr Comp Phys 280(1):R241–R247

Clarke G, Grenham S, Scully P, Fitzgerald P, Moloney RD, Shanahan F et al (2012) The microbiome-gut-brain axis during early life regulates the hippocampal serotonergic system in a sex-dependent manner. Mol Psychiatry 18(6):666–673

Cortamira NO, Seve B, Lebreton Y, Ganier P (1991) Effect of dietary tryptophan on muscle, liver and whole-body protein synthesis in weaned piglets: relationship to plasma insulin. Br J Nutr 66(3):423–435

Duanmu Q, Tan B, Wang J, Huang B, Li J, Kang M et al (2021) The amino acids sensing and utilization in response to dietary aromatic amino acid supplementation in LPS-induced inflammation piglet model. Front Nutr 8:819835

Eder K, Peganova S, Kluge H (2001) Studies on the tryptophan requirement of piglets. Arch Tierernahr 55(4):281–297

Eder K, Nonn H, Kluge H, Peganova S (2003) Tryptophan requirement of growing pigs at various body weights. J Anim Physiol Anim Nutr (Berl) 87(9–10):336–346

Goodarzi P, Habibi M, Roberts K, Sutton J, Shili CN, Lin D, Pezeshki A (2021) Dietary tryptophan supplementation alters fat and glucose metabolism in a low-birthweight piglet model. Nutrients 13(8):2531

Guzik AC, Matthews JO, Kerr BJ, Bidner TD, Southern LL (2006) Dietary tryptophan effects on plasma and salivary cortisol and meat quality in pigs1,2. J Anim Sci 84(8):2251–2259

He L, Yang H, Hou Y, Li T, Fang J, Zhou X et al (2013) Effects of dietary l-lysine intake on the intestinal mucosa and expression of CAT genes in weaned piglets. Amino Acids 45(2):383–391

Htoo JK, Araiza BA, Sauer WC, Rademacher M, Zhang Y, Cervantes M, Zijlstra RT (2007) Effect of dietary protein content on ileal amino acid digestibility, growth performance, and formation of microbial metabolites in ileal and cecal digesta of early-weaned pigs. J Anim Sci 85(12):3303–3312

Hu Z, Feng L, Jiang Q, Wang W, Tan B e, Tang X, Yin Y (2023) Intestinal tryptophan metabolism in disease prevention and swine production. Anim Nutr 15:364–374

Hubbard TD, Murray IA, Perdew GH (2015) Indole and tryptophan metabolism: endogenous and dietary routes to Ah receptor activation. Drug Metab Dispos 43(10):1522–1535

Jansman AJM, Cirot O, Corrent E, Lambert W, Ensink J, van Diepen JTM (2019) Interaction and imbalance between indispensable amino acids in young piglets. Animal 13(5):941–949

Jiangmin (2015) The influence of tryptophan to antioxidant capacity and amino acid transporter. (master), Nanchang University

Jiao X, Ma W, Chen Y, Li Z (2016) Effects of amino acids supplementation in low crude protein diets on growth performance, carcass traits and serum parameters in finishing gilts. Anim Sci J 87(10):1252–1257

Jung ID, Lee JS, Lee C-M, Noh KT, Jeong Y-I, Park WS et al (2010) Induction of indoleamine 2,3-dioxygenase expression via heme oxygenase-1-dependant pathway during murine dendritic cell maturation. Biochem Pharmacol 80(4):491–505

Kerkaert HR, Cemin HS, Woodworth JC, DeRouchey JM, Dritz SS, Tokach MD et al (2021) Improving performance of finishing pigs with added valine, isoleucine, and tryptophan: validating a meta-analysis model. J Anim Sci 99(1)
Koopmans SJ, Ruis M, Dekker R, Diepen H v, Korte M, Mroz Z (2005) Surplus dietary tryptophan reduces plasma cortisol and noradrenaline concentrations and enhances recovery after social stress in pigs. Physiol Behav 85(4):469–478
Kwon WB, Soto JA, Stein HH (2021) Effects of dietary leucine and tryptophan on serotonin metabolism and growth performance of growing pigs. J Anim Sci 100(1):skab356
Lay DC Jr, Enneking SA, Anderson NC, Richert BT, Sapkota A (2021) Effect of short-term high tryptophan diet fed to sows on their subsequent piglet behavior. Transl Anim Sci 5(3):txab127
Le Floc'h N, Melchior D, Sève B (2008) Dietary tryptophan helps to preserve tryptophan homeostasis in pigs suffering from lung inflammation1. J Anim Sci 86(12):3473–3479
Le Floc'H N, Otten W, Merlot E (2011) Tryptophan metabolism, from nutrition to potential therapeutic applications. Amino Acids 41(5):1195–1205
Li YZ, Baidoo SK, Johnston LJ, Anderson JE (2011) Effects of tryptophan supplementation on aggression among group-housed gestating sows. J Anim Sci 89(6):1899–1907
Liang H, Dai Z, Liu N, Ji Y, Chen J, Zhang Y et al (2018) Dietary L-tryptophan modulates the structural and functional composition of the intestinal microbiome in weaned piglets. Front Microbiol 9:1736
Lin Y, Jiang Z, Yu D, Zhengli W, W, Liu Y, Zhang Z (1999) Study on tryptophan requirement parameters of weaned piglets. Animal Nutr 11(3):44–50
Liu W, Mi S, Ruan Z, Li J, Shu X, Yao K et al (2017) Dietary tryptophan enhanced the expression of tight junction protein ZO-1 in intestine. J Food Sci 82(2):562–567
Liu J, Zhang Y, Li Y, Yan H, Zhang H (2019a) L-tryptophan enhances intestinal integrity in diquat-challenged piglets associated with improvement of redox status and mitochondrial function. Animals (Basel) 9(5)
Liu JB, Yan HL, Cao SC, Liu J, Li ZX, Zhang HF (2019b) The response of performance in grower and finisher pigs to diets formulated to different tryptophan to lysine ratios. Livest Sci 222:25–30
Liuyan L, X, Li C (2019) The research of tryptophan nutrition. Chin Feed (5):50–53
Luo J, Dujian W, Xiangbing M, Hejun Y, Chen D (2019) Effects of porcine transmissible gastroenteritis virus infection on intestinal barrier function and nutrition strategy. J Nanjing Agric Univ 42(5):787–792
Ma X, Yu M, Liu Z, Deng D, Cui Y, Tian Z, Wang G (2020) Effect of amino acids and their derivatives on meat quality of finishing pigs. J Food Sci Technol 57(2):404–412
Mao X, Lv M, Yu B, He J, Zheng P, Yu J et al (2021) Correction to: the effect of dietary tryptophan levels on oxidative stress of liver induced by diquat in weaned piglets. J Anim Sci Biotechnol 12(1):116
Marion J, Biernat M, Thomas F, Savary G, Le Breton Y, Zabielski R et al (2002) Small intestine growth and morphometry in piglets weaned at 7 days of age. Effects of level of energy intake. Reprod Nutr Dev 42(4):339–354
Miao J, Xu J, Wang Z, Yin Y (2018) The effect of dietary supplementation of tryptophan on milk composition and hormones related to lactation of sow. J Fujian Agri Forest Univ 47(4):461–466
Miao J, Adewole D, Liu S, Xi P, Yang C, Yin Y (2019) Tryptophan supplementation increases reproduction performance, milk yield, and milk composition in lactating sows and growth performance of their piglets. J Agric Food Chem 67(18):5096–5104
Moehn S, Pencharz PB, Ball RO (2012) Lessons learned regarding symptoms of tryptophan deficiency and excess from animal requirement studies. J Nutr 142(12):2231S–2235S
Mosnier E, Le Floc'h N, Etienne M, Ramaekers P, Seve B, Pere MC (2010) Reduced feed intake of lactating primiparous sows is associated with increased insulin resistance during the peripartum period and is not modified through supplementation with dietary tryptophan. J Anim Sci 88(2):612–625

Munn AL, Weaver AC, van Wettere W (2021) Supplementary tryptophan fed to sows prior to and after farrowing to improve piglet growth and survival. Animals (Basel) 11(9)

Naatjes M, Htoo JK, Walter K, Tölle KH, Susenbeth A (2014) Effect of dietary tryptophan to lysine ratio on growth of young pigs fed wheat–barley or corn based diets. Livest Sci 163:102–109

Poletto R, Kretzer FC, Hotzel MJ (2014) Minimizing aggression during mixing of gestating sows with supplementation of a tryptophan-enriched diet. Physiol Behav 132:36–43

Sato H, Miura M, Fujieda T, Taciak M, Kikusato M, Sato K, Toyomizu M (2021) Growth performance responses to increased tryptophan supplementation in growing barrows fed three different very low crude protein corn and soybean meal-based diets fortified with essential amino acids. Anim Sci J 92(1):e13605

Shen YB, Voilque G, Odle J, Kim SW (2012a) Effects of elevating tryptophan intake on growth and hypothalamic serotonin secretion and stress hormone secretion in nursery pigs. J Anim Sci 90:2264–2275

Shen YB, Voilque G, Odle J, Kim SW (2012b) Dietary L-tryptophan supplementation with reduced large neutral amino acids enhances feed efficiency and reduces stress hormone secretion of nursery pigs under social-mixing stress. J Nutr 142:1540–1546

Shen YB, Coffey MT, Kim SW (2015) Effects of short term supplementation of L-tryptophan and reducing large neutral amino acid along with L-tryptophan supplementation on growth and stress response in pigs. Anim Feed Sci Technol 207:245–252

Simongiovanni A, Corrent E, Le Floc'h N, van Milgen J (2012) Estimation of the tryptophan requirement in piglets by meta-analysis. Animal 6(4):594–602

Sterndale SO, Miller DW, Mansfield JP, Kim JC, Pluske JR (2020) Increasing dietary tryptophan in conjunction with decreasing other large neutral amino acids increases weight gain and feed intake in weaner pigs regardless of experimental infection with enterotoxigenic Escherichia coli. J Anim Sci 98(8)

Tang M, Chen X, Liang F, Jicheng MQ, Hexin (2008) Study on the balance model of digestible Lys, Met+Cys, Thr, and Trp in low protein diet for pigs. Animal Nutr 20(04):397–403

Tang X, Xiong K, Fang R, Li M (2022) Weaning stress and intestinal health of piglets: a review. Front Immunol 13

Tao W, Zhao D, Sun M, Wang Z, Lin B, Bao Y et al (2018) Intestinal absorption and activation of decitabine amino acid ester prodrugs mediated by peptide transporter PEPT1 and enterocyte enzymes. Int J Pharm 541(1–2):64–71

Tossou MCB, Liu H, Bai M, Chen S, Cai Y, Duraipandiyan V et al (2016) Effect of high dietary tryptophan on intestinal morphology and tight junction protein of weaned pig. Biomed Res Int 2016:1–6

Trevisi, Melchior, Mazzoni, Casini, De Filippi S, Minier L et al (2009) A tryptophan-enriched diet improves feed intake and growth performance of susceptible weanling pigs orally challenged with Escherichia coli K88. J Anim Sci 87(1):148–156

Valini GA d C, Arnaut PR, França I, Ortiz MT, Oliveira MJK d, Melo ADB et al (2023) Increased dietary Trp, Thr, and met supplementation improves growth performance and protein deposition of salmonella-challenged growing pigs under poor housing conditions. J Anim Sci 101:skad141

Wangjun ZY, Fang Z, Linyan, Che L, Yangmin, Wude (2012) Effects of dietary threonine and tryptophan on immune response of growing pigs inoculated porcine reproductive and respiratory syndrome modified live virus vaccine. Animal Nutr 25(6):1189–1198

Wenfeng M, Guoliang WZ, Zhao F (2020) Effects of dietary tryptophan levels on organ index and small intestine related gene expression in weaning pigs. Chin J Agric Biotechol 29(08):1538–1545

Willemen SA, Che L, Dewilde S, Van Hauwaert ML, De Vos M, Huygelen V et al (2014) Enteric and serological distribution of serotonin and its precursor tryptophan in perinatal low and normal weight piglets. Animal 8(5):792–799

Wu G (2009) Amino acids: metabolism, functions, and nutrition. Amino Acids 37(1):1–17

Xia Y (2012) Effects of different protein diets supplement with different tryptophan levels on performance and serum biochemical assays for piglets. Yangzhou University, Yangzhou

Xiao-lan L, Zheng R, Shu-mei M, Min J, Xin W, Kang Y et al (2016) Effect of Trp on the intestinal barrier function in pigs. Modern Food Sci Technol 32(3):42–47

Xu K, Bai M, Bin P, Duan Y, Wu X, Liu H, Yin Y (2019) Negative effects on newborn piglets caused by excess dietary tryptophan in the morning in sows. J Sci Food Agric 99(6):3005–3016

Yano JM, Yu K, Donaldson GP, Shastri GG, Ann P, Ma L et al (2015) Indigenous bacteria from the gut microbiota regulate host serotonin biosynthesis. Cell 161(2):264–276

Yanzhi Z, Wenfeng M, Xiaochen C, Xiangfang Z (2020) Tryptophan catabolism and its application. Animal Nutr 32(03):1019–1024

Yin Y-L (2011) Tryptophan metabolism in animals important roles in nutrition and health. Front Biosci S3(1):286–297

Yin H, Zhang S, Sun J, Yi X, Qiao S (2008) Effects of low protein diets with different net energy levels on growth performance and nutrient digestibility of pigs. Nutr Feedstuffs 44(13):25–28

Youjian S, Defa L, Jianjun X, Zhiguo C (2005) Effects of low protein diet supplemented with tryptophan on performance, serum free amino acids and urea nitrogen of piglets. Chin J Animal Sci 41(1):26–28

Zelante T, Iannitti RG, Cunha C, De Luca A, Giovannini G, Pieraccini G et al (2013) Tryptophan catabolites from microbiota engage aryl hydrocarbon receptor and balance mucosal reactivity via interleukin-22. Immunity 39(2):372–385

Zhang H, Yin J, Li D, Zhou X, Li X (2007) Tryptophan enhances ghrelin expression and secretion associated with increased food intake and weight gain in weanling pigs. Domest Anim Endocrinol 33(1):47–61

Zhang R, Huang G, Ren Y, Wang H, Ye Y, Guo J et al (2022) Effects of dietary indole-3-carboxaldehyde supplementation on growth performance, intestinal epithelial function, and intestinal microbial composition in weaned piglets. Front Nutr 9:896815

Zhao J, Zhang X, Liu H, Brown MA, Qiao S (2018) Dietary protein and gut microbiota composition and function. Curr Prot Pept Sci 20(2):145–154

Chapter 4
Tryptophan Nutrition in Poultry and Ruminant Animals

Changji Lu, Yuanfan Deng, Weiqing Ma, Wence Wang, Pingping Li, Panpan Shi, Tianhai Yan, Yulong Yin, and Peng Huang

Abstract Tryptophan is an essential amino acid that plays an important role in protein synthesis and serves as a precursor for important neurotransmitters and hormones such as serotonin and melatonin. Its involvement in various physiological processes including immune response, growth, and reproduction implies the importance of tryptophan nutrition in farmed animals. Understanding tryptophan nutrition is crucial for meeting animal needs, promoting growth performance, and reducing stress levels. Although the tryptophan requirement may vary among different animal species, the significance of a balanced tryptophan diet is generally recognized. This review provides insights into the importance of tryptophan nutrition in poultry and ruminant production, examining the

C. Lu
Fujian Sunner Development Limited, Nanping City, Fujian Province, China
e-mail: snlcj@sunner.cn.com

Y. Deng
CJ International Trading Corporation Limited, Guangzhou, China
e-mail: yuanfan.deng@cj.net

W. Ma · W. Wang (✉)
Guangdong Provincial Key Laboratory of Animal Nutrition and Regulation, College of Animal Science, South China Agricultural University,
Guangzhou, Guangdong, Province, China
e-mail: maweiqing@stu.scau.edu.cn; wangwence@scau.edu.cn

P. Li · P. Shi · P. Huang (✉)
College of Animal Science and Technology, Hunan Agricultural University, Changsha, China
e-mail: 488162468@stu.hunau.edu.cn; 1140110690@stu.hunau.edu.cn
huangpeng@hunau.edu.cn

T. Yan (✉)
Ruminant Nutrition Unit, Agri-Food and Biosciences Institute, Hillsborough, UK
e-mail: tianhai.yan@afbini.gov.uk

Y. Yin (✉)
Tianjin Institute of Industrial Biotechnology, Chinese Academy of Sciences, National Center of Technology Innovation for Synthetic Biology, Tianjin, China

Institute of Subtropical Agriculture, Chinese Academy of Science, Changsha, China
e-mail: yinyulong@isa.ac.cn

Y. Yin et al. (eds.), *Tryptophan in Animal Nutrition and Human Health*,
https://doi.org/10.1007/978-981-97-4719-1_4

effects of dietary tryptophan concentrations on growth, stress response, and immunity. The results underscore the need for appropriate tryptophan supplementation to optimize the overall health and performance of farmed animals.

Abbreviations

5-HT	5-Hydroxytryptamine
5-HTP	5-Hydroxytryptophan
ASD	Autism spectrum disorder
BCAA	Branched-chain amino acids
CORT	Corticosterone
CYS	Cysteine
DDGS	Distiller's dried grains with soluble
DL-Trp	DL-Tryptophan
FAO	Food and Agriculture Organization
FCR	Feed conversion ratio
FFA	Free fatty acid
GBA	Gut-brain axis
GI	Gastrointestinal
HSP70	Heatshockprotein70
IDO	Indoleamine 2,3-dioxygenase
IgA	Immunoglobulin A
IgG	Immunoglobulin G
IgY	Immunoglobulin Y
IL-10	Interleukin-10
IL-1β	Interleukin-1β
IL-6	Interleukin-6
LNAA	Large molecule neutral amino acids
L-Trp	L-Tryptophan
Lys	Lysine
MET	Methionine
MTNR1A	Melatonin receptor 1A
MTNR1C	Melatonin receptor 1C
SID	Standardized ileal digestibility
TDO	Tryptophan 2,3-dioxygenase
TNF-α	Tumor necrosis factor-α

4.1 Tryptophan Nutrition in Poultry

Tryptophan is an essential amino acid that has several roles in poultry production performance. One of the critical functions of tryptophan is its regulation of animal immune response. Tryptophan can directly participate in the regulation of animal humoral immunity and promote the differentiation of bone marrow T-lymphocyte precursors into mature T lymphocytes, enhancing disease resistance. Additionally, tryptophan metabolites, such as 5-hydroxytryptamine, can bind to different receptors to exert corresponding immune functions. Tryptophan also regulates animal feeding behavior, mainly by acting on the feeding center of the hypothalamus and affecting the secretion of peptide gastric hunger, thereby enhancing feed conversion efficiency and promoting growth and development. Tryptophan is also involved in regulating protein synthesis by promoting the synthesis of hepatic nucleoproteins and enhancing the activity of nuclear membrane nucleoside triphosphatase. Many studies have been conducted on the role of tryptophan in poultry production performance. For example, adding tryptophan to the diet can increase the egg production rate and average egg weight of laying hens and improve the feed conversion ratio of broilers. However, tryptophan utilization is influenced by the competitive carrier system it shares with other large neutral amino acids, and the beneficial effects of tryptophan on cognitive behavior are only observed when its concentration in the blood surpasses certain thresholds.

In addition to its role in poultry production performance, tryptophan is also associated with the regulation of several behavioral and physiological processes through the stimulation of serotonergic activity. Tryptophan is the immediate precursor for serotonin synthesis, and its uptake and utilization by the brain are influenced by the presence of large neutral amino acids. When the concentration of tryptophan in blood surpasses certain thresholds, it can have beneficial effects on cognitive behavior. Low levels of tryptophan in the diet can result in aggressive behavior in chickens. The level of tryptophan uptake by the brain also has a significant impact on animals experiencing certain types of stress.

Finally, the altered metabolism of tryptophan is one of the core biological signatures of prenatal maternal stress. Tryptophan plays a critical role in developing social-emotional behaviors in offspring, and its altered metabolism due to maternal stress can have long-lasting effects on the offspring's behavior and physiology. Therefore, ensuring an adequate level of tryptophan in the diet of poultry and other animals is crucial for their growth, development, and immune response, as well as their behavioral and physiological processes.

4.1.1 Tryptophan Nutrition in Chickens

4.1.1.1 Overview of Tryptophan Nutrition

Tryptophan is an essential amino acid that plays an important nutritional role not only in chickens but also in humans and animals. It is considered a limiting amino acid, which means that its deficiency limits the growth and productive performance of chickens. Tryptophan is a precursor for the synthesis of serotonin and methemoglobin in living organisms, and these compounds play important physiological functions in living organisms.

4.1.1.2 Applications of Tryptophan in Chickens

1. Promote Growth

The growth of chickens is influenced by the dietary supply of amino acids. Tryptophan is a limiting amino acid, and its deficiency will limit the growth rate of chickens. Therefore, it is necessary to add enough tryptophan to the feed to ensure the health and growth development of chickens.

2. Improve Muscle Tissue Quality

Muscle growth is the most important quantification of production performance in chickens. Tryptophan can promote muscle growth and repair and improve the quality and elasticity of muscle tissue. In addition, tryptophan can increase the protein content and nutritional value of chicken meat.

3. Improve Egg Production Performance

Adding an appropriate amount of tryptophan to feed can improve the egg-laying performance of chickens. Tryptophan is a precursor of serotonin, which plays an important role in regulating ovarian function in chickens. Therefore, adding tryptophan to feed can promote the synthesis of serotonin in chickens, thus improving egg production performance.

4.1.1.3 The Role of Tryptophan in Feed Formulation

The level of tryptophan in feed formulations needs to be adjusted according to the growth and production needs of chickens. Inadequate levels of amino acids in feed can limit the growth and production performance of chickens, while excess amino acids can increase feed costs and negatively affect the feed composition balance. Therefore, it is important to add sufficient tryptophan to the feed.

1. Amino Acid Ratio in Compound Feeds

The ratio of amino acids in feed is one of the key factors affecting the growth and performance of chickens. When adding tryptophan to the feed, the content of other amino acids in the feed needs to be considered to ensure a balanced amino acid ratio. This can be achieved by matching the ratio of different amino acids in the feed.

2. Optimize Feed Formulation

To ensure the balance of amino acid contents in the feed, the feed formulation needs to consider a range of factors such as the growth and production requirements of chickens and the feed cost. Depending on the growth and production stages of chickens, the tryptophan content in the feed formula needs to be adjusted to ensure a balance of various nutrients in the feed.

3. Use High-Quality Raw Materials

The quality of ingredients in feed has an important impact on the growth and production performance of chickens. The use of high-quality ingredients can improve the amino acid content of the feed and ensure a balance of various nutrients in the feed. This can be achieved by using high-quality protein sources and adding suitable amino acid supplements.

4.1.1.4 Application of Tryptophan Nutrition in Broilers

The synthesis amount of tryptophan in broilers is very low, which cannot meet the nutritional needs of broilers, and broilers need to obtain tryptophan from diets. Tryptophan is easy to be deficient in the corn-soybean meal diet, so tryptophan is considered to be the third limiting amino acid of broilers after methionine (MET) and lysine (Lys). The following are the trials from 1999 to 2005 included in CVB (Table 4.1) and the supplementary reports on tryptophan requirement of broilers from 2005 to 2020 (Table 4.2).

Table 4.1 Experimental summary of tryptophan requirement of broilers included in CVB (2018)

Trial No.	Estimate *R*	Std. Err. *R*	R^2	Source
1a	0.191	0.0107	0.981	Corzo et al. (2005)
1b	0.175	0.002	0.999	Corzo et al. (2005)
2a	0.238	0.0337	0.949	Fatufe et al. (2005)
2b	0.18	0.0178	0.973	Fatufe et al. (2005)
3a	0.17	0.0104	0.988	Shan et al. (2003): trial 1
3b	0.156	0.0108	0.982	Shan et al. (2003): trial 1
4a	0.153	0.0094	0.987	Shan et al. (2003): trial 2
4b	0.14	0.0074	0.985	Shan et al. (2003): trial 2
5a	0.166	0.0103	0.988	Rosa et al. (2001): trial 1
5b	0.148	0.0034	0.998	Rosa et al. (2001): trial 1
6a	0.156	0.0042	0.998	Rosa et al. (2001): trial 2
6b	0.151	0.0081	0.987	Rosa et al. (2001): trial 2
7a	0.149	0.0068	0.993	Rosa et al. (2001): trial 3
7b	0.135	.	0.976	Rosa et al. (2001): trial 3
8a	0.149	0.011	0.982	Rosa et al. (2001): trial 4
8b	0.129	.	0.959	Rosa et al. (2001): trial 4
9	0.133	.	0.434	Castro et al. (2000)
10	0.142	.	0.524	Castro et al. (2000)
11	0.167	0.0363	0.643	Mack et al. (1999)
12a	0.15	0.0131	0.987	Baker et al. (2002)
12b	0.125	0.0095	0.984	Baker et al. (2002)

Note: "a" represents the evaluation value of the original test data (0.163), and "b" represents the evaluation value after data processing of CVB (2018)

Table 4.2 Studies on tryptophan requirement of broilers from 2005 to 2020

Breed	Phase	Gender	Key points of the report	Source
Hubbard	35–56 d	Male	The tryptophan requirement of broilers is about 0.198%	Hsia et al. (2005)
Ross	8–21 d	Male	A dietary tryptophan content of 0.15 g/MJ ME_N can be considered adequate for broilers aged 2 to 3 weeks	Fatufe et al. (2005)
Peterson×arbor acres	0–14 d	Male	The dietary digestible tryptophan requirement of broilers aged 0–14 days is 1.8 g/kg	Blair et al. (2008)
Ross 708	1–18 d	.	Dietary tryptophan/Lys of 18% can meet the growth requirements of high-yield broilers	Corzo (2012)
Cobb 500	22–42 d	Male	A digestible tryptophan level of 0.2255% can improve feed conversion, and digestible tryptophan level of 0.1919% can significantly improve the slaughter rate	Duarte et al. (2013)
Cobb 500	22–42 d	Male	According to performance and meat quality, the AID tryptophan requirement of male broilers aged from 22 to 42 days is about 0.15%	Wang et al. (2015)
		Female	For female broilers, the AID tryptophan requirement was equal to or not greater than 0.11%	
Cobb 400	1–21 d	Male	The digestible tryptophan level in the basal diet was 0.209% and 0.187%, respectively, and the ratio of tryptophan to Lys was 16%, which could meet the requirements of broilers	Borges et al. (2016)
Arbor acre	1–56 d		Under the cold temperatures in winter, the tryptophan requirement of chicks was 0.24% and that of large chickens was 0.23%	Opoola et al. (2016)
Arbor acres	1–56 d		Under the hot temperatures in summer, the tryptophan requirement of chicks was 0.24% and that of large chickens was 0.21%	Opoola et al. (2016)
Cobb 500	0–42 d	Male	Male broilers at 0–14 d, 14–28 d, and 28–42 d required 86, 264, and 441 mg standardized ileal digestibility (SID) tryptophan/d, respectively	Acácio et al. (2020)
		Female	Female broilers at 0–14 d, 14–28 d, and 28–42 d required 76.5, 237, and 278 mg SID tryptophan/d, respectively	
Ross 308	1–10 d	Male	According to the requirements of average daily gain and feed/gain, SID tryptophan and SID tryptophan/Lys were 1.58 and 1.36 g/kg and 0.13, respectively	Lee et al. (2020)

There is a large variation in tryptophan requirements of chickens as presented in Table 4.3. The influencing factors may be as follows:

CVB (2018) used BL models with lower evaluation values than QL models (0.17 vs. 0.19). BL models represent a population requirement, while QL models produce the average individual requirement (better uniformity for broilers, worse for piglets) (Corzo et al. 2005).

Table 4.3 Tryptophan has a variety of physiological functions in broilers, related reports are as follows

Functions	Key points of the report	Source
Intestinal	The injection of tryptophan into fertilized eggs can promote the intestinal development of broilers in the early stages, thus affecting the growth and weight of broilers in later stages.	Nayak et al. (2015)
	Tryptophan alleviates the harmful effects of ochratoxin A on intestinal mucosa (changes in intestinal morphology, immunoglobulin A (IgA)/ immunoglobulin Y (IgY) antibody levels, and IgA+ cells.	Ricci et al. (2021)
	Increasing the dietary tryptophan/LNAA value can increase feed intake, improve meat quality (drop loss and shear force), adjust the ratio of pathogenic/non-pathogenic bacteria in the intestinal tract, and make the intestinal tract more healthy.	Bello (2016)
Immunity	The combination of arginine and tryptophan can regulate the growth performance. In addition, it regulates the systemic immune response to infectious bursal disease.	Emadi et al. (2011)
	Tryptophan levels higher than those recommended by the NRC improved the antioxidant capacity and humoral and cellular immunity of broilers from 7 to 21 days.	Mund et al. (2020)
	The addition of 0.03–0.12% tryptophan to the NRC (1994) standard may improve the immune function and antioxidant capacity of white-feathered broilers under continuous light by promoting the duodenal synthesis of melatonin. In addition, 0.26% dietary tryptophan level may be an appropriate dietary tryptophan level to improve immune function and antioxidant performance of male Arbor Acres broilers aged from 1 to 21 days.	Zhang et al. (2022)
Carcass	Increasing the supplemental level of tryptophan (145% tryptophan) can improve the growth performance, feed efficiency, meat yield, and profit production of commercial broilers.	Khatun et al. (2020)
Stress	Dietary supplementation of tryptophan 1.5 times higher than the standard supplemental level can alleviate oxidative stress and improve the welfare and feed efficiency of broilers raised in high stocking density cages.	Wang (2014)
	Tryptophan level and stocking density had a significant interaction with plasma cholesterol level. The results showed that supplementation of 0.25 g/kg tryptophan, up to 114% of the NRC-recommended value, had a positive effect on reducing plasma cholesterol levels.	Gogary and Azzam (2014)
	The performance and health status of broilers can be improved by increasing dietary tryptophan levels under hot summer temperatures.	Helal et al. (2020)
	Dietary higher level of tryptophan can improve the welfare status of broilers before and after transport stress (decrease serum corticosterone (CORT) and HSP70, increase 5-hydroxytryptamine (5-HT)). The balance of pathogenic bacteria/non-pathogenic bacteria in the intestinal tract can be achieved by increasing dietary tryptophan levels.	Bello et al. (2017)
	Dietary tryptophan can relieve oxidative stress damage in the liver and intestine caused by heat stress and reduce inflammation by inhibiting NLRP3 inflammatory and downstream cascade reactions.	Ouyang et al. (2022)
	Tryptophan alleviates stress-induced barrier structure and immune disorders in animals by decreasing the secretion of stress hormones and blocking the binding of stress hormones to their receptors	Yue (2017)
	Appropriately increasing dietary tryptophan level (dietary tryptophan level is 0.24%) can enhance plasma 5-HT metabolic pathway, increase feed intake, reduce neurotoxicity, and alleviate the occurrence of stressful behaviors such as feather pecking, to improve the performance and feather coverage of breeding hens.	Liu et al. (2022)

1. Gender

The gender of broilers may influence tryptophan requirements. For example, Wang et al. (2015) reported that the tryptophan requirement of hens at 22–42 days of age may not exceed 0.11% (tryptophan/Lys = 0.16) while that of cocks is between 0.18 and 0.20% (tryptophan/Lys = 0.20–0.22) According to Rosa et al. (2001), taking feed conversion as an indicator, male and female broilers need 0.16% and 0.17% tryptophan, respectively.

2. Breed

Rosa et al. (2001) studied the difference in tryptophan requirement of white-feathered broilers with different genotypes and concluded that the tryptophan requirement of classic Ross broilers and high-yield Ross broilers was 0.18% and 0.17%, respectively, based on body weight.

According to the research results of Xi et al. (2011) on yellow-feathered broilers, the tryptophan requirement of females was 0.203% (tryptophan/Lys = 0.18) and that of males was 0.193% (tryptophan/Lys = 0.17) based on weight gain (quadratic curve model, 95% confidence). Using feed conversion ratio (FCR) as the evaluation index (quadratic curve model, 95% confidence), the tryptophan requirement was 0.210% (tryptophan/Lys = 0.19) for female and 0.190% (tryptophan/Lys = 0.17) for male. The study of Ma et al. (2011) showed that the weight of newborn chicks of yellow-feathered broiler breeders increased by 3.99% when 0.18% tryptophan was added to the basal diet.

3. Stocking Density

The tryptophan requirement of broiler chickens varies at different stocking densities. Wang (2014) reported that different levels of tryptophan (0.18 and 0.27%) had no effect on the growth performance of broilers at low stocking density (11.0 feathers/m^2), but high tryptophan (0.27%) could significantly improve the growth performance of broilers at high stocking density (15.4 feathers/m^2).

According to the study of Gogary and Azzam (2014), there is a significant interaction between tryptophan level and stocking density on plasma cholesterol level. Adding 0.25 g/kg of tryptophan reaches 114% of the NRC-recommended value, which has a positive effect on reducing plasma cholesterol levels, indicating that the requirement for tryptophan will increase at high stocking density.

However, Goo et al. (2019) reported that broilers aged 21 to 42 days raised at high density (18 broilers/m^2) had poor growth performance compared with broilers raised at low density (9 broilers/m^2), while dietary addition of 0.16% tryptophan had no significant effect on broilers raised at either low or high density.

4. Balance Between Tryptophan and Large Neutral Amino Acids

Tryptophan enters the brain in competition with other large neutral amino acids (LNNAs, valine, isoleucine, leucine, tyrosine, phenylalanine, and methionine) through a common transporter protein. Thus, the amount of tryptophan entering the brain depends on the plasma concentrations of tryptophan in relation to the other

LNAAs, tryptophan/LNAA. When the diet contains more LNNAs such as Leu, Val, Ile, and Met, the decrease of tryptophan/LNAA will cause the lack of tryptophan entering the brain, thus affecting production performance. Therefore, the level of LNNAs in the diet is too high, which will increase the requirement for tryptophan in broilers.

Bello (2016) reported that increasing tryptophan/LNAA in the diet could increase the feed intake of broilers, improve meat quality (drip loss and shear stress), adjust the ratio of pathogenic/non-pathogenic bacteria in the gut, and make the gut more healthy. Broilers fed a diet with a higher ratio of tryptophan/LNAA had lower heat shock protein Heatshockprotein70 (HSP70) values, higher serotonin, and better body condition. These results suggested that the tryptophan level should be increased in a high tryptophan/LNAA diet, and the balance between tryptophan and LNAA should be reached.

CVB (2018) also reported that SID tryptophan/Lys increased significantly with the SID-Met+cysteine (CYS)/SID-Lys level of the diet ($P = 0.004$), using weight gain as the evaluation index.

5. Functions of Tryptophan in Broilers

6. Intestinal-Related Function

Tryptophan is an essential nutritious amino acid for broilers, and a diet rich in this amino acid can affect intestinal epithelial cells by inducing anti-inflammatory cytokines and enhancing immune system defense by increasing blood immunoglobulin levels (Bai et al. 2017). In the basal or low crude protein diet, the ratio of villus height to crypt depth of broilers can be increased by adding tryptophan (Fatemi and Toghyani 2018). Wang et al. (2015) showed that treatment with 5-hydroxytryptophan (5-HTP), a precursor of tryptophan, could increase IgA concentration in the duodenum and ileum of chickens Nayak et al. (2015) reported that injecting tryptophan into 18-day-old fertilized eggs could increase the villus height, middle villus width, and crypt depth in the early stage of broilers, promote intestinal development, and thus affect the later growth and field weight of broilers.

Ricci et al. (2021) reported that tryptophan alleviated the harmful effects of ochratoxin A on intestinal mucosa (changes in intestinal morphology, IgA/IgY antibody levels, and IgA+ cells).

Bello et al. (2017) showed that increasing dietary tryptophan levels could achieve a favorable balance of pathogenic bacteria/non-pathogenic bacteria in the intestine.

7. Immune-Related Function

Tryptophan, its metabolites, and key enzymes play an important role in regulating animal immune system response especially humoral immunity (Yao et al. 2011). By reducing the secretion of stress hormones and blocking the binding of stress hormones to its receptors, tryptophan alleviates the intestinal barrier structure and immune disorders caused by stress in animals (Yue 2017). Harden et al. (2016) reported that dietary tryptophan deficiency inhibited the immune function of poultry, resulting in a significant increase in morbidity and mortality of poultry, and these symptoms were alleviated after tryptophan supplementation. Wang (2011) showed that dietary tryptophan supplementation significantly increased the serum IgG and IgA contents of broilers aged 1–3 weeks, which reached the highest level when the tryptophan supplemental level was 0.06% and then showed a downward trend.

Emadi et al. (2011) showed that supplementation of tryptophan with twice NRC level in the basal diet had positive immunoregulatory effects on IFNa, IFNg, and immunoglobulin G (IgG) immune responses of broilers challenged by infectious bursa disease vaccine, which can be used as a practical method to reduce vaccine-related stress. It can enhance the innate immunity and adaptive immune response of chickens and improve the health status of chickens.

Mund et al. (2020) reported that dietary tryptophan levels of 0.3% and 0.5% increased overall antibody response and IgG levels of broilers compared with 0.2% tryptophan ($p < 0.05$) which did not affect IgM levels.

Zhang et al. (2022) showed that the addition of 0.03–0.12% tryptophan to the NRC (1994) standard might improve the immune function and antioxidant capacity of white-feathered broilers under continuous long light by promoting the duodenal synthesis of melatonin. In addition, 0.26% dietary tryptophan level might be an appropriate dietary tryptophan level to improve immune function and antioxidant performance of male Arbor Acres broilers aged from 1 to 21 days. With the increase of tryptophan concentration in the basal diet, the plasma concentrations of interleukin-1β (IL-1β), defective interleukin-6 (IL-6), and tumor necrosis factor-α (TNF-α) of broilers were significantly decreased, while the plasma IgG concentration was significantly increased.

8. Carcass-Related Function

Khatun et al. (2020) reported that the yield of chicken thigh and drumstick supplemented with 145% tryptophan was significantly higher than that of the control group ($P < 0.01$).

However, Kerr et al. (2005) conducted a study on broilers at the age of 49 days before the threshing, which showed that no matter whether directly using tablets (14 h before the threshing), through drinking water (14 h before the threshing), or by adding tryptophan into feed (3 days before the threshing, 46–49 days), the carcass performance of broilers was not affected, but the water intake was reduced.

9. Stress-Related Function

Tryptophan alleviates inflammation caused by stress. Indoleamine 2,3-dioxygenase (IDO) activity is induced by cytokines IFNa, IFNb, and TNF-α. Tryptophan depletion induced by IDO reduces the proliferation of microorganisms and regulates the activity of T cells. The production of tryptophan metabolites also helps to inhibit the proliferation of T cells and exert antioxidant activity (Le Floc'h et al. 2011). When an inflammatory response occurs, the catabolism of tryptophan is enhanced, and the requirement of tryptophan in poultry is increased (Li et al. 2016).

Ouyang et al. (2022) reported that dietary tryptophan supplementation could alleviate oxidative stress damage in the liver and intestine caused by heat stress and reduce inflammation by inhibiting NLRP3 inflammasome and downstream cascade reaction.

Helal et al. (2020) showed that increasing the dietary tryptophan level (tryptophan/Lys increased from 0.16 to 0.24) under hot summer temperatures could improve the performance and health status of broilers.

As reported by Birkl et al. (2017) tryptophan deficiency reduced plasma tryptophan levels, affected the synthesis of the neurotransmitter serotonin, and thus reduced feed intake of poultry and even caused feather pecking and other behaviors. Hierden et al. (2004) suggested that feather pecking was triggered by low serumergic neurotransmission and that mild feather pecking was reduced by increasing Trp in the diet as the intensity of serumergic neurotransmission increased.

Bello et al. (2017) reported that dietary supplementation of higher levels of tryptophan can improve the welfare status of broilers before and after transport stress (reducing serum CORT and HSP70 and increasing 5-HT).

4.1.1.5 Precautions for Tryptophan

Although tryptophan has an important role in poultry feeds, the following points need to be considered when using tryptophan in poultry feeds:

1. Appropriate Supplementation

Although tryptophan has an important nutritional role for poultry, adding excess quantity can have adverse effects on the health of poultry. Therefore, when adding tryptophan, the appropriate amount should be determined according to the breed and growth stage and the content of other nutrients in the feed formula.

2. Shelf Life

Tryptophan is a substance that is easily oxidized and can be easily affected by light and high temperatures and lose its activity. Therefore, when storing and using tryptophan, care should be taken to avoid light and high temperatures and to strictly control its shelf life.

3. Balance with Other Nutrients

Tryptophan is an amino acid that interacts with other amino acids, minerals, vitamins, and other nutrients. Therefore, the interaction between tryptophan and other nutrients should be taken into consideration when designing feed formulations to ensure the comprehensive utilization of nutrients by poultry.

4. Feed Production

During feed production, the mixing time and mixing speed of the feed should be controlled to ensure the uniform distribution of tryptophan in the feed. In addition, attention should be paid to protecting the stability of tryptophan during feed processing to avoid its destruction by heat or oxidation.

4.1.1.6 Conclusion

Tryptophan is an essential amino acid for the growth and production of chickens. Adding an appropriate amount of tryptophan to feed can promote chicken growth, improve muscle tissue quality, and increase egg production performance. The tryptophan content in feed formulations needs to be adjusted according to the growth and production needs of chickens to ensure the balance of amino acids in feed. The balance and sufficiency of nutrients in feed can be achieved by optimizing feed formulations, matching the ratio of amino acids in feed, and using high-quality ingredients.

4.1.2 Tryptophan Nutrition in Ducks

4.1.2.1 Current Situation of Duck Breeding in China

It is estimated that the demand for animal-derived food will double by the middle of this century, due to the increase in global economy and population, compared to the beginning of this century (Costa et al. 2021). As an important part of the poultry industry, the duck industry scale will be further expanded. China is the country with the largest number of ducks in the world, with meat ducks accounting for 69% of the world's total (Hou and Liu 2022), and duck meat, eggs, and feathers can be used as important commodities for primary processing and deep processing. Therefore, there is still much room for development in the economic benefits of the duck breeding industry.

Currently, there are two primary methods for raising ducks: traditional outdoor free-range breeding and indoor flat breeding or cage breeding (Bai et al. 2022; Zhou et al. 2023). Outdoor free-range farming is advantageous because it saves on food costs by allowing ducks to consume fish, shrimp, and aquatic plants in the water. However, this method of raising ducks requires a high amount of water resources. Additionally, due to environmental protection concerns and other factors, duck breeding technology has gradually shifted toward environmentally friendly indoor breeding. Indoor feeding requires a high standard of feed. On one hand, high protein levels in feed can lead to gout in ducks during the brooding period (Choi et al. 2004; Xi et al. 2020). On the other hand, commercial feed for ducks primarily consists of corn and soybean meal, which are largely imported from China and are subject to significant price fluctuations due to international factors (Wang et al. 2020). Therefore, a low-protein diet with balanced D-amino acid supplies has gradually gained acceptance. This type of diet can reduce the dependence on soybean meal while simultaneously decreasing environmental pollution and nitrogen emissions (Such et al. 2021). Table 4.4 lists the tryptophan content of commonly used feed ingredients for ducks.

Tryptophan, as an essential amino acid, plays a significant role in the low-protein diet. Compared to other poultry, there is currently less research on low-protein diets

Table 4.4 Content of tryptophan in commonly used feed ingredients for ducks

Feed name	Tryptophan (%)
Corn grain	0.06–0.09
Soybean meal	0.57
Broken rice	0.12
Paddy	0.09
Wheat bran	0.18–0.26
Corn gluten meal	0.29–0.35
Distiller's dried grains with soluble (DDGS)	0.20

in ducks, which is also an area where further research is needed on the use of tryptophan in duck production. Additionally, tryptophan plays a role in regulating energy metabolism and promoting feed energy utilization. Adding 0.9% or 2% tryptophan to feed for young zebrafish led to better use of energy in the feed, as determined by measuring energy and nitrogen retention values. The content of serotonin in the brain was positively correlated with the amount of tryptophan added (Teixeira et al. 2021). Tryptophan is involved in the synthesis of body proteins and has important physiological significance in regulating body growth and immune function via its metabolites, including 5-hydroxytryptophan, melatonin, kynurenine, NAD^+, and indole.

The current work mainly summarizes the demand for tryptophan in ducks, and summarizes the application of tryptophan and its metabolites in duck production performance, meat, eggs, feathers and other animal-derived products. The purpose was to provide a basis for better application of tryptophan in duck breeding industry.

4.1.2.2 Demand and Utilization of Tryptophan in Ducks

The amount of tryptophan required for different animals varies depending on factors such as animal type, breed, age, feeding environment, and health status. Within a certain range, tryptophan can improve the growth and production performance of animals, but both excessive and insufficient amounts can have adverse effects. Table 4.5 shows the optimal requirements for tryptophan in different duck breeds at different stages, and in general, the demand for tryptophan in ducks follows a trend of increasing first and then decreasing. Additionally, in inflammatory states, the demand for tryptophan in ducks increases. The demand for tryptophan in ducks can also be affected by dietary additives, as shown by the decreased demand for tryptophan in Pekin ducks supplemented with niacin (Xie et al. 2012). It is important to note that evaluating an individual's demand for tryptophan using different indicators may yield different values, so the actual demand for production should be taken into consideration when selecting the optimal requirement (Nguyen et al. 2019). Based on studies where body weight gain and feed conversion rate were dependent variables and tryptophan was an independent variable, the bioavailability of tryptophan in soybean meal for ducks was found to be 94% and 92%, respectively (Adeola 1998).

Table 4.5 Optimum requirements for tryptophan in ducks

Breed	Age	Optimum level of tryptophan (%)	Reference
Pekin duck	1–14	0.20	Xie et al. (2014)
Xichang duck	1–28	0.30	Wang (2014)
Jinding duck	1–28	0.30	Liu (2013)
Longyan duck	22–37	0.29–0.31	Zhang et al. (2021)
Jinding duck	84–118	0.247–0.250	Zhang (2013))
Shanma duck	120–260	0.20	Zhang et al. (2016)

4.1.2.3 Effects of Tryptophan on Duck Performance

Under high-density feeding, the addition of 0.48% or 0.78% tryptophan to the Pekin duck diet can improve average daily weight gain and feed conversion rate (Liu et al. 2015). The feed intake of Pekin ducks showed a trend of first increasing and then decreasing with a gradual increase in the amount of tryptophan added, indicating that the amount of tryptophan added should be appropriate (Han et al. 2009). 5-Hydroxytryptophan, which is involved in the regulation of feeding, is distributed in the Purkinje cell layer of the cerebellum of ducks. After treatment with 100–200 mg/kg tryptophan, 5-HT significantly increased (Liu et al. 2008). Additionally, indole, a tryptophan metabolite, has a certain promoting effect on growth performance. Moreover, 5-hydroxytryptophan can stimulate the contraction of gastrointestinal (GI) smooth muscle, promote GI peristalsis, and enhance nutrient digestion and absorption. However, excessive tryptophan may compete with other neutral amino acid transporters, resulting in reduced absorption of other neutral amino acids and decreased production performance. Adding 250 mg/kg indole to the diet of Gaoyou ducks increased body weight by 0.54%, decreased food intake by 21.38%, reduced production costs at the fourth week of the experiment, and improved the villus height of the intestine (Li et al. 2022).

4.1.2.4 Effects of Tryptophan on Duck Meat, Eggs, and Feather Production

Animal protein is the primary source of protein for humans, and tryptophan plays an essential role in protein synthesis and metabolism. The quality of duck meat, eggs, and feathers depends on protein deposition. Protein deposition, in turn, relies on protein synthesis and metabolism from a cellular metabolism perspective (Zhao et al. 2020). Supplementing 0.04% tryptophan to the diet of Xichang ducks can improve nutritional metabolism, increase serum total protein and albumin levels, and promote muscle protein deposition. This offers an opportunity to meet people's demand for animal protein (Bomfim et al. 2020; Li et al. 2020; Wang 2014).

Duck eggs are a crucial animal product for laying ducks and an essential indicator of breeding duck reproductive performance. Egg quality is determined by its quantity and quality, with protein height and Haugh unit being critical indicators.

Higher protein height and Haugh unit indicate fresher commercial egg quality or higher reproductive performance of breeding eggs. Adding tryptophan to diets can increase the weight of fallopian tubes and ovaries, as well as the number and weight of follicles, thereby increasing egg production. Melatonin, another compound that increases estradiol levels in female ducks, stimulates follicular growth and development (He 2003; Zhang et al. 2021). In contrast, nicotinic acid, another metabolite of tryptophan, added to the diet of Shanma duck from 0–90 mg/kg, had no significant impact on egg production and egg quality (Wang et al. 2023). However, with the increase of the added amount, the fat content in the egg yolk gradually decreased, providing a basis for the production of special duck eggs. Excessive tryptophan supplementation can lead to a decrease in protein height and Haugh unit. Therefore, the supplementation of tryptophan in the diet of laying ducks should not exceed 0.4% (Zhang et al. 2021). Additionally, 5-hydroxytryptophan can increase intracellular calcium levels, promoting intestinal transport of electrolytes and improving eggshell quality (Khattak and Helmbrecht 2019).

Feather pecking is a common and thorny problem in modern duck breeding. To some extent, it is an aggressive behavior, ranging from causing damage to feathers to causing individual death (Jiao et al. 2019). It has a serious impact on animal welfare and product quality, especially on duck feathers. Generally speaking, feather pecking is related to breeding management such as breeding density and light, as well as genetic factors (Zhang 2011). In recent years, studies have found that feather pecking is related to the level of 5-hydroxytryptophan, which is lower in ducks with feather pecking. Increased feather pecking behavior was also found in the tryptophan depletion model (Birkl et al. 2019; Mindus et al. 2021; Wang 2022). When tryptophan is deficient in feed, ducks attempt to supplement tryptophan by pecking their feathers. Of course, supplementation with tryptophan is not the only way to alleviate feather pecking. Transplanting fecal bacteria from non-feather pecking ducks can still improve feather pecking behavior, which may be due to the presence of bacteria involved in tryptophan metabolism in fecal bacteria (Wang 2022).

4.1.2.5 Effects of Tryptophan on Other Aspects of Ducks

Early Chinese and foreign publications have identified several key physiological characteristics. Specifically, melatonin is known to play a crucial role in the regulation of various physiological processes in ducks. For example, injecting exogenous melatonin during duck embryo hatching altered heart rate (Hochel and Nichelmann 2001), and the secretion of endogenous melatonin is essential for proper development during hatching. Furthermore, melatonin can slow down the excretion of hypertonic sodium chloride from salt glands by binding to melatonin receptors (Ching et al. 1999). This regulatory mechanism may have important implications for raising ducks in seawater. Studies have also demonstrated that melatonin receptors melatonin receptor 1A (MTNR1A) and melatonin receptor 1C (MTNR1C) are related to the selection of early-maturing traits in ducks. The expression of MTNR1A mRNA in the ovaries of Shaoxing ducks in the late-maturing group is significantly higher than in the early-maturing group, with MTNR1C showing the opposite trend (Feng et al. 2018). Moreover, melatonin has been found to have positive effects on duck immune performance, such as alleviating the decrease in serum total egg and globulin levels caused by LPS and improving the immune function of Shanma ducks (Jiang et al. 2018). Melatonin also has been shown to alleviate liver inflammation caused by OTA-induced intestinal injury by restoring the diversity of intestinal microorganisms and improving the tight junction proteins in the jejunum and cecum (Xia et al. 2021). Overall, these findings highlight the important role that melatonin plays in duck physiology and may have important implications for the poultry industry.

In summary, tryptophan and its metabolites play an important role in improving the growth and production performance of ducks and have been proven to improve immune performance and alleviate toxin damage. Further research needs to investigate the appropriate requirements of tryptophan in low-protein diets of ducks and the physiological regulation of tryptophan on duck body metabolism.

4.2 Tryptophan Nutrition in Ruminants

4.2.1 Tryptophan Nutrition in Sheep

Tryptophan is an essential amino acid in human and animal body metabolism and is a precursor to the synthesis of the neurotransmitter serotonin. Sheep is one of the most important domestic animals and one of the most important meat sources in China. Lamb quality, safety, and hygiene are major consumer concerns. Appropriate dietary supplementation of tryptophan during sheep rearing can improve the growth performance and meat quality of sheep and reduce stress response and disease incidence (Nicholls et al. 1999; Wang et al. 2023). This chapter focuses on the use of tryptophan in sheep, including a detailed discussion on the nutritional function of tryptophan and the effects of dietary addition of tryptophan.

4.2.1.1 Nutritional Function of Tryptophan

1. Synthetic Neurotransmitter

Tryptophan is an important precursor to the synthesis of the neurotransmitter serotonin, which can regulate the physiological functions of animals such as nervous, endocrine, immune, and digestive.

2. Improve Appetite

Tryptophan can promote animal appetite, increase food digestion and absorption, and improve feed conversion rate.

3. Promote Growth

Tryptophan is an essential amino acid, which can promote protein synthesis in animals and accelerate growth rate.

4. Anti-stress

Tryptophan can alleviate the stress response of animals, reduce the damage of stress to the body, and improve immunity and disease resistance. In addition to the nutritional functions already mentioned, tryptophan has several other important functions.

As a biosynthetic substance: Tryptophan is a biosynthetic substance that can synthesize a variety of important physiological active substances, such as serotonin, zeaxanthin, melanin, and thyroxine. These substances are important for sustaining vital activities and promoting growth and development.

Effects on the nervous system: Serotonin is an important neurotransmitter that regulates the excitability of the nervous system, affecting mood, behavior, cognition, and sleep. As a precursor to serotonin, tryptophan affects the synthesis and release of serotonin, which in turn has effects on the nervous system.

Immune system regulation: Tryptophan can regulate the functions of the immune system to maintain body homeostasis. It affects the proliferation and differentiation of T lymphocytes, regulates the balance between cellular and humoral immunity, and enhances the body's resistance to pathogens.

Effect on the cardiovascular system: Tryptophan can lower blood pressure and improve cardiovascular function. It can reduce the level of angiotensin II by inhibiting the activity of angiotensin-converting enzyme, thereby dilating blood vessels and lowering blood pressure.

Impact on gut microbes: Tryptophan affects the type and number of gut microbes, thus influencing gut health. It promotes the growth and metabolism of probiotics, inhibits the growth and metabolism of harmful bacteria, and maintains the balance of the intestinal flora.

In summary, tryptophan, as an important nutrient, has a variety of important functions and effects. In animal husbandry, scientific and rational use of tryptophan can promote the growth and development of livestock and poultry, increase egg production rates and meat quality, reduce feed costs, and improve economic benefits.

4.2.1.2 Tryptophan Requirements in Sheep Production

The tryptophan requirements of sheep vary with different stages of growth and physiological states. In addition, sheep demand is affected by other factors such as the amount of tryptophan in feed, protein quality and digestibility, ambient temperature and humidity, and the estrus period.

1. Supplemental Amount

The supplemental amount of tryptophan should be determined according to the breed, age, weight, growth stage of sheep, and other factors. Overall, the amount of tryptophan added is 0.1–0.2 g/kg feed DM.

2. Adding Method

Tryptophan can be added to feed in the form of feed additives. Common feed additives are tryptophan salts, tryptophan peptides, etc. Alternatively, tryptophan can be added to feed in natural feed such as soybeans and fish meal.

3. Addition Time

The addition time of tryptophan should be determined according to the growth stage and feeding management of sheep. In general, tryptophan can be added to the feed throughout the growing period. Under certain circumstances, such as in sheep with a strong stress response, tryptophan may be added in appropriate amounts during stress.

4.2.1.3 Matters Needing Attention

1. Appropriate Additional Amount

Excessive addition of tryptophan will affect the growth and development of sheep and their health, leading to nutritional imbalance and disease.

2. Dietary Nutrient Balance

The addition of tryptophan should be matched with other nutrients in the feed to maintain the balance and comprehensiveness of the feed.

3. Feed Quality

The addition of tryptophan should choose high-quality feed and qualified feed additives and avoid using inferior feed and unqualified feed additives.

4.2.1.4 Effect of Application of Tryptophan in Sheep

1. Promote Growth

Tryptophan can promote the growth and development of sheep, increase weight and muscle mass, and improve meat quality. Studies have shown that sheep fed a diet supplemented with tryptophan gained about 10% more daily weight than control animals.

2. Improve the Resistance to Disease

Tryptophan can promote the proliferation and differentiation of immune cells, enhance the activity of immune cells, and improve the resistance to disease. It showed that sheep fed tryptophan were less likely to develop the disease than controls.

3. Improve Meat Quality

Tryptophan can increase the content of carnitine and creatine in muscle, increase the brightness and water retention of meat, and improve meat quality. At the same time, tryptophan also promotes the formation and differentiation of muscle fibers and improves firmness and tenderness in meat.

4. Reduce Stress Response

Sheep is affected by various stress factors in the process of raising, such as environmental changes, feed discomfort, and diseases, which affect the health and growth of sheep. Tryptophan relieves the stress response and reduces stress-related damage to sheep (Brown et al. 2001).

In summary, the use of tryptophan in sheep can significantly improve the production performance and economic benefits of sheep production, meat quality, immunity, and adaptability and reduce stress response. Therefore, an appropriate dietary addition of tryptophan i is an effective production management measure.

4.2.1.5 Supply of Tryptophan

Tryptophan is not produced in sheep itself and must be obtained through dietary feeding. Sheep feeds contain some amounts of natural tryptophan and normally need to be supplemented with tryptophan. Common tryptophan supplements currently on the market include L-tryptophan (L-Trp) and DL-tryptophan (DL-Trp). L-Trp is a naturally occurring L-Trp with high bioavailability and biological activity. D-tryptophan, a synthetic hybrid made from a mixture of L-Trp and D-tryptophan, is less bioavailable but relatively inexpensive (Fenderson and Bergen 1972).

4.2.1.6 Regulation of the Nutrition of Tryptophan

The level of tryptophan in sheep is influenced not only by the supply routes of tryptophan but also by some regulatory factors. These include the following:

1. The Content of Other Amino Acids in the Feed

The proportion and content of amino acids in feed can affect the absorption and utilization of amino acids in sheep and therefore the supply of tryptophan.

2. Protein Quality and Digestibility in Feed

Differences in protein quality and digestibility affect the utilization of protein in sheep and thus the supply of tryptophan.

3. Ambient Temperature and Humidity

High temperatures and humidity affect the sheep's appetite and ability to digest and absorb tryptophan, which in turn affects the supply of tryptophan.

4. Physiological State of Sheep

For example, sheep in heat have an increased demand for tryptophan, and a suitable increase in tryptophan supply is required. In general, ensuring the supply of tryptophan in the feed and adjusting the amount of tryptophan in the feed according to the growth stage and physiological state of the sheep can meet the need for tryptophan in the sheep to the maximum extent and promote the growth and healthy development of the sheep.

4.2.1.7 Tryptophan Nutrition and the Immune System of Sheep

Tryptophan, as an important nutrient, not only plays an important role in the growth and healthy development of sheep but is also closely related to the sheep's immune system. Some products of the tryptophan metabolism pathway play an important role in regulating the function of immune cells and can regulate the proliferation, differentiation, function, and metabolism of immune cells, thus affecting the immune function of sheep. The relationship between tryptophan nutrition and the immune system in sheep is discussed below in terms of tryptophan metabolites.

1. The Regulatory Effects of Tryptophan Metabolites on the Proliferation and Differentiation of Immune Cells

Some products of the tryptophan metabolism pathway play an important role in regulating the proliferation and differentiation of immune cells. For example, the tryptophan metabolites cortisol and oxytocin promote T-lymphocyte proliferation and differentiation, whereas serotonin and melatonin inhibit T-lymphocyte proliferation and differentiation. In addition, fentanyl and imidazolide, the metabolites of tryptophan, also promote the proliferation and differentiation of T lymphocytes.

2. The Regulatory Effect of Tryptophan Metabolites on Immune Cell Function

Some products of the tryptophan metabolism pathway can regulate the function of immune cells and affect immune responses. For example, serotonin and melatonin inhibit the activity of monocytes and macrophages, thus affecting the inflammatory response. However, serotonin and quinolinone can promote the activity of T and B lymphocytes and enhance the immune response.

3. Regulation of Tryptophan Metabolites on Immune Cell Metabolism

Some products of the tryptophan metabolism pathway can regulate the metabolism of immune cells and affect the immune response. For example, serotonin and melatonin promote the release of cytochrome c, thus enhancing the apoptotic effect of immune cells. Uridine and uracil affect the activity and function of immune cells by affecting the metabolism of nucleic acids.

The above regulatory effects indicate that tryptophan metabolites have an important regulatory effect on the immune system of sheep and can affect the proliferation, differentiation, function, and metabolism of immune cells. Therefore, ensuring an adequate supply of tryptophan in sheep not only meets the nutritional needs of sheep but also maintains the normal functioning of the immune system and thus improves resistance and immunity (Wang et al. 2013).

4.2.1.8 Conclusion

Tryptophan is an essential amino acid, and its use in sheep promotes growth and development and improves meat quality, immunity, disease resistance, and stress response. The dietary supplementation amount of tryptophan should be determined based on the breed, age, weight, and growth stage of the sheep, and attention should be taken to select an appropriate combination of feed ingredients and quality and feed additives.

In actual production, tryptophan should be added reasonably according to the growth and development of the sheep and the management of the feed to achieve the best results. At the same time, the research and application of tryptophan should be strengthened to further improve the production performance and economic benefits of sheep and contribute more to the development of animal husbandry.

4.2.2 Tryptophan Nutrition in Cattle

4.2.2.1 Importance and Impact of Cattle Farming

Cattle are one of the most important livestock in the world. Products such as beef, milk, dairy, and leather have high demand in the global market. In addition, cattle are used for plowing, transportation, and material storage. According to information from the Food and Agriculture Organization of the United Nations (FAO), the global cattle inventory is about 1.4 billion, of which India has the highest inventory with more than 320 million heads. Brazil, China, and the USA follow closely behind, each with inventories exceeding 100 million heads. Overall, global cattle farming plays an important role in human livelihoods and diets and is important in many countries and regions.

4.2.2.2 Tryptophan Deficiency in Cattle

L-Tryptophan (Trp; L-α-aminoindole-3-propionic acid) is a nutritionally essential amino acid for monogastric animals (e.g., humans, pigs, dogs, rats, mice, and chickens) and preweaning ruminants (e.g., calves and lambs) due to the lack of endogenous synthesis. It is a white powder and a neutral amino acid with a pI value of 5.96 [pKa (α-COOH) = 2.46; pKa (α-NH3+) = 9.41)]. Trp was first isolated from casein in 1902 by F.G. Hopkins using base hydrolysis. Like some other amino acids (e.g., homocysteine and CYS), Trp binds non-covalently with serum albumin. Ruminant animals, including cows, often cannot obtain sufficient tryptophan from their diet (Kollmann et al. 2008; Mclaren et al. 1965). This is because the protein in their feed is often deficient in tryptophan, and ordinary feed cannot provide enough tryptophan to meet the maximum requirement of ruminant animals (Li and Song 2020). Additionally, due to the degradation by rumen microorganisms, a large amount of supplemented dilute tryptophan can also be broken down, affecting the effectiveness of tryptophan supplementation (Tan et al. 2021).

4.2.2.3 Tryptophan Plays an Important Role in the Health of Cows

Tryptophan is an essential amino acid for mammals, but during the course of evolution, mammals have lost the ability to synthesize tryptophan. Therefore, most animals need to obtain tryptophan through their diets. In agricultural animals, tryptophan is widely used as a food supplement to meet their nutritional requirements (Yao et al. 2011). Tryptophan plays a significant role in cow health and production. It has important effects on feed intake behavior, growth, reproduction, immune function, and stress responsiveness in dairy cows (Wu 2010). Research has shown that tryptophan deficiency could lead to reduced nitrogen deposition, delayed growth, low feed intake, stress sensitivity, and rough coat (Coşkun et al. 2006; Cubero et al. 2006; Henry et al. 1992).

1. Effect on the Function of the Intestinal Tract

A study on Hanwoo cattle demonstrated that L-treatment increased melatonin synthesis and might have increased the secretion of intestinal hormones (Lee et al. 2019). Although the mechanisms behind the modulation of blood ghrelin by L-Trp remain unclear, tryptophan is the sole precursor of serotonin, a key neurotransmitter that participates in the modulation of central neurotransmission and enteric physiological function. Moreover, gut microbiota can influence tryptophan metabolism, which in turn affects the gut-brain axis (GBA) (Gao et al. 2020). Intestinal microbiota metabolize tryptophan into indole derivatives (Roager and Licht 2018; Smith and Macfarlane 1996), which regulate bacterial motility and biofilm formation. Biofilms are multicellular structures formed by cell-expressed adhesins, which can prevent the invasion of non-indole-producing pathogens such as *Salmonella* and *Pseudomonas aeruginosa* (Ellermann and Sartor 2018). Tryptophan metabolism plays a central role in the GBA. The major pathways of tryptophan metabolism include serotonin synthesis, the kynurenine pathway, and microbial degradation pathways, which are implicated to varying degrees in disorders such as depression and irritable bowel syndrome. Impaired tryptophan metabolism has been linked to various diseases such as GI functional disorders, depression, and autism spectrum disorder (ASD), highlighting the potential role of tryptophan metabolism in the pathophysiology of these diseases (Gao et al. 2020). Peripheral 5-HT acts as a signaling molecule to regulate intestinal function, and by activating 5-HT receptors, the signal is transmitted to neurons, which in turn regulate a series of physiological activities such as intestinal secretion, peristalsis, vasodilation, and nutrient absorption (Mohammad-Zadeh et al. 2008).

2. Enhancing Stress Resistance

The physiological stress response is designed to promote survival, as hormones and neurotransmitters act as signals to prepare the body for trauma and infection.

However, if the heightened immune response is directed against self-antigens or becomes dysregulated after prolonged activation, the system may contribute to pathological changes. L-Trp is an essential amino acid for animals, which is not only involved in the regulation of physiological functions such as nociception, sleep, feeding, and body temperature, but also in the alleviation of stress response. Numerous studies have confirmed that L-Trp can alleviate behavioral changes induced by chronic stress and improve the stress resistance of humans and animals (Guzik et al. 2006; Koopmans et al. 2005). Tryptophan metabolites, as signaling molecules in the microbiota-GBA, play a crucial role in regulating stress. Tryptophan, as a precursor of 5-HT, can adapt to dynamic stress and modulate anxiety and fear by increasing its metabolism pathways toward 5-HT. Additionally, tryptophan can act as a prebiotic to promote the growth of beneficial bacteria, regulate gut microbiota, and exert an anti-stress effect. As weaning represents a significant stressor for calves, it is necessary to strengthen related research.

3. Effect on Milk Production

Tryptophan promotes mammary gland development and milk production in lactating animals, increasing milk production. Studies have shown that adding rumen-protected tryptophan to the diet of lactating cows could increase milk production, blood growth hormone levels, and antioxidant capacity of cows. In addition, the addition of rumen-protected tryptophan can increase melatonin levels in milk (Yao et al. 2020). The effect of adding tryptophan and per-rumen tryptophan to the diet on the production performance and blood index of lactating cows was found. The experiment showed that the addition of tryptophan and per-rumen tryptophan could improve milk yield, blood growth hormone content, and antioxidant ability of cows, and the effect of adding per-rumen tryptophan was better.

4. Improving Immune Function

L-tryptophan has two outcomes in the body: one is as a substrate for the synthesis of tissue proteins and the other is catabolized to produce bioactive metabolites. In the animal body, there are three main pathways for tryptophan catabolism: (1) the kynurenine pathway metabolism, accounting for about 95% of the entire metabolic pathway, tryptophan in the indole 2,3-dioxygenase (IDO), and tryptophan 2,3-dioxygenase (TDO) to generate kynurenine, after which most of the kynurenine is catalyzed hydrolysis into 3-hydroxy-octopamine benzoic acid, and finally, through a multistage enzymatic reaction; generation of active molecules such as pyridine carboxylic acids, quinolinic acid, and nicotinamide adenine dinucleotide; and only a small portion of kynurenine to generate kynurenine quinolinic acid; (2) tryptophan through hydroxylation and decarboxylation and ultimately the generation of melatonin; and (3) tryptophan metabolism to generate indole acetic acid discharged from the stomach.

Tryptophan and its metabolites have the function of regulating the immunity of livestock and poultry, mainly by regulating the production of lymphocytes and immunoglobulins, so that the body's immunity is improved. Tryptophan is directly involved in the regulation of humoral immunity, promoting the differentiation of bone marrow T-lymphocyte precursors (Weinlich et al. 2006; Widner et al. 2000) into mature T lymphocytes and enhancing disease resistance, while L-Trp deficiency leads to reduced humoral immunity. IDO and TDO catalyze tryptophan catabolism in the body. IDO is expressed at low levels in the thymus, lung, and intestine and is upregulated when the body is infected or inflamed. Activation of IDO predisposes to tryptophan depletion in the cellular microenvironment. If IDO activity expression is abnormally high, it can lead to excessive loss of tryptophan in the cellular microenvironment, leaving some important functional cells in a state of "tryptophan starvation," cells in a rapidly dividing phase or pathogenic microorganisms, such as tumor cells, as well as viruses and bacteria, are sensitive. These phenomena suggest that tryptophan is closely related to immune function. Tryptophan has a certain influence on the phagocytic ability of macrophages, the proliferation of splenic lymphocytes, and the proliferation and differentiation processes of T cells. The precursor of tryptophan, 5-HT, can bind to various receptors and exert its corresponding immunological functions. 5-HT can increase the production of IL-6 and interleukin-10 (IL-10), both of which can promote humoral immune responses and regulate adaptive immune responses. In addition, tryptophan-derived metabolites related to immune function include melatonin, which can promote the enlargement of immune organs and increase the proliferation of lymphocytes and splenic cells (Wu 2009). Melatonin can also induce macrophages to produce interferon and interleukin, exerting immunoregulatory effects and significantly increasing the release of immune factors by immune cells (Reiter 1998). When the body undergoes inflammation, the demand for tryptophan increases, and supplementing tryptophan can significantly enhance the immune system of animals, leading to enhanced tryptophan catabolism. Conversely, insufficient tryptophan in feed can affect the nutritional balance of livestock and poultry, decrease immunity, increase the likelihood of disease, and correspondingly increase mortality rates (Wu 2009). Kolodziej found that during the process of immune regulation, the production of the pro-inflammatory cytokine interleukin-17 (IL-17) is inhibited by kynurenine and its products, leading to the inhibition of the maturation and differentiation of helper T-cell 17 (Th17) and the transformation of CD4+ T cells into CD4 + CD25 + FOXP3+ regulatory T cells, which helps alleviate the inflammatory response in the body (Kolodziej 2013). Studies have also shown that the addition of tryptophan to the diet can increase the total immunoglobulin content in the blood. Numerous studies have confirmed the important role of tryptophan in enhancing animal immune function.

5. Regulating Growth Function

Tryptophan mainly regulates growth performance by affecting animal feeding behavior. It mainly regulates animal feeding through the following ways: (a) 5-Hydroxytryptamine (5-HT, also known as serotonin) acts on the hypothalamic feeding center to achieve the regulatory effect on animal feeding. 5-HT is a derivative of tryptophan. Tryptophan is hydroxylated to form 5-hydroxytryptophan and then decarboxylated to form 5-HT; (b) the ratio of branched-chain amino acids (BCAA), LNAA, and tryptophan can also affect animal feeding behavior; (c) hormonal regulation of animal feeding behavior; and (d) 5-HT binds to different receptors to regulate animal feeding behavior.

6. Regulating Protein Synthesis

Tryptophan is a component of proteins, and some nucleic acid tryptophan receptors bind to tryptophan (Conejos et al. 2021). Their binding has a good promoting effect on liver nuclear protein synthesis and promotes protein synthesis. When studying the mechanism of tryptophan regulating protein synthesis, it was found that tryptophan could affect the metabolism of proteins in the liver, promote cytoplasmic poly(A)-RNA synthesis, promote the release of labeled RNA from the nucleus, and enhance the activity of nuclear membrane nucleoside triphosphatase (Herschel et al. 1995). Studies have shown that tryptophan could increase protein synthesis in the liver and muscles by stimulating insulin release. If animals lack amino acids in their daily diet, the synthesis rate of proteins in the liver and muscles may decrease, which may be related to a decreased nutrient absorption rate (Cortamira et al. 1991). In the body, changes in LNAA and plasma-free fatty acid (FFA) concentrations can affect whether tryptophan exists in the blood in a free or bound state. Tryptophan also participates in regulating protein synthesis by promoting the synthesis of liver nuclear proteins and enhancing the activity of nuclear membrane nucleoside triphosphatase.

4.2.2.4 Tryptophan Supplementation Methods and Effectiveness Assessment

Cows primarily obtain amino acids through the digestion of protein derived from microorganisms and plants. Therefore, the quality and quantity of absorbed amino acids have a significant impact on cow health, milk production, and milk quality. In recent years, researchers have been working on finding methods to improve the efficiency of tryptophan supplementation in cows. One approach is to prepare rumen-protected tryptophan through high-temperature treatment and extrusion using a pelletizer and adding other ingredients to reduce its degradation in the rumen. This process is expected to enhance the bioavailability of tryptophan supplementation in cows.

4.2.2.5 Conclusion

In summary, the application of tryptophan in cattle is of great significance for maintaining cattle health, improving milk production, and enhancing the quality of milk. In terms of tryptophan supplementation, it is crucial to choose appropriate supplementation methods to improve the biological utilization of tryptophan. Future studies should focus on exploring more effective tryptophan supplementation strategies to meet the nutritional requirements of cows and improve their production performance.

References

Acácio RM, Sakomura NK, Viana GS, Malheiros EB, Fernandes JBK, Suzuki RM (2020) Optimum standardized ileal digestible tryptophan intake for broiler chickens. Livest Sci 240

Adeola O (1998) Bioavailability of tryptophan in soybean meal and tryptophan retention in the carcasses of four-week-old ducks. Poult Sci 77(9):1312–1319

Bai M, Liu H, Xu K, Oso AO, Wu X et al (2017) A review of the immunomodulatory role of dietary tryptophan in livestock and poultry. Amino Acids 49(1):67–74. https://doi.org/10.1007/s00726-016-2351-8

Bai H, Yang B, Dong Z, Li X, Song Q et al (2022) Research note: effects of cage and floor rearing systems on growth performance, carcass traits, and meat quality in small-sized meat ducks. Poult Sci 101(1)

Baker D, Batal A, Parr T, Augspurger N, Parsons C (2002) Ideal ratio (relative to lysine) of tryptophan, threonine, isoleucine, and valine for chicks during the second and third weeks posthatch. Poult Sci 81(4):485–494

Bello UAH (2016) Influence of dietary tryptophan to large neutral amino acids ratios on performance and well-being of broiler chickens subjected to stressful conditions

Bello UA, Idrus Z, Yong Meng G, Awad EA, Soleimani Farjam A (2017) Gut microbiota and transportation stress response affected by tryptophan supplementation in broiler chickens. Ital J Anim Sci:1–7

Birkl P, Kjaer JB, Szkotnicki W, Forsythe P, Harlander-Matauschek A (2017) Acute tryptophan depletion: the first method validation in an avian species (Gallus gallus domesticus). Poult Sci 96(9):3021–3025

Birkl P, Chow J, McBride P, Kjaer JB, Kunze W, Forsythe P, Harlander-Matauschek A (2019) Effects of acute tryptophan depletion on repetitive behavior in laying hens. *Front Vet Sci* 6

Blair K, Finger E, Marsh A, Morton J, Mondillo K, Buzas B et al (2008) The role of 5-HTTLPR in choosing the lesser of two evils, the better of two goods: examining the impact of 5-HTTLPR genotype and tryptophan depletion in object choice. Psychopharmacology 196(1):29–38

Bomfim MAD, Marchao RS, Ribeiro FB, de Siqueira JC, Silva LR, Takishita SS (2020) Digestible tryptophan requirement for tambaqui (*Colossoma macropomum*) fingerlings. Rev Ciênc Agron 51(2)

Borges B, Samuel NRDC, Strighini JH, Mascarenhas AG, Mello HHDC, Martins JMDS et al (2016) Digestible tryptophan levels for male broilers in pre-starter and starter diets. Semina Ciencias Agrarias 37(4):2529–2538

Brown J, Howie SE, Entrican G (2001) A role for tryptophan in immune control of chlamydial abortion in sheep. Vet Immunol Immunopathol 82(1–2):107–119

Castro AJD, Gomes PC, Pupa JMR, Rostagno HS, Albino LFT, Nascimento AH (2000) Tryptophan requirement for broilers from 1 to 21 days of age. Revista Brasileira de Zootecnia 29:1743–1749

Ching AC, Hughes MR, Poon AM, Pang SF (1999) Melatonin receptors and melatonin inhibition of duck salt gland secretion. Gen Comp Endocrinol 116(2):229–240

Choi HK, Atkinson K, Karlson EW, Willett W, Curhan G (2004) Purine-rich foods, dairy and protein intake, and the risk of gout in men. N Engl J Med 350(11):1093–1103

Conejos JRV, Nejad JG, Kim J-E, Moon J-O, Lee J-S, Lee H-G (2021) Supplementing with L-tryptophan increases medium protein and alters expression of genes and proteins involved in milk protein synthesis and energy metabolism in bovine mammary cells. Int J Mol Sci 22(5)

Cortamira NO, Seve B, Lebreton Y, Ganier P (1991) Effect of dietary tryptophan on muscle, liver and whole-body protein synthesis in weaned piglets: relationship to plasma insulin. Br J Nutr 66(3):423–435

Corzo A (2012) Determination of the arginine, tryptophan, and glycine ideal-protein ratios in high-yield broiler chicks. J Appl Poult Res 21(1):79–87. https://doi.org/10.3382/japr.2011-00362

Corzo A, Kidd MT, Thaxton JP, Kerr BJ (2005) Dietary tryptophan effects on growth and stress responses of male broiler chicks. Br Poult Sci 46(4):478–484

Coşkun S, Ozer C, Gönül B, Take G, Erdoğan D (2006) The effect of repeated tryptophan administration on body weight, food intake, brain lipid peroxidation and serotonin immunoreactivity in mice. Mol Cell Biochem 286:133–138

Costa M, Cardoso C, Afonso C, Bandarra NM, Prates JAM (2021) Current knowledge and future perspectives of the use of seaweeds for livestock production and meat quality: a systematic review. J Anim Physiol Anim Nutr 105(6):1075–1102

Cubero J, Valero V, Narciso D, Rivero M, Marchena J, Rodríguez A, Barriga C (2006) L-tryptophan administered orally at night modifies the melatonin plasma levels, phagocytosis and oxidative metabolism of ringdove (Streptopelia roseogrisea) heterophils. Mol Cell Biochem 293(0):79–85

Duarte KF, Otto MJ, Rosemeire da SF, Siqueira JC, Puzotti MM, Garcia EA et al (2013) Digestible tryptophan requirements for broilers from 22 to 42 days old. Rev Bras Zootec 42(10):728–733

Ellermann M, Sartor RB (2018) Intestinal bacterial biofilms modulate mucosal immune responses. J Immunol Sci 2(2):13–18

Emadi M, Jahanshiri F, Kaveh K, Hair-Bejo M, Ideris A, Alimon AR (2011) Nutrition and immunity: the effects of the combination of arginine and tryptophan on growth performance, serum parameters and immune response in broiler chickens challenged with infectious bursal disease vaccine. Avian Pathol 40(1):63–72

Fatemi, M., Toghyani, M. (2018). Effect of tryptophan supplementation in protein deficient diets on performance, gut development and immune responses in broiler chickens. Islamic Azad University *Rasht Branch*(1)

Fatufe AA, Hirche F, Rodehutscord M (2005) Estimates of individual factors of the tryptophan requirement based on protein and tryptophan accretion responses to increasing tryptophan supply in broiler chickens 8–21 days of age. Arch Anim Nutr 59(3):181–190

Fenderson CL, Bergen WG (1972) Effect of ration composition and protein level on plasma free tryptophan and ruminal microbial tryptophan content in sheep. J Anim Sci 35(4):896

Feng P, Zhao W, Xie Q, Zeng T, Lu L, Yang L (2018) Polymorphisms of melatonin receptor genes and their associations with egg production traits in Shaoxing duck. Asian Australas J Anim Sci 31(10):1535–1541

Gao K, Mu CL, Farzi A, Zhu WY (2020) Tryptophan metabolism: a link between the gut microbiota and brain. Adv Nutr 11(3):709–723

Gogary EMR, Azzam MM (2014) Effects of dietary tryptophan levels and stocking density during the growing-finishing phase on broiler performance and immunity. Asian J Animal Vet Adv 9(9):568–577

Goo D, Kim JH, Park GH, Delos Reyes JB, Kil DY (2019) Effect of stocking density and dietary tryptophan on growth performance and intestinal barrier function in broiler chickens. Poult Sci 98(10):4504–4508

Guzik AC, Matthews JO, Kerr BJ, Bidner TD, Southern LL (2006) Dietary tryptophan effects on plasma and salivary cortisol and meat quality in pigs. J Anim Sci 84(8):2251–2259

Han XF, Gao YP, Hou SS (2009) Study on tryptophan requirement of Peking duckings in 1~14 days. Chin Feed 0(22):22–23

Harden JL, Lewis SM, Lish SR, Suárez-Fariñas M, Gareau D, Lentini T et al (2016) The tryptophan metabolism enzyme L-kynureninase is a novel inflammatory factor in psoriasis and other inflammatory diseases. J Allergy Clin Immunol 137(6):1830–1840

He LH (2003) Effects of melatonin passive immunization on the follicle development and reproductive hormone in female ducks. Hunan Agricultural University

Helal A, Abdelazeem F, Thabet H (2020) Influence of dietary threonine and tryptophan supplementations on broiler productivity traits under egyptian summer conditions. Arab Univ J Agric Sci (2)

Henry Y, Seve B, Colleaux Y, Ganier P, Saligaut C, Jego P (1992) Interactive effects of dietary levels of tryptophan and protein on voluntary feed intake and growth performance in pigs, in relation to plasma free amino acids and hypothalamic serotonin. J Anim Sci 70(6):1873–1887

Herschel S, Rabinder NK et al (1995) Tryptophan binding to nuclei of rat liver and hepatoma. J Nutr Biochem 6(2):73–79

Hierden Y, Boer S, Koolhaas J, Korte S (2004) The control of feather pecking by serotonin. Behav Neurosci 118(3):575–583
Hochel J, Nichelmann M (2001) Ontogeny of heart rate responses to exogenous melatonin in Muscovy duck and chicken embryos. Life Sci 69(19):2295–2309
Hou S, Liu L (2022) The current situation, future development trend and suggestions of waterfowl industry in 2021. Chin Animal Ind 58(03):227–231
Hsia LC, Hsu JH, Liao CT (2005) The effect of varying levels of tryptophan on growth performance and carcass characteristics of growing and finishing broilers. Asian Aust J Animal Sci 18(2)
Jiang D, Lao Y, Wu C, Wang J, Bao H, Lin Z et al (2018) Effects of exogenous melatonin and lipopolysaccharide on the growth and immune function of Shanma ducks. J South Chin Agric Univ 39(5):47–52
Jiao H, Yan Z, Ma Q, Li X, Jiang Y, Liu Y, Chen J (2019) Influence of Xiaoyaosan on depressive-like behaviors in chronic stress-depressed rats through regulating tryptophan metabolism in hippocampus. Neuropsychiatr Dis Treat 15:21–31
Kerr BJ, Moran E, Kidd MT (2005). Effect of supplementary tryptophan prior to marketing on carcass quality in broilers1
Khattak F, Helmbrecht A (2019) Effect of different levels of tryptophan on productive performance, egg quality, blood biochemistry, and caecal microbiota of hens housed in enriched colony cages under commercial stocking density. Poult Sci 98(5):2094–2104
Khatun A, Chowdhury S, Roy B, Sultana F, Krishnan P, Channarayapatna G (2020) Effect of increasing levels of dietary tryptophan on growth performance, meat yields and cost benefit of commercial broiler chickens. J Bangl Agric Univ (0)
Kollmann M, Locher M, Locher F, Eder K, Meyer H, Bruckmaier R (2008) Effects of tryptophan supplementation on plasma tryptophan and related hormone levels in heifers and dairy cows. Domest Anim Endocrinol 34(1):14–24
Kolodziej L (2013) Systemic metabolism of tryptophan and its catabolites, kynurenine and 3-HAA, in mice with inflammatory arthritis. Gene 512(1):23–27
Koopmans SJ, Ruis M, Dekker R, van Diepen H, Korte M, Mroz Z (2005) Surplus dietary tryptophan reduces plasma cortisol and noradrenaline concentrations and enhances recovery after social stress in pigs. Physiol Behav 85(4):469–478
Lee SB, Lee KW, Wang T, Lee JS, Jung US et al (2019) Intravenous administration of L-tryptophan stimulates gastrointestinal hormones and melatonin secretions: study on beef cattle. J Anim Sci Technol 61(4):239–244
Lee JY, Sung YK, Kong C (2020) Ideal ratios of standardized ileal digestible methionine, threonine, and tryptophan relative to lysine for male broilers at the age of 1 to 10 days. Anim Feed Sci Technol 262
Le Floc'h, N, Otten W, Merlot E (2011) Tryptophan metabolism, from nutrition to potential therapeutic applications. Amino Acids 41:1195–1205.
Li, Song R (2020) The regulation of zein biosynthesis in maize endosperm. Theor Appl Genet 133(5):1443–1453
Li HW, Zhu Q, Wu L, Yin Y, Kong X (2016) Physiological function and dietary application of tryptophan in livestock and poultry. *Chin J Animal Nutr 28*(3):659–664
Li YL, Li J-M, Ning L-J, Lu D-L, Luo Y et al (2020) Mitochondrial fatty acid β-oxidation inhibition promotes glucose utilization and protein deposition through energy homeostasis remodeling in fish. J Nutr 150(9):2322–2335
Li YJ, Yan W, Ling ZP, Tan J, Ding XL et al (2022) Effects of dietary indole level on growth performance, serum biochemistry and index and trace element content of Gaoyou ducks. Feed Res 45(21):44–49
Liu X (2013) Effects of levels of tryptophan on growth and blood biochemical indexes of 0–4 week-old duking. Northeast Agricultural University
Liu ZH, Li XY, Tong JJ, Qiu ZY, Zhan HC et al (2008) Duck cerebellum participates in regulation of food intake via the neurotransmitters serotonin and neuropeptide Y. Nutr Neurosci 11(5):200–206

Liu Y, Yuan J, Zhang L, Zhang Y, Cai S et al (2015) Effects of tryptophan supplementation on growth performance, antioxidative activity, and meat quality of ducks under high stocking density. Poult Sci 94(8):1894–1901

Liu M, Huang C, Chen Y, Wang D, Shi L et al (2022) Effects of dietary tryptophan level on feather damage performance and kynurenine metabolism of laying breeders in natural mating cage. Chin J Animal Nutr 34(12):7770–7780

Ma Ye, Zhan X, Zhu Q, Liu W, Xia L (2011) Dietary tryptophan level affects performance, antioxidant function and serum biochemical indices of yellow-feathered broiler breeders. Chin J Animal Nutr 23(12):2177–2182

Mack S, Bercovici D, De Groote G, Leclercq B, Lippens M, Pack M et al (1999) Ideal amino acid profile and dietary lysine specification for broiler chickens of 20 to 40 days of age. Br Poult Sci 40(2):257–265

Mclaren GA, Anderson GC, Barth KM (1965) Influence of methionine and tryptophan on nitrogen utilization by lambs fed high levels of non-protein nitrogen. J Anim Sci 24(1):231–234

Mindus C, van Staaveren N, Fuchs D, Gostner JM, Kjaer JB, Kunze W et al (2021) L. rhamnosus improves the immune response and tryptophan catabolism in laying hen pullets. Sci Rep 11(1)

Mohammad-Zadeh L, Moses LC, Gwaltney-Brant SM (2008) Serotonin: a review. J Vet Pharmacol Ther 31(3):187–199

Mund MD, Riaz M, Mirza MA, Rahman Zu, Mahmood T, Ahmad F, Ammar A (2020) Effect of dietary tryptophan supplementation on growth performance, immune response and antioxidant status of broiler chickens from 7 to 21 days. Vet Med Sci 6(1):48–53

Nayak J, Ezhilvalavan, Kirubaharan, John, Ranjan, Manimaran (2015) Effect of in ovo feeding of arginine and/or tryptophan on hatchability and small intestinal morphology in broiler chicken. Ind J Poultry Sci

Nguyen L, Salem SMR, Salze GP, Hieu D, Davis DA (2019) Tryptophan requirement in semi-purified diets of juvenile Nile tilapia Oreochromis niloticus. Aquaculture 502:258–267

Nicholls T, Nitsos I, Walker DW (1999) Tryptophan metabolism in pregnant sheep: increased fetal kynurenine production in response to maternal tryptophan loading. Am J Obstet Gynecol 181(6):1452–1460

Opoola, Bawa, Agubosi (2016) Determination of optimum tryptophan requirement of broiler chickens reared in the cool season under tropical environment

Ouyang G, Li Q, Zhou H, Zheng J, Guo S, Liu S (2022) Effects of dietary tryptophan supplementation on antioxidant capacity and inflammatory response in liver and intestine of broilers subjected to heat stress. Chin J Vet Sci 42(06):1005–4545

Reiter RJ (1998) Oxidative damage in the central nervous system: protection by melatonin. Prog Neurobiol 56(3):359–384

Ricci FG, Terkelli LR, Venancio EJ, Justino L, dos Santos BQ, Baptista AAS et al (2021) Tryptophan attenuates the effects of OTA on intestinal morphology and local IgA/IgY production in broiler chicks. Toxins 13(1)

Roager HM, Licht TR (2018) Microbial tryptophan catabolites in health and disease. Nat Commun 9(1):3294. https://doi.org/10.1038/s41467-018-05470-4

Rosa A, Pesti GM, Edwards HM, Bakalli R (2001) Tryptophan requirements of different broiler genotypes. Poult Sci 80(12):1718–1722

Shan A, Sterling K, Pesti G, Bakalli R, Driver J, Tejedor A (2003) The influence of temperature on the threonine and tryptophan requirements of young broiler chicks. Poult Sci 82(7):1154–1162

Smith E, Macfarlane G (1996) Enumeration of human colonic bacteria producing phenolic and indolic compounds: effects of pH, carbohydrate availability and retention time on dissimilatory aromatic amino acid metabolism. J Appl Bacteriol 81(3):288–302

Such N, Pal L, Strifler P, Horvath B, Koltay IA, Rawash MA et al (2021) Effect of feeding low protein diets on the production traits and the nitrogen composition of excreta of broiler chickens. Agric Basel 11(8)

Tan P, Liu H, Zhao J, Gu X, Wei X, Zhang X et al (2021) Amino acids metabolism by rumen microorganisms: nutrition and ecology strategies to reduce nitrogen emissions from the inside to the outside. Sci Total Environ 800:149596

Teixeira C, Rodrigues P, Serrão P, Figueira L, Guimarães L, Teles LO et al (2021) Dietary tryptophan supplementation does not affect growth but increases brain serotonin level and modulates the expression of some liver genes in zebrafish (Danio rerio). Fish Physiol Biochem 47(5):1541–1558

Wang P (2011) Effects of dietary tryptophan on growth and biochemical indexes of 0-3weeks AA broilers. Northeast Agricultural University

Wang FM (2014) Effect of tryptophan on growth performance and nutrition of Xichang duck. Heilongjiang Animal Sci Vet Med (13):92–94

Wang J (2022) Effects of fecal bacteria transplantation on intestinal microbiota, brain neurotransmitters and feather-pecking behavior in muscovy duck. Guizhou University

Wang B, Qu Y, Liu T, Bao X, Yang L et al (2013) Immunohistochemical evidence for the presence of tryptophan hydroxylase in the brains of insects as revealed by sheep anti-tryptophan hydroxylase polyclonal antibody

Wang, Min, Yuan (2015) Apparent ileal digestible tryptophan requirements of 22- to 42-day-old broiler chicks. J Appl Poult Res pfv061

Wang H, Long W, Chadwick D, Velthof GL, Oenema O et al (2020) Can dietary manipulations improve the productivity of pigs with lower environmental and economic cost a global meta-analysis. Agric Ecosyst Environ 289

Wang J, Ding L, Yu X, Wu F, Zhang J et al (2023) Tryptophan improves antioxidant capability and meat quality by reducing responses to stress in nervous Hu sheep. Meat Sci 204:109267

Weinlich G, Murr C, Richardsen L, Winkler C, Fuchs D (2006) Decreased serum tryptophan concentration predicts poor prognosis in malignant melanoma patients. Dermatology 214(1):8–14

Widner B, Leblhuber F, Walli J, Tilz GP, Demel U, Fuchs D (2000) Tryptophan degradation and immune activation in Alzheimer's disease. J Neural Transm (Vienna) 107(3):343–353

Wu G (2009) Amino acids: metabolism, functions, and nutrition. Amino Acids 37(1):1–17

Wu G (2010) Functional amino acids in growth, reproduction, and health. Adv Nutr (Bethesda, Md) 1(1):31–37

Xi P, Ling Y, Zheng C (2011) Effects of dietary tryptophan levels on growth, whole-body nutrient retention and hypothalamic serotonin concentration in yellow-feathered broilers aged 1 to 21 days. Chin J Animal Nutr 23(001):43–52

Xi Y, Huang Y, Li Y, Yan J, Shi Z (2020) Fermented feed supplement relieves caecal microbiota dysbiosis and kidney injury caused by high-protein diet in the development of gosling gout. Animals 10(11)

Xia D, Yang L, Li Y, Chen J, Zhang X, Wang H et al (2021) Melatonin alleviates ochratoxin A-induced liver inflammation involved intestinal microbiota homeostasis and microbiota-independent manner. J Hazard Mater 413

Xie M, Han X, Hou S, Huang W, Yu J (2012) Interrelationship between tryptophan and niacin of Pekin ducks aged from 1 to 14 days. Chin J Animal Nutr 24(12):2335–2341

Xie M, Han XF, Huang W, Hon SS (2014) Effects of niacin status on tryptophan requirements of starter white Pekin ducks. Livest Sci 159:75–78

Yao K, Fang J, Yin YL, Feng ZM, Tang ZR, Wu G (2011) Tryptophan metabolism in animals: important roles in nutrition and health. Front Biosci (Schol Ed) 3(1):286–297

Yao S, Wu H, Ma H, Fu Y, Wei W et al (2020) Effects of rumen bypass melatonin feeding (RBMF) on milk quality and mastitis of Holstein cows. PeerJ 8:e9147

Yue Y (2017) Mechanism of tryptophan on intestinal epithelial barrier and immune functions in chronic unpredictable stress exposed animals. China Agricultural University

Zhang SF (2011) Nutritional factors and improvement measures that affect the growth of meat duck feather feathers. Chin Poultry 33(16):50–51

Zhang K (2013) Effects of different levels tryptophan on growth biochemical indexes of reserve ducks. Northeast Agricultural University

Zhang H, Chen W, Yan L, Wang S, Ruan D et al (2016) Tryptophan optimal requirement for laying ducks in peak laying period. Chin J Animal Nutr 28(4):1060–1067

Zhang YN, Ruan D, Wang S, Huang XB, Li KC et al (2021) Estimation of dietary tryptophan requirement for laying duck breeders: effects on productive and reproductive performance, egg quality, reproductive organ and ovarian follicle development and serum biochemical indices. Poult Sci 100(8)

Zhang Y, Ma D, Yu M, Feng J, Zhang M (2022) Effects of dietary tryptophan level on intestinal melatonin synthesis, immune function and antioxidant capacity of broilers under continuous long light exposure. China Animal Industry

Zhao Y, Li J-Y, Jiang Q, Zhou X-Q, Feng L, Liu Y et al (2020) Leucine improved growth performance, muscle growth, and muscle protein deposition through AKT/TOR and AKT/FOXO3a signaling pathways in hybrid catfish Pelteobagrus vachelli x Leiocassis longirostris. Cells 9(2)

Zhou Y, Yan C, Chen D, Zhang C, Zhao X (2023) Integration of grape-duck production pattern boosts duck behavior, meat quality, fecal microbiota and soil microorganisms. Agric Basel 13(1)

Chapter 5
Tryptophan Metabolism in Human Diseases

Yadong Zhao, Libin Wang, Xiong Xiong, Yi-Zhou Gao, and Hou-De Zhou

Abstract Tryptophan (Trp) is an essential amino acid needed in the human body. When metabolized in the body, it produces various bioactive compounds such as serotonin, melatonin, and indoles. These compounds play important roles in regulating neural activities, energy, and nutrition metabolism. However, abnormalities in tryptophan metabolism have also been associated with a variety of human diseases such as neurological diseases, cardiovascular diseases, obesity, and diabetes. This chapter describes the biological connection between intestinal Trp metabolism and human diseases and also discusses the changes in Trp metabolism under disease models.

Y. Zhao (✉)
School of Food and Pharmacy, Zhejiang Ocean University,
Zhoushan, Zhejiang Province, China
e-mail: yadong@kth.se

L. Wang (✉)
College of Horticulture, Nanjing Agricultural University, Nanjing, Jiangsu, China
e-mail: wanglibin@njau.edu.cn

X. Xiong (✉)
College of Food Science and Light industry, Nanjing Tech University, Nanjing, China
e-mail: xiongx0624@njtech.edu.cn

Y.-Z. Gao (✉)
The Center for Microbes, Development and Health, Shanghai Institute of Immunity and Infection, Chinese Academy of Sciences, Shanghai, China
e-mail: gaoyizhou@yeah.net

H.-D. Zhou (✉)
National Clinical Research Center for Metabolic Diseases, The Second Xiangya Hospital of Central South University, Changsha, Hunan, China
e-mail: houdezhou@csu.edu.cn

Y. Yin et al. (eds.), *Tryptophan in Animal Nutrition and Human Health*,
https://doi.org/10.1007/978-981-97-4719-1_5

Abbreviations

5-HT	5-Hydroxytryptamine
AhR	Aryl hydrocarbon receptor
BBB	Blood-brain barrier
CD	Crohn's disease
CNS	Central nervous system
CRWLD	Calorie-restricted weight-loss diet
CVD	Cardiovascular disease
FOXP3	Forkhead box p3
HIOMT	Hydroxyindole-O-methyltrasferase
HIV	Human immunodeficiency virus
HPLC	High-performance liquid chromatography
IAA	Indole acetic acid
IBD	Inflammatory bowel disease
IDO1	Indoleamine 2,3-dioxygenase 1
IPA	Indole propionic acid
KYN	Kynurenine
KYNA	Kynurenic acid
MetS	Metabolic syndrome
MLT	Melatonin
NAD^+	Nicotinamide adenine dinucleotide
NAS	N-Acetylserotonin
NAT	5-Serotonin-N-acetyltransferase
NDD	Neurodegenerative diseases
NMDA	N-methyl-d-aspartate
ox-LDL	Oxidized low-density lipoprotein
PDA	Pancreatic ductal adenocarcinoma
pDCs	Plasmacytoid dendritic cells
PXR	Pregnane X receptor
QA	Quinolinic acid
QPRT	Quinolinic acid phosphoribosyl transferase
TDO	Tryptophan 2,3-dioxygenase
Trp	Tryptophan
UC	Ulcerative colitis

5.1 Introduction

In addition to being a substrate for protein synthesis, Trp can produce biologically important compounds in the host through several metabolic pathways (Fig. 5.1). Trp and its metabolites play critical roles in maintaining human health, preventing and treating various diseases, and developing new therapeutic targets.

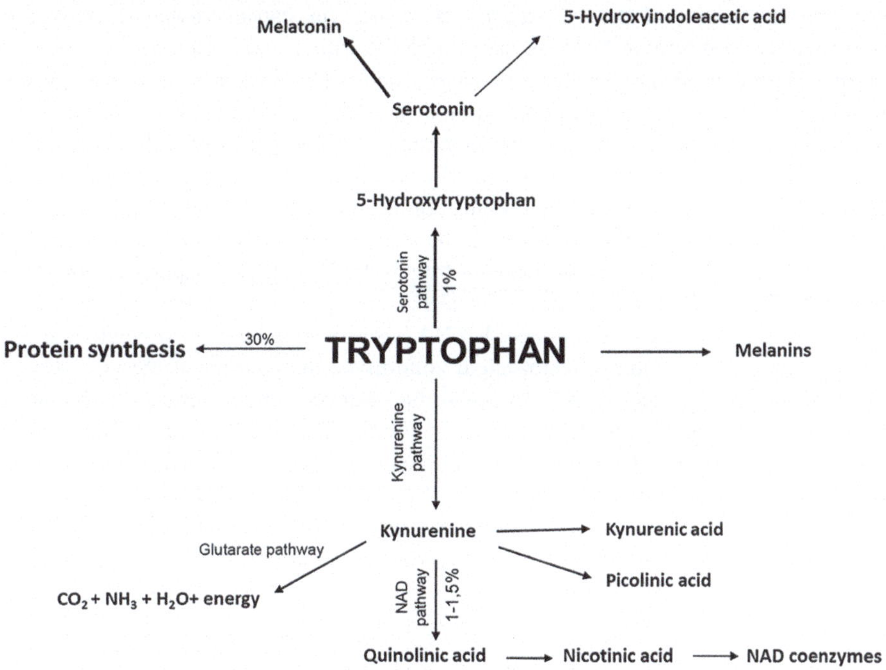

Fig. 5.1 Main pathways of Trp metabolism in humans (Comai et al. 2020)

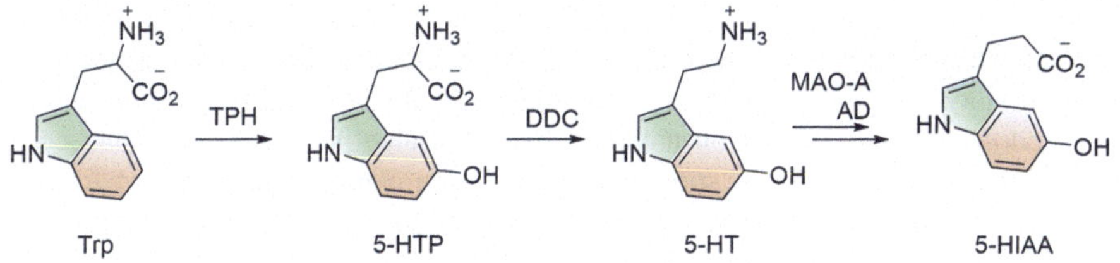

Fig. 5.2 Trp metabolism through the serotonin pathway

Trp can be utilized to synthesize the neurotransmitter serotonin through the 5-hydroxytryptamine (5-HT) pathway (Fig. 5.2). In some organs such as the intestine and pineal gland, 5-HT can be then changed into N-acetylserotonin (NAS) through 5-serotonin-N-acetyltransferase (NAT). NAS is further converted into melatonin (MLT) by the catalyst of hydroxyindole-O-methyltrasferase (HIOMT). The neurotransmitter 5-HT plays important roles in the regulation of memory, appetite and digestion, sexual behavior, mood, and sleep (Lucki 1998). The neuromodulator MLT is not only an effective antioxidant due to the presence of two G-protein-coupled receptors MT1 and MT2, but also a critical regulator in many processes, such as pain, circadian rhythm, mood, and sleep (Gabriella and Comai 2019; Gobbi and Comai 2019). More particularly, only the free Trp component could cross the blood-brain barrier (BBB) to reach the brain via carrier-mediated transport (Fernstrom and Wurtman 1971; Fernstrom and Wurtman 1972; Gessa et al. 1974).

In various central and peripheral diseases, such as Huntington's disease, acquired immunodeficiency syndrome, depression, schizophrenia, Alzheimer's disease, suicide, and cancer, the activation of the kynurenine (KYN) pathway and the resultant imbalance in the generation of downstream metabolites are always involved (Chen et al. 2009). Among the KYN metabolites, quinolinic acid (QA) and kynurenic acid (KYNA) have been largely employed in the study of neuropsychiatric and neurodegenerative disorders because of their binding to the N-methyl-d-aspartate (NMDA) receptors. Take QA for example, it works as an agonist of NMDA receptors to induce excitotoxic neuronal damage; KYNA has proven to be neuroprotective as an NMDA antagonist. Under physiological conditions, the amount of the neurotoxic QA and the neuroprotective KYNA has to be balanced. In contrast, the imbalance between QA and KYNA is characteristic of some pathological events, while higher levels of QA increased neurotoxicity. For example, pro-inflammatory cytokines involved in neuroinflammatory diseases are released to boost IDO and then increase Trp degradation through the KYN pathway, thus resulting in an increase in QA and/or a decrease in KYNA (Maes et al. 2011; Manchia et al. 2017; Myint et al. 2007).

5.2 Central Nervous System (CNS) Disease

5.2.1 Tryptophan Alteration in Mental Diseases

Mental disease refers to a wide range of mental health disorders, including depression, anxiety, and schizophrenia. These mental health disorders not only have a tremendous negative impact on patients' lives, but also impose a significant economic and social burden on society. Accumulating evidence in recent years has revealed that the occurrence of mental diseases is closely related to the disorder of Trp metabolism in the human body.

Trp is a precursor for the neurotransmitters serotonin and melatonin and also acts as the substrate for the synthesis of Niacin. Trp in the human body is mainly metabolized through the kynurenine (KYN) and serotonin pathways, with the KYN pathway mainly producing the 5-hydroxytryptamine (5-HT) and the serotonin pathway producing serotonin and melatonin (Maes et al. 2002). Disturbances of Trp metabolism are closely related to the occurrence of mental diseases (Karu et al. 2016). For instance, in individuals suffering from depression, there is a reduction in plasma Trp levels, leading to decreased 5-HT synthesis and subsequently resulting in symptoms such as low mood and diminished interest. Among patients with anxiety disorders, however, an increase in 5-HT synthesis may lead to excessive tension and fear (Ramos-Chávez et al. 2018). Moreover, disturbances in Trp metabolism, such as elevated serotonin levels, are also present in patients with schizophrenia. In addition to endogenous Trp metabolism, gut microbes are also involved in the production of specific Trp metabolites and indirectly affect host physiology. Gut microbes metabolize Trp mainly through the production of indole-like compounds, such as indole propionic acid (IPA) and indole acetic acid (IAA) (Sonner et al. 2019). These compounds can influence the structure and function of the intestinal flora, thereby affecting the immune and nervous systems of the host.

5.2.2 *Tryptophan Alteration in Neurological Diseases*

Studies in recent years showed that the occurrence of neurological disorders is closely linked to the imbalance of Trp metabolism. Alterations in Trp metabolism may be an important mechanism that triggers neurological disorders. It is found that Trp levels in serum are reduced in patients with cerebral infarction and cerebral hemorrhage, possibly due to impaired Trp transport caused by disruption of the blood-brain barrier and cerebral ischemia and hypoxia (Ko et al. 2018). Moreover, alterations in serotonin levels may also be involved in the occurrence of cerebrovascular diseases (Barcelon et al. 2019). For patients with encephalitis and epilepsy, researches indicated that the Trp metabolic pathway is altered, resulting in reduced levels of Trp and its metabolites. This may be related to the inflammatory response of the patient's neurons and glial cells, as well as to the role of Trp in the regulation of neuronal excitatory and inhibitory processes (González Esquivel et al. 2017). In patients with brain tumors, studies have shown alterations in Trp metabolic pathways and changes in the levels of Trp metabolites (Fiore and Murray 2021). These may be related to abnormalities in tumor cell metabolism and the role of Trp and its metabolites in cell proliferation and apoptosis.

5.2.3 *Tryptophan Alteration in Neurodegenerative Diseases*

Neurodegenerative diseases (NDD) are a class of disorders, including Alzheimer's disease, Parkinson's disease, and Huntington's disease, which are usually associated with neuronal degeneration. Recent research found that Trp and its metabolites are related to the occurrence and development of neurodegenerative diseases. Firstly, Trp and its metabolites play essential roles in the growth, development, and function of neurons (Kondo et al. 2013). Trp-deficiency will cause abnormal growth and development of neurons, which may increase the risk of neuronal degeneration (Delgado 2000). Secondly, Trp and its metabolites can affect the redox status and metabolism in neurons (Hajsl et al. 2020). Oxidative stress is a major pathological mechanism of neurodegenerative diseases, while Trp and its metabolites are capable of reducing the damage caused by oxidative stress, exhibiting protective effects on neurons (Platten et al. 2019). Moreover, Trp and its metabolites can also regulate the synthesis and release of neurotransmitters, including dopamine, serotonin, norepinephrine, and other neurotransmitters, which play an important role in neuronal degeneration (Gibson 2018). Finally, Trp and its metabolites also play critical roles in neuroinflammation and neuropathological responses. It has been found that neuroinflammation is an important pathological mechanism in neurodegenerative diseases, and Trp and its metabolites have been shown to reduce the neuroinflammatory responses in neurons (Zhang et al. 2021). Together, Trp and its metabolites play essential functions in the treatment of neurodegenerative diseases, especially in the growth, development, and function of neurons as well as the regulation of oxidative stress and neurotransmitters.

5.2.4 Pharmacological Effects of Tryptophan in Neuropsychiatric Diseases

Trp has important pharmacological effects in neuropsychiatric disorders. Serotonin is a neurotransmitter synthesized through the Trp metabolic pathway, which plays a key role in inhibiting mood disorders and controlling behavior. Therefore, by increasing the uptake and metabolism of Trp, serotonin levels can be elevated, thereby improving the symptoms of mental disorders such as anxiety, depression, and other emotional disorders. In a study examining the use of Trp as an adjunctive treatment for depression, the results showed that Trp supplementation manifested better positive effects compared to traditional antidepressant medications (Sarris et al. 2016). Trp supplements (4 or 8 mg/day) reduced the incidence of intervention (Morand et al. 1983). In a double-blind, placebo-controlled study where patients were categorized into groups treated either with Trp (up to 6 g/day) or with placebo, Trp treatment did not affect the number of violent incidents; however, it was able to significantly reduce the use of injections of sedatives and antipsychotics needed to control aggressive behaviors (Volavka et al. 1990). Moreover, Trp intake improves sleep efficiency and total sleep time and reduces sleep fragmentation index and sleep latency at the same time (Bravo et al. 2013). It is found that Trp can be used as a hypnotic for patients with chronic insomnia, which can shorten the sleep onset. Importantly, the dose of therapeutic Trp did not show side effects, such as typical sedative-hypnotic effects including impaired psychomotor and mental performance, intolerance, and physiologic sleep architecture changes.

5.2.5 Limitations of Tryptophan Being a Biomarker of Neuropsychiatric Disorders

Despite that Trp has been widely applied as an important biomarker in the study of neuropsychiatric diseases, there are still some limitations. Firstly, the relationship between Trp and neuropsychiatric disorders is not unidirectional. Although some studies have shown that disturbances in Trp metabolism are associated with the occurrence and development of neuropsychiatric disorders (Miller et al. 2013), neuropsychiatric disorders themselves may also affect the metabolic processes of Trp (Modoux et al. 2021). Therefore, there may be some errors in judging the occurrence and development of neuropsychiatric diseases merely by using Trp as a biomarker.

Secondly, the metabolic pathway of Trp is very complex, and its metabolites have different effects on neuropsychiatric disorders. For example, some metabolites like serotonin and melatonin are deemed helpful for treating diseases like depression and anxiety. Other metabolites such as vanillic acid and anterior pituitary hormones, however, may not be associated with neuropsychiatric diseases, and they may even have negative effects on these diseases (Myint et al. 2012). Therefore, relying solely on the total Trp level to assess the severity of neuropsychiatric disorders and treatment effectiveness may overlook the bioactive differences in metabolites, leading to inaccurate judgments.

Thirdly, there are also some limitations in the measurement methods and standardization of Trp. To date, high-performance liquid chromatography (HPLC) and gas chromatography are commonly used methods for Trp measurement. These measurements, however, differ in sensitivity and specificity, which may result in biased results (Lesniak et al. 2013). In addition, there is a lack of a well-established standardization for Trp, and there are variations in the units of measurement and standards used in different studies, making it difficult to compare and validate the results from these studies (Bray et al. 1994).

5.3 Cancer and Tumors

5.3.1 Overexpression of IDO1 in Tumor Microenvironment

Indoleamine 2,3-dioxygenase 1 (IDO1) is a rate-limiting metabolizing enzyme which converts Trp into kynurenine (Kyn). It inhibits the immune function of T cells and natural killer cells by metabolizing Trp. As a key regulator, IDO1 plays an important role in regulating immune escape, promoting tumor growth and metastasis, and enhancing drug resistance to tumor chemotherapy and immunotherapy. High expression of IDO1 is coupled with the occurrence, development, and prognosis of most tumors. Studies have shown that in the microenvironment of malignant tumors, the expression of IDO1 is significantly increased, and the expression level of IDO1 is positively correlated with the malignant degree of tumors (Liang et al. 2019). For example, IDO1 is expressed in primary colon cancer cells. A number of studies have found that Trp concentration decreased and kynurenine metabolites increased in patients with colon cancer, indicating an increase in IDO1 activity. The high expression of IDO1 often means a high probability of metastasis and poor prognosis of colorectal cancer. The discovery of a large number of cells expressing IDO1 in adjacent lymph nodes often means a lower 5-year survival rate in patients suffering from colorectal cancer. The expression of IDO1 seems to be an important signal in the early stage of colorectal cancer, which may be driven by colonic dysplasia and related gene expression. In some tumor cell lines, IDO1 is constitutively expressed, such as HT-29 and HCT-116 cell lines, but not in the case of Caco-2. Since this expression pattern happens without the inflammatory environment, suggesting that IDO1 expression might be associated with the genetic background that drives CRC.

Cancer cells are resistant to apoptosis because of the quinolinic acid–dependent production of nicotinamide adenine dinucleotide (NAD^+) in these cells. A unique characteristic of cancer cells is their ability to resist survival pressures. For example, cancer cells may not only survive but also exhibit high proliferative activity due to the high level of NAD^+ even under hypoxic conditions. NAD^+ is a cofactor of many enzymes and is essential for maintaining cell integrity. Therefore, in addition to the nicotinamide pathway, cancer cells also need to meet their rapid consumption of NAD^+ through other synthetic pathways to maintain their anti-apoptotic properties. The quinolinic acid (QA) produced by the kynurenine metabolic pathway is a substitute, and it can synthesize NAD^+ by quinolinic acid phosphoribosyl transferase (QPRT). Colon cancer cells also express QPRT. If colon cancer cells produce NAD^+ through quinolinic acid, then the combination of IDO1 and QPRT inhibitors may have an even stronger therapeutic effect on colon cancer.

The overexpression of IDO1 was found in 90% of gastric cancer patients. It has been also reported that the expression of IDO1 is not only closely associated with the depth of invasion of gastric tumor cells and metastasis of lymph nodes, but also reduces the infiltration of $CD4^+$ and $CD8^+$ T cells. Gastric tumor cells may promote immune escape through IDO1 activity. These studies have found that Forkhead box p3 (FOXP3) is closely related to the expression of IDO1 in gastric cancer cells. Additionally, increased expression of IDO1 has also been reported in human pancreatic cancer. In metastatic pancreatic ductal adenocarcinoma (PDA), the overexpression of IDO1 is associated with an increase in the number of T cells. However, in patients having lymph node metastasis, IDO1 expression was observed in both primary and metastatic tumor cells. This indicates that PDA survives in lymph nodes through IDO1-related immune escape. High expression of IDO1 has also been found in esophageal cancer, and its high expression is linked to less lymphocyte infiltration and a lower survival rate. In the p53-knockout mouse model of esophageal cancer, the IDO1 expression in metastatic tumors is closely related to the expression of the periosteal protein. This protein induces IDO1 expression by activating the signal transducer and transcription 1, suggesting the connection between inflammatory signaling pathways and esophageal cancer.

It is worth noticing that IDO1 displays double effects on tumor treatment. On the one hand, IDO1 can inhibit the activity of immune cells, thereby reducing their immune recognition and attack on tumor cells and decreasing the effectiveness of tumor treatment (Shao et al. 2021). On the other hand, overexpression of IDO1 in tumor cells can also enhance their resistance to chemotherapy and immunotherapy (Li et al. 2017). The overexpression of IDO1 in malignant tumors leads to the depletion of Trp and the accumulation of kynurenine metabolites, resulting in the immune escape of tumor cells (Tang et al. 2022). Therefore, inhibiting the expression or activity of IDO1 has become a potential and effective way in the treatment of tumors. Currently, various IDO1 inhibitors have been developed including small molecule compounds and antibodies, such as navoximod (Jung et al. 2019), NLG919 (Nayak et al. 2014), and 1-methyl Trp (1-MT) (Qian et al. 2009). These inhibitors can be utilized along with or in combination with chemotherapy, radiotherapy, and immunotherapy to improve the efficacy of clinic treatment (Le Naour et al. 2020).

5.3.2 *Adaptation of Tumor Cells to Tryptophan Depletion*

With in-depth research on the tumor microenvironment, it has been discovered that tumor cells possess a unique adaptive mechanism known as an adaption to Trp depletion (Platten et al. 2012). Trp is one of the essential amino acids in the human body, serving not only as a fundamental substrate for protein synthesis but also as a precursor for the synthesis of neurotransmitters and various bioactive molecules. Tumor cells exhibit high metabolic activity, and they meet the needs of rapid growth and proliferation by promoting their own metabolism (Kim and Tomek 2021). In this process, tumor cells consume Trp through different pathways, including enzymatic degradation or converting it into other metabolites (Amobi et al. 2017). Two key enzymes in the Trp metabolic pathway are tryptophan 2,3-dioxygenase (TDO) and indoleamine 2,3-dioxygenase (IDO) (Klar et al. 2020). These two enzymes catalyze the degradation of Trp by converting it into other metabolites such as kynurenine and quinolinic acid. Among them, IDO expression is increased in tumor cells and is involved in various biological processes such as tumorigenesis, development, and immune escape (Sari et al. 2019). On the one hand, high expression of IDO inhibits the activity of immune cells and reduces their immune recognition and attack capacity, thereby increasing the survival of tumor cells (He et al. 2020). On the other hand, IDO degrades Trp to provide the energy necessary for tumor cell growth, thus meeting the demands of rapid growth and proliferation of tumor cells (Favre et al. 2010).

5.4 The Role of Tryptophan Metabolism in Other Diseases

5.4.1 *Intestinal Inflammatory Diseases*

Inflammatory bowel disease (IBD) represents a class of inflammatory disorders in the gastrointestinal tract (Cosnes et al. 2011). The chance of IBD is increasing and become prevalent in most countries around the world. Based on clinical features, IBD could be categorized into ulcerative colitis (UC) and Crohn's disease (CD). In CD, inflammation can occur along with the entire gastrointestinal tract and penetrate the intestinal wall. However, UC is confined to be present in the rectum and colon, with inflammation occurring only in the inner layer of the intestinal wall. The natural incidence of IBD is often unpredictable, and its symptoms can be characterized by relapse and remission stages. Although the pathogenesis of IBD has not been fully elaborated, it has been demonstrated to be associated with both endogenous (immune system and genetic susceptibility) or exogenous (drugs, food, psychological stress, and microbiota) factors.

In the metabolism of food-derived Trp, the intrinsic gut microbiota is actively involved in healthy people. In fact, Trp could be converted to different metabolites in the gut lumen; for example, bacterial decarboxylases can produce serotonin, while bacterial tryptophanases are able to generate indole derivatives. In addition to the conversion of some into indoleacetate and indolepropionate, the others can be excreted as indoxyl sulfate through the urinary system. Certain indole derivatives including indole-3-acrylic acid, indole-3-propionic acid, indole-3-acetaldehyde, indole-3-acetic acid, and indole-3-aldehyde have been found to be binding with the aryl hydrocarbon receptor (AhR) (Alexeev et al. 2018). AhR is a crucial intestinal immune regulator; therefore, these indole derivatives are beneficial to not only keep the gut environment stable and integrative but also regulate intestinal epithelial immune function, thus prohibiting pathogen invasion. Furthermore, AhR in peripheral immune cells maintains the homeostasis required for the tolerance of symbiotic microbiota (Sarris et al. 2016). Another Trp derivative, indolyl-3-propionic acid, binds to the pregnane X receptor (PXR) to maintain mucosal homeostasis and the function of the intestinal epithelial barrier.

Although it is known that the gut microbiota regulates the human endogenous Trp metabolism, the detailed metabolic pathways are still not fully understood due to the complexity of catalytic enzymes in symbiotic bacteria (Agus et al. 2018). In addition, Trp catabolism has been found to be strain-specific, meaning that not all symbiotic bacteria produce AhR ligands (Zelante et al. 2013; Alexeev et al. 2018). Therefore, the balance between the host and bacterial Trp metabolism has to be established to maintain local gut function, immune homeostasis via AhR activity, and the gut-brain axis connection via serotonin signaling. Conversely, disrupted Trp metabolism causes an ecological imbalance in the gut (Agus et al. 2018). The impact of gut dysbiosis on IBD has been clearly demonstrated. IBD patients are often accompanied by a depletion of the gut mucosal microbiota and microbial infiltration (Yu 2018), with a decreased production of AhR and PXR ligands (Wlodarska et al. 2017). The AhR deficiency reduces the production of IL-22, which is crucial for intestinal mucosal regeneration and protection, that in turn damages tight junctions in intestinal epithelial cells, thus increasing permeability and activating the pro-inflammatory pathways (Tarun et al. 2010). The imbalance in a gut ecosystem as a pathogenic factor is involved in chronic inflammatory episodes, T-cell activation, and the mucosal immune system. The level of ecological disruption is significantly correlated with imbalanced plasma pro-inflammatory cytokines and increased Trp catabolism (Guan and Zhang 2017; Nikolaus et al. 2017). The accumulation of QA in active IBD with increased expression of IDO1 is a marker of activation of the KYN pathway (Nikolaus et al. 2017). Although Trp metabolism exerts its pathogenic effect on the alteration in microbial communities and host-specific genetic contributions of IBD (Bruno et al. 2016, 2018), whether Trp deficiency is a result or cause of IBD should be further elucidated.

5.4.2 Cardiovascular Disease (CVD)

CVD is one of the major world-leading causes of death, in which myocardial infarction and coronary heart disease are two of the most common types. Forming atherosclerotic plaque and inducing thrombosis are the common causes of these two types of CVD. In coronary heart disease, the blocking and reduction of blood flow going through the myocardium result in heart failure, heart attacks, or angina. Hyperlipidemia and hypertension are considered to be reasons for chronic endothelial damage in atherosclerosis. During this process, the lipoproteins accumulate in the arterial intima, which induces cholesterol deposition, activates endothelial cells, and infiltrates leukocytes (Rafieian-Kopaei et al. 2014). Oxidative damage induced by high blood glucose and high blood lipids attacks lipids and proteins, causing them to enter the lipoprotein metabolism which is recognized by scavenger receptors on macrophage surfaces. Macrophages are transformed into foam cells which are accumulated in the subendothelial compartment to clean oxidized low-density lipoprotein (ox-LDL) through phagocytosis.

The risk factors causing atherosclerosis are well identified, including diabetes, hypertension, obesity, smoking, and dyslipidemia. The relation between these risk factors and Trp metabolism has been investigated (Myint et al. 2007; Marie et al. 2016; Hiraishi et al. 2018). Plasma Trp is negatively correlated with the incidence of CVD (Edward et al. 2017). However, diabetes, dyslipidemia, obesity, insulin resistance, and mild inflammation are related to the increase of IDO1 activity and KYN (Myint et al. 2007; Marie et al. 2016; Zhang et al. 2017; Florence et al. 2018). Trp metabolites through the 5-HT and KYN pathways might be relevant to cardiovascular events, as they have a direct impact on the vascular system (Wang et al. 2010; Pechanova et al. 2014; Manchia et al. 2017).

In the pathogenesis of atherosclerosis, chronic inflammation plays an important role. When CVD develops, several cytokines, such as interferons, chemokines, and interleukins, are produced due to the stimulation by the intrinsic cells in the atherosclerotic plaque (Ramji and Davies 2015). In particular, the coordination between pro-inflammatory cytokine INF-γ and the activation of macrophages and T lymphocytes can induce the release of reactive oxygen species and promote the IDO1 activation of macrophages in the atherosclerotic plaques. In addition, INF-γ may promote thrombosis via the upregulation of tissue factors in activated macrophages (Yun et al. 2016; Yuki et al. 2018; Baumgartner et al. 2019). Moreover, IDO expression and activity can be differentially regulated by inflammatory pathways in different types of immune cells. IDO1 expression located in the atherosclerotic intima has been found to occur in plasmacytoid dendritic cells (pDCs), which play critical roles in balancing the pro-inflammatory environment (Yun et al. 2016). The pDCs have direct immunomodulatory effects via local induction of regulatory T cells that contribute to building an immunosuppressive microenvironment. However, the protective effect of pDCs on atherosclerosis and whether they help to stabilize atherosclerotic plaques await further investigation.

CVD is always accompanied by a decrease in plasma Trp level and an increase in KYN/Trp ratio, which are regarded as crucial clinical predictors (Ringdal et al. 2011; Ringdal et al. 2013; Edward et al. 2017). Though neopterin and INF-γ are well-known systemic inflammatory biomarkers, the plasma KYN/Trp ratio can accurately predict cardiovascular events and mortality (Ringdal et al. 2011; Sulo et al. 2013). According to large-scale epidemiological studies, Trp is associated with coronary artery disease (Wirleitner et al. 2003), as well as mortality risk for cardiovascular incidents (Hui et al. 2016). Finally, elevated plasma KYN levels can be used to predict an increased risk of acute myocardial infarction in those patients having stable angina pectoris (Ringdal et al. 2015).

Several studies have indicated that Trp should be related to the development of emotional disorders which are associated with CVD. In fact, the plasma Trp level can be influenced by the presence of chronic low-grade inflammation in CVD, thus affecting the synthesis of brain 5-HT (Leathwood 1987). The interrelationship between severe depression and CVD is evident by (1) an increased incidence of heart disease in patients with depression and (2) a higher incidence of severe depression in patients with CVD than in healthy populations (Maes et al. 2010).

5.4.3 Chronic Kidney Disease

An interesting inverse relation between circulating Trp level and chronic kidney disease stages ($P < 0.0001$) was demonstrated (Debnath et al. 2017). It has been found that the downstream bioactive Trp metabolites QA, KYNA, KYN, and the kidney disease severity show a robustly positive correlation ($P < 0.0001$). The further multiple linear regression suggests that neither TNF-α nor IL-6 was relevant to the KYN/Trp ratio after adjusting for estimated glomerular filtration rate (eGFR), but TNF-α is found to be independently related to KYN when the effect of eGFR was taken into account. All these findings indicate that both inflammation and impaired kidney function cause the accumulation of toxic Trp metabolites, thus associating with chronic kidney disease. Schulman (2011) described the molecular events that occur when the drug AST-120, a charcoal compound, inhibits intestinal absorption of Trp derivatives. The drug is expected to slow the progression of chronic kidney disease in humans by inhibiting the synthesis of transforming growth factor-beta (TGF-β) in the kidney. In addition, the drug showed more beneficial effects beyond preventing renal fibrosis. In fact, elevated indoxyl sulfate (IS) levels interfere with tryptophan binding to albumin, leading to higher free tryptophan in plasma. This in turn leads to increased catabolism of Trp.

5.4.4 Diabetes

It has been suggested that glucose consumption can be caused by a high-protein/low-carbohydrate diet, thus increasing the risk of type 2 diabetes for patients. Inubushi et al. (2012) studied the effects of Trp on glucose absorption in the mouse intestine, which observed that Trp inhibited serum glucose, glucose absorption in the intestine, and insulin levels. This suggests that Trp inhibits blood glucose levels and reduces adverse reactions associated with insulin secretion by pancreatic β-cells. Chen et al. (2016) evaluated the possible relationship between Trp and the type 2 diabetes development of 213 individuals, including 51 diabetes patients and 162 healthy individuals. They utilized ultra-high-performance liquid chromatography-triple quadrupole mass spectrometry to measure Trp levels and found that (a) serum Trp levels were positively correlated with the diabetes development risk; (b) the positive correlation between Trp levels and insulin resistance, triglyceride secretion, and blood pressure levels of patients was observed; and (c) Trp may represent a new biomarker associated with the risk of diabetes, but this needs to be validated in other types of diabetes and in larger populations. In terms of diabetes, Imahori et al. (2018) isolated 4-quinlylaldoxime and indole-3-aldehyde from the in vitro reaction between Trp and glucose. Indole-3-aldehyde exhibited mutagenic effects in a *Salmonella typhimurium* YG1024 infection mouse model. Because the 4-quinolinealdoxime was only detected in the plasma while it was absent in the diets of the diabetic rats, it was interpreted that these isolated compounds under diabetic conditions were derived from the plasma and they could generate genetic damage as indicated by their mutagenicity toward *Salmonella typhimurium* YG1024.

5.4.5 Human Immunodeficiency Virus and Lung Infection

The Trp oxidation due to the immune induction of indoleamine 2,3 dioxygenase is considered to be the major cause of Trp depletion in patients infected by human immunodeficiency virus (HIV). Thus, Priyesh et al. (2016) examined the plasma Trp level in 105 HIV-infected patients and 60 healthy individuals without HIV infection. They found that the plasma Trp levels in HIV patients were 44.1% lower than those in healthy individuals. The Trp level is correlated with C-reactive protein, interleukin-6, and neopterin. The researchers concluded that the causes of the low Trp level in HIV-infected patients are most likely due to food insecurity and high inflammatory activity. Moreover, inflammation-induced Trp consumption in HIV-infected patients has a broader pro-inflammatory effect on the nutritional status of patients with HIV infection.

Strasser et al. (2016) found that compared with the placebo control group, serum Trp levels were 11% lower in trained athletes after 12 weeks of daily administration of probiotics. They also showed a 2.2-fold increase in the proportion of upper respiratory tract infections in the placebo control group compared to the probiotics-treated group, suggesting that probiotics may not only improve exercise-induced decrease in Trp levels but also reduce the incidence of infection while no effect on exercise performances was observed. The Trp catabolism may improve the immune system function and then prevent infections. There are also some other studies in terms of the complex relationships between Trp metabolism and inflammation-related depression, weight loss, and exercise (Barbara et al. 2015; Barbara and Dietmar 2016). Nayak et al. (2014) found that human chlamydia pneumonia pathogenic microorganisms are more sensitive to Trp availability than animal strains, reflecting the chronic nature of human host infection. This finding suggests that Trp may be useful in treating patients with pneumonia (Lieben et al. 2018).

5.4.6 Overweight and Obesity

Excessive food intake may lead to excessive consumption of Trp precursors and dietary Trp. In addition, Trp also plays the role of regulating heat intake in the body (Harald et al. 2014). A recent study has suggested that obesity is related to metabolic disorders of tyrosine and Trp (Barbara et al. 2015). Besides, these compounds have also been reported to display effects on the treatment of neuropsychiatric disorders (André et al. 2014). Barbara et al. (2015) studied the effect of a 2-week trial of a calorie-restricted weight-loss diet (CRWLD) on the circulating level of leptin and then studied the short-term dietary effects of inflammatory biomarkers and Trp on adults with overweight. They found that the biosynthesis of a natural precursor was affected, which may be related to mood disorders and increased demand for carbohydrates. Leptin as a peptide hormone regulating the fat homeostasis and satiety is produced by adipocytes, and it could tune energy balance and food intake (Doorn et al. 2017). Most obesity patients are characterized by high levels of leptin (Lönnqvist et al. 1995; Hundal et al. 2000). Recent papers have shown the dramatic effect of weight loss on leptin circulating levels (Klempel 2011; Musil et al. 2015). The study by Barbara et al. (2015) showed that reducing circulating leptin level could improve lipid levels, blood pressure, and insulin sensitivity. In the low-calorie diet group, Kyn and Trp levels were significantly reduced by 15% and 17%, separately. In the extremely low-calorie diet group, their concentrations reduced by 21% and 16%, respectively, and leptin level also decreased by 46% (Barbara et al. 2015).

Weight loss caused by increased metabolism and accelerated satiety may involve 5-hydroxytryptamine energy-driven mechanisms. Therefore, dietary supplementing Trp in calorie-restricted diets may be beneficial for treating obesity (Edward et al. 2017). Furthermore, the increased supply of Trp promotes the production of 5-HT and reduces the depressive symptoms of overweight patients (Musil et al. 2015). Therefore, dietary supplementing Trp may be an effective approach to treat the uncontrolled weight increase or prevent neuropsychiatric symptoms (Barbara et al. 2015).

Weight loss in obesity patients shows that we can prevent obesity by improving the proportion of Trp intake and providing other obesity-related signals that activate the inflammation and immune system. However, for the obese patients who had bariatric surgery intervention, no decrease in the Trp intake or the expression of immune biomarkers was found (Myung et al. 2016). Current literatures reported the connection between Trp metabolism and obesity, which is found to be closely associated with immune-mediated inflammation (Brandacher et al. 2007; Harald et al. 2014; Reininghaus et al. 2014).

Harald et al. (2014) analyzed 527 participants aged between 10 and 65 years and found that both the serum Kyn level and the Kyn/Trp ratios were significantly higher in overweight/obese adults (aged between 18 and 65 years) than in the control group. While in the overweight and obese juvenile male group (age $\leq$ 18 years old), the Kyn/Trp ratio was lower than that of the control group. In addition, the Kyn level and the Kyn/Trp ratios of adults with metabolic syndrome (MetS) were significantly increased compared to the control. Serum Trp levels were decreased in the overweight/obese adult female group, but there was no significant difference between the normal-weight patients and the overweight/obese group with or without metabolic syndrome (Harald et al. 2014).

These researches showed that there were significant differences in obesity and Trp metabolism between adolescents and adults. For example, adolescents seemed to be prone to suffer from Th2-mediated immune response, while obese adults are likely relying on Th1-mediated immune mechanisms. These differences provide potential clinical significance as follows: firstly, during the childhood and adolescence period, conservative treatment of obesity by changing their lifestyles including appropriate physical activities could prevent irreversible clinical damage before immune activation; secondly, in obese patients, the age-independent presence of Th-1 markers can be verified by the analysis of the Kyn/Trp ratio (Harald et al. 2014).

Moreover, in obese patients, a potential pathogenic link between IDO-mediated Trp reduction, chronic immune activation, and 5-HT levels has been pointed out by many researchers (Raheja et al. 2015; Yvonne et al. 2015). Obesity syndromes are related to systemic inflammation and immune activation and are also related to the synthesis and release of pro-inflammatory cytokines in adipose tissue such as the hormone leptin, INF-γ, and TNF-α. IDO can be induced by IFN-γ, and it is involved in regulating immune responses and degrading Trp to form N-formylkynurenine, which is then converted into nicotinic acid. Changes in Trp metabolism may reduce the production of 5-hydroxytryptamine, which then leads to reduced satiety, mood disorders, and depression, resulting in an increased calorie intake and ultimately obesity.

Obesity has been proven to be related to a decreased plasma Trp level, which is not associated with dietary intake or weight loss (Jun et al. 2015; Ritze et al. 2016). As mentioned above, Trp is a precursor to biosynthesize 5-HT, while 5-HT is a biochemical and neurotransmitter regulator maintaining the balance between satiety and hunger. Similarly, specific neurons in the brain control appetite (Zhang et al. 2015). Overeating and obesity are caused by suppressing satiety (Brandacher et al. 2007). In this case, 5-HT acts as a satiety signal and neurotransmitter to control food intake (Halford and Blundell 2000). This process also inhibits the expression of neuropeptide Y in the hypothalamus via suppressing hunger and controlling obesity (Wurtman and Wurtman 1995; Manousopoulou et al. 2016). Furthermore, carbohydrate or fat intake can be specifically regulated by serotonin (Blundell and Lawton 1995; Bray 2001).

5.5 Conclusion and Outlook

Trp as a nutritionally essential amino acid is an important contributor to protein synthesis, and it is also a precursor for the vitamin niacin and the neurotransmitter serotonin, thus regulating various physiological processes. Therefore, it is crucial to accurately and sensitively measure the Trp levels. Currently, a number of approaches have been developed including the non-hydrolytic ninhydrin method and non-hydrolytic near-infrared spectroscopy, which are widely utilized for plant breeding.

A major challenge is to develop new strategies to generate cereals and legumes with high-Trp/lysine/methionine through plant-based molecular genetic engineering methods. The protein quality of this genetically modified plant food may be similar to that of meat products; thus, they are likely to be a potential way to alleviate malnutrition at an affordable cost. Besides, it has been reported that proper dietary Trp intake can alleviate the course of a variety of chronic diseases, so it is of great importance to initiate more clinical investigations to exploit the therapeutic potential of cheap and high-Trp-enriched foods. In addition, the role of Trp metabolites through the Trp-gut microbiota interactions in the etiology should be further studied, and its prevention effects on human diseases need to be clearly elucidated.

References

Agus A, Planchais J, Sokol H (2018) Gut microbiota regulation of tryptophan metabolism in health and disease. Cell Host Microbe 23:716–724

Alexeev EE, Lanis JM, Kao DJ, Campbell EL, Kelly CJ, Battista KD, Gerich ME, Jenkins BR, Walk ST, Kominsky DJ, Colgan SP (2018) Microbiota-derived indole metabolites promote human and murine intestinal homeostasis through regulation of interleukin-10 receptor. Am J Pathol 188:1183–1194

Amobi A, Qian F, Lugade AA, Odunsi K (2017) Tryptophan catabolism and cancer immunotherapy targeting IDO mediated immune suppression. Adv Exp Med Biol 1036:129–144

André C, Dinel A-L, Ferreira G, Layé S, Castanon N (2014) Diet-induced obesity progressively alters cognition, anxiety-like behavior and lipopolysaccharide-induced depressive-like behavior: focus on brain indoleamine 2,3-dioxygenase activation. Brain Behav Immu 41:10–21

Barbara S, Dietmar F (2016) Diet versus exercise in weight loss and maintenance: focus on tryptophan. Int J Tryptophan Res 9:9–16

Barbara S, Ken B, Dietmar F (2015) Effects of a caloric restriction weight loss diet on tryptophan metabolism and inflammatory biomarkers in overweight adults. Eur J Nutr 54:33–42

Barcelon E, Mukaino T, Yokoyama J, Uehara T, Ogata K, Kira J, Tobimatsu S (2019) Grand total EEG score can differentiate Parkinson's disease from Parkinson-related disorders. Front Neurol 10:398

Baumgartner R, Forteza MJ, Ketelhuth DFJ (2019) The interplay between cytokines and the kynurenine pathway in inflammation and atherosclerosis. Cytokine 122:154148

Blundell JE, Lawton CL (1995) Serotonin and dietary fat intake: effects of dexfenfluramine. Metabolism 44:33–37

Brandacher G, Hoeller E, Fuchs D, Weiss HG (2007) Chronic immune activation underlies morbid obesity: is IDO a key player? Curr Drug Metab 8:289–295

Bravo R, Matito S, Cubero J, Paredes SD, Franco L, Rivero M, Rodríguez AB, Barriga C (2013) Tryptophan-enriched cereal intake improves nocturnal sleep, melatonin, serotonin, and total antioxidant capacity levels and mood in elderly humans. Age (Dordr) 35:1277–1285

Bray GA (2001) Drug treatment of obesity. Rev Endocr Metab Disord 2:402–407

Bray MR, Carriere AD, Clarke AJ (1994) Quantitation of tryptophan and tyrosine residues in proteins by fourth-derivative spectroscopy. Anal Biochem 221:278–284

Bruno L, Richard ML, Valentin L, Hang-Phuong P, Marie-Laure M, Gregory DC, Chantal B, Sarah J, Hoffmann TW, Natividad JM, Loic B, Soraya T, Aurélie C-M, Isabelle N-L, Fatiha M, Philippe S, Anne B, Jacques C, Bernhard R, Laurent B, Jean-Marie L, Philippe L, Xavier RJ, Harry S (2016) CARD9 impacts colitis by altering gut microbiota metabolism of tryptophan into aryl hydrocarbon receptor ligands. Nat Med 22:598–609

Bruno L, Natividad JM, Harry S (2018) Aryl hydrocarbon receptor and intestinal immunity. Mucosal Immunol 11:1024–1038

Chen T, Zheng X, Ma X, Bao Y, Ni Y, Hu C, Rajani C, Huang F, Zhao A, Jia W, Jia W (2016) Tryptophan predicts the risk for future type 2 diabetes. PLoS One 11:e0162192

Chen Y, Joshi YV, Raju JS, Zhang ZJ (2009) A theory of combative advertising. Mark Sci 28:1–19

Comai S, Bertazzo A, Brughera M, Crotti S (2020) Chapter five – Tryptophan in health and disease. In: Makowski GS (ed) Advances in clinical chemistry. Elsevier, pp 165–218

Cosnes J, Gower–Rousseau C, Seksik P, Cortot A (2011) Epidemiology and natural history of inflammatory bowel diseases. Gastroenterology 140:1785–1794

Debnath S, Velagapudi C, Redus L, Thameem F, Kasinath B, Hura CE, Lorenzo C, Abboud HE, O'Connor JC (2017) Tryptophan metabolism in patients with chronic kidney disease secondary to type 2 diabetes: relationship to inflammatory markers. Int J Tryptophan Res 10:1178646917694600

Delgado PL (2000) Depression: the case for a monoamine deficiency. J Clin Psychiatry 61(Suppl 6):7–11

Doorn CV, Macht VA, Grillo CA, Reagan LP (2017) Leptin resistance and hippocampal behavioral deficits. Physiol Behav 176:207–213

Edward Y, Miguel R-C, Marta G-F, Yan Z, Estefania T, Clish Clary B, Jordi S-S, Liming L, Wang Dong D, Dolores C, Montse F, Enrique G-G, José L, Ramón E, Emilio R, Montserrat C, Fernando A, Dora R, Lluis S-M, Sorlí Jose V, H. F. B, Martinez-Gonzalez Miguel A (2017) Increases in plasma tryptophan are inversely associated with incident cardiovascular disease in the prevención con dieta mediterránea (PREDIMED) study. J Nutr Biochem 147:314–322

Favre D, Mold J, Hunt PW, Kanwar B, Loke P, Seu L, Barbour JD, Lowe MM, Jayawardene A, Aweeka F, Huang Y, Douek DC, Brenchley JM, Martin JN, Hecht FM, Deeks SG, McCune JM (2010) Tryptophan catabolism by indoleamine 2,3-dioxygenase 1 alters the balance of TH17 to regulatory T cells in HIV disease. Sci Transl Med 2:32ra36

Fernstrom JD, Wurtman RJ (1971) Brain serotonin content: physiological dependence on plasma tryptophan levels. Science 173:149–152

Fernstrom JD, Wurtman RJ (1972) Brain serotonin content: physiological regulation by plasma neutral amino acids. Science 178:414–416

Fiore A, Murray PJ (2021) Tryptophan and indole metabolism in immune regulation. Curr Opin Immunol 70:7–14

Florence A, Giada M, Nathaly G, Teresa RC, Jose ZG, Lars K, Knut D-J, Van den Eynde B, Ciriana O, Ursula G, von Herrath MG (2018) Loss of IDO1 expression from human pancreatic β-cells precedes their destruction during the development of type 1 diabetes. Diabetes 67:171281

Gabriella G, Comai S (2019) Differential function of melatonin MT1 and MT2 receptors in REM and NREM sleep. Front Endocrinol 10:87

Gessa GL, Biggio G, Fadda F, Corsini GU, Tagliamonte A (1974) Effect of the oral administration of tryptophan-free amino acid mixtures on serum tryptophan, brain tryptophan and serotonin metabolism. J Neurochem 22:869–870

Gibson EL (2018) Tryptophan supplementation and serotonin function: genetic variations in behavioural effects. Proc Nutr Soc 77:174–188

Gobbi G, Comai S (2019) Sleep well. Untangling the role of melatonin MT1 and MT2 receptors in sleep. J Pineal Res 66:e12544

González Esquivel D, Ramírez-Ortega D, Pineda B, Castro N, Ríos C, Pérez de la Cruz V (2017) Kynurenine pathway metabolites and enzymes involved in redox reactions. Neuropharmacology 112:331–345

Guan Q, Zhang J (2017) Recent advances: the imbalance of cytokines in the pathogenesis of inflammatory bowel disease. Mediat Inflamm 2017:4810258

Hajsl M, Hlavackova A, Broulikova K, Sramek M, Maly M, Dyr JE, Suttnar J (2020) Tryptophan metabolism, inflammation, and oxidative stress in patients with neurovascular disease. Meta 10:208

Halford JCG, Blundell JE (2000) Separate systems for serotonin and leptin in appetite control. Ann Med 32:222–232

Harald M, Summers KL, Andreas M, Sieglinde Z, Gunter A, Ruth P, Schnedl WJ, Eva R, Katharina P, Daniel W, Dietmar F (2014) Obesity-related dysregulation of the tryptophan-kynurenine metabolism: role of age and parameters of the metabolic syndrome. Obesity 22:195–201

He JG, Li BB, Zhou L, Yan D, Xie QL, Zhao W (2020) Indoleamine 2,3-dioxgenase-transfected mesenchymal stem cells suppress heart allograft rejection by increasing the production and activity of dendritic cells and regulatory T cells. J Investig Med 68:728–737

Hiraishi MN, Silva LE, Pritesh L (2018) Dysregulation of tryptophan catabolism in metabolic syndrome. Metab Syndr Relat Disord 16:135–142

Hui Z, Ueland PM, Arve U, Eussen SJPM, Vollset SE, Ottar N, Øivind M, Despoina T, Klaus M, Tell GS (2016) Plasma biomarkers of inflammation, the kynurenine pathway, and risks of all-cause, cancer, and cardiovascular disease mortality: the Hordaland health study. Am J Epidemiol 183:249–258

Hundal RS, Krssak M, Dufour S, Laurent D, Lebon V, Chandramouli V, Inzucchi SE, Schumann WC, Petersen KF, Landau BR, Shulman GI (2000) Mechanism by which metformin reduces glucose production in type 2 diabetes. Diabetes 49:2063–2069

Imahori D, Matsumoto T, Kojima N, Hasei T, Sumii M, Sumida T, Yamashita M, Watanabe T (2018) Chemical structures of novel Maillard reaction products under hyperglycemic conditions. Chem Pharm Bull 66:363–367

Inubushi T, Kamemura N, Oda M, Sakurai J, Nakaya Y, Harada N, Suenaga M, Matsunaga Y, Ishidoh K, Katunuma N (2012) L-tryptophan suppresses rise in blood glucose and preserves insulin secretion in type-2 diabetes mellitus rats. J Nutr Sci Vitaminol 58:415–422

Jun N, Hail K, Sangkyu P (2015) Peripheral serotonin: a new player in systemic energy homeostasis. Mol Cells 38:1023–1028

Jung KH, LoRusso P, Burris H, Gordon M, Bang YJ, Hellmann MD, Cervantes A, Ochoa de Olza M, Marabelle A, Hodi FS, Ahn MJ, Emens LA, Barlesi F, Hamid O, Calvo E, McDermott D, Soliman H, Rhee I, Lin R, Pourmohamad T, Suchomel J, Tsuhako A, Morrissey K, Mahrus S, Morley R, Pirzkall A, Davis SL (2019) Phase I study of the indoleamine 2,3-dioxygenase 1 (IDO1) inhibitor navoximod (GDC-0919) administered with PD-L1 inhibitor (atezolizumab) in advanced solid tumors. Clin Cancer Res 25:3220–3228

Karu N, Mckercher C, Nichols D, Jose M (2016) Tryptophan metabolism, its relation to inflammation and stress markers and association with psychological and cognitive functioning: Tasmanian chronic kidney disease pilot study. BioMed Central 17:171

Kim M, Tomek P (2021) Tryptophan: a rheostat of cancer immune escape mediated by immunosuppressive enzymes IDO1 and TDO. Front Immunol 12:636081

Klar R, Michel S, Schell M, Hinterwimmer L, Zippelius A, Jaschinski F (2020) A highly efficient modality to block the degradation of tryptophan for cancer immunotherapy: locked nucleic acid-modified antisense oligonucleotides to inhibit human indoleamine 2,3-dioxygenase 1/ tryptophan 2,3-dioxygenase expression. Cancer Immunol Immunother 69:57–67

Klempel MC, Varady KA (2011) Reliability of leptin, but not adiponectin, as a biomarker for diet-induced weight loss in humans. Nutr Rev 69:145–154

Ko M, Choi-Kwon S, Jun SE, Kim JH, Cho KH, Nah HW, Song H, Kim JS (2018) Poststroke emotional disturbances and a tryptophan hydroxylase 2 gene polymorphism. Brain Behavior 8:e00892

Kondo M, Nakamura Y, Ishida Y, Yamada T, Shimada S (2013) The 5-HT3A receptor is essential for fear extinction. Learn Mem 21:1–4

Le Naour J, Galluzzi L, Zitvogel L, Kroemer G, Vacchelli E (2020) Trial watch: IDO inhibitors in cancer therapy. Onco Targets Ther 9:1777625

Leathwood PD (1987) Tryptophan availability and serotonin synthesis. Proc Nutr Soc 46:143–156

Lesniak WG, Jyoti A, Mishra MK, Louissaint N, Romero R, Chugani DC, Kannan S, Kannan RM (2013) Concurrent quantification of tryptophan and its major metabolites. Anal Biochem 443:222–231

Li F, Zhang R, Li S, Liu J (2017) IDO1: an important immunotherapy target in cancer treatment. Int Immunopharmacol 47:70–77

Liang C, Peng L, Zeng S, Zhao Q, Tang L, Jiang X, Zhang J, Yan N, Chen Y (2019) Investigation of indoleamine 2,3-dioxygenase 1 expression in uveal melanoma. Exp Eye Res 181:112–119

Lieben CK, Blokland A, Deutz NE, Jansen W, Han G, Hupperts RM (2018) Intake of tryptophan-enriched whey protein acutely enhances recall of positive loaded words in patients with multiple sclerosis. Clin Nutr 37:321–328

Lönnqvist F, Arner P, Nordfors L, Schalling M (1995) Overexpression of the obese (ob) gene in adipose tissue of human obese subjects. Nat Med 1:950–953

Lucki I (1998) The spectrum of behaviors influenced by serotonin. Biol Psychiatry 44:151–162

Maes M, Verkerk R, Bonaccorso S, Ombelet W, Bosmans E, Scharpé S (2002) Depressive and anxiety symptoms in the early puerperium are related to increased degradation of tryptophan into kynurenine, a phenomenon which is related to immune activation. Life Sci 71:1837–1848

Maes M, Ruckoanich P, Chang YS, Mahanonda N, Berk M (2010) Multiple aberrations in shared inflammatory and oxidative & nitrosative stress (IO&NS) pathways explain the co-association of depression and cardiovascular disorder (CVD), and the increased risk for CVD and due mortality in depressed patients. Prog Neuro-Psychopharmacol Biol Psychiatry 35:769–783

Maes M, Leonard BE, Myint AM, Kubera M, Verkerk R (2011) The new '5-HT' hypothesis of depression: cell-mediated immune activation induces indoleamine 2,3-dioxygenase, which leads to lower plasma tryptophan and an increased synthesis of detrimental tryptophan catabolites (TRYCATs), both of which contribute to the onset of depression. Prog Neuro-Psychopharmacol Biol Psychiatry 35:702–721

Manchia M, Carpiniello B, Valtorta F, Comai S (2017) Serotonin dysfunction, aggressive behavior, and mental illness: exploring the link using a dimensional approach. ACS Chem Neurosci 8:961–972

Manousopoulou A, Koutmani Y, Karaliota S, Woelk CH, Manolakos ES, Karalis K, Garbis SD (2016) Hypothalamus proteomics from mouse models with obesity and anorexia reveals therapeutic targets of appetite regulation. Nutr Diabetes 6:e204

Marie F, Benjamin H, Robert C, Audrey L, Loic Y, Marie V, Arredouani A, Michel M (2016) The kynurenine pathway is activated in human obesity and shifted toward kynurenine monooxygenase activation. Obesity 24:2066–2074

Miller AH, Haroon E, Raison CL, Felger JC (2013) Cytokine targets in the brain: impact on neurotransmitters and neurocircuits. Depress Anxiety 30:297–306

Modoux M, Rolhion N, Mani S, Sokol H (2021) Tryptophan metabolism as a pharmacological target. Trends Pharmacol Sci 42:60–73

Morand C, Young SN, Ervin FR (1983) Clinical response of aggressive schizophrenics to oral tryptophan. Biol Psychiatry 18:575–578

Musil F, Blaha V, Ticha A, Hyspler R, Haluzik M, Lesna J, Smahelova A, Sobotka L (2015) Effects of body weight reduction on plasma leptin and adiponectin/leptin ratio in obese patients with type 1 diabetes mellitus. Physiol Res 64:221–228

Myint A-M, Kim YK, Verkerk R, Scharpé S, Steinbusch H, Leonard B (2007) Kynurenine pathway in major depression: evidence of impaired neuroprotection. J Affect Disord 98:143–151
Myint AM, Schwarz MJ, Müller N (2012) The role of the kynurenine metabolism in major depression. J Neural Transm 119:245–251
Myung OC, Sangkyu P, Hail K (2016) Serotonin as a new therapeutic target for diabetes mellitus and obesity. Diabetes Metab J 40:89–98
Nayak A, Hao Z, Sadek R, Vahanian N, Ramsey WJ, Kennedy E, Mautino M, Link C, Bourbo P, Dobbins R, Adams K, Diamond A, Marshall L, Munn DH, Janik J, Khleif SN (2014) A phase I study of NLG919 for adult patients with recurrent advanced solid tumors. J Immunother Cancer 2:P250
Nikolaus S, Schulte B, Al-Massad N, Thieme F, Schulte DM, Bethge J, Rehman A, Tran F, Aden K, Häsler R, Moll N, Schütze G, Schwarz MJ, Waetzig GH, Rosenstiel P, Krawczak M, Szymczak S, Schreiber S (2017) Increased tryptophan metabolism is associated with activity of inflammatory bowel diseases. Gastroenterology 153:1504–1516
Pechanova O, Paulis L, Simko F (2014) Peripheral and central effects of melatonin on blood pressure regulation. Int J Mol Sci 15:17920–17937
Platten M, Wick W, Van den Eynde BJ (2012) Tryptophan catabolism in cancer: beyond IDO and tryptophan depletion. Cancer Res 72:5435–5440
Platten M, Nollen EAA, Röhrig UF, Fallarino F, Opitz CA (2019) Tryptophan metabolism as a common therapeutic target in cancer, neurodegeneration and beyond. Nat Rev Drug Discov 18:379–401
Priyesh B, Levay PF, Margaretha V (2016) Tryptophan depletion in context of the inflammatory and general nutritional status of a low-income South African HIV-infected population. J Health Popul Nutr 35:5
Qian F, Villella J, Wallace PK, Mhawech-Fauceglia P, Tario JD Jr, Andrews C, Matsuzaki J, Valmori D, Ayyoub M, Frederick PJ, Beck A, Liao J, Cheney R, Moysich K, Lele S, Shrikant P, Old LJ, Odunsi K (2009) Efficacy of levo-1-methyl tryptophan and dextro-1-methyl tryptophan in reversing indoleamine-2,3-dioxygenase-mediated arrest of T-cell proliferation in human epithelial ovarian cancer. Cancer Res 69:5498–5504
Rafieian-Kopaei M, Setorki M, Doudi M, Baradaran A, Nasri H (2014) Atherosclerosis: process, indicators, risk factors and new hopes. Int J Prev Med 5:927–946
Raheja UK, Dietmar F, Ina G, Brenner LA, Rovner SF, Iqra M, Daniel W, Harald M, Dan R, Postolache TT (2015) In psychiatrically healthy individuals, overweight women but not men have lower tryptophan levels. Pteridines 26:79–84
Ramji DP, Davies TS (2015) Cytokines in atherosclerosis: key players in all stages of disease and promising therapeutic targets. Cytokine Growth Factor Rev 26:673–685
Ramos-Chávez LA, Roldán-Roldán G, García-Juárez B, González-Esquivel D, Pérez de la Cruz G, Pineda B, Ramírez-Ortega D, García Muñoz I, Jiménez Herrera B, Ríos C, Gómez-Manzo S, Marcial-Quino J, Sánchez Chapul L, Carrillo Mora P, Pérez de la Cruz V (2018) Low serum tryptophan levels as an indicator of global cognitive performance in nondemented women over 50 years of age. Oxidative Med Cell Longev 2018:8604718
Reininghaus EZ, McIntyre RS, Reininghaus B, Geisler S, Bengesser SA, Nina L, Karen H, Armin B, Fabian K, Renate U, Hans-Peter K, Sieglinde Z, Dietmar F, Harald M (2014) Tryptophan breakdown is increased in euthymic overweight individuals with bipolar disorder: a preliminary report. Bipolar Disord 16:432–440
Ringdal PE, Øivind M, Magne UP, Hall S-H, Reinhard S, Jannicke I, Erik NJ, Marta E, Gard S, Øyvind B, Rolf B, Ottar N (2011) Systemic markers of interferon-γ-mediated immune activation and long-term prognosis in patients with stable coronary artery disease. Arterioscler Thromb Vasc Biol 31:698–704
Ringdal PE, Tveitevåg SGF, Hall S-H, Magne UP, Marta E, Erik NJ, Jannicke I, Reinhard S, Miodini NR, Ottar N (2013) Urinary excretion of kynurenine and tryptophan, cardiovascular events, and mortality after elective coronary angiography. Eur Heart J 34:2689–2696

Ringdal PE, Nora T, Eussen SJPM, Magne UP, Elin S, Tveitevåg SGF, Øivind M, Klaus M, Gunnar M, Arve U, Erik NJ, Nilsen DW, Ottar N (2015) Associations of plasma kynurenines with risk of acute myocardial infarction in patients with stable angina pectoris. Arterioscler Thromb Vasc Biol 35:455–462

Ritze Y, Schollenberger A, Sinno MH, Bühler N, Böhle M, Bárdos G, Sauer H, Mack I, Enck P, Zipfel S, Meile T, Königsrainer A, Kramer M, Bischoff SC (2016) Gastric ghrelin, GOAT, leptin, and leptinR expression as well as peripheral serotonin are dysregulated in humans with obesity. Neurogastroenterol Motil 28:806–815

Sari S, Tomek P, Leung E, Reynisson J (2019) Discovery and characterisation of dual inhibitors of tryptophan 2,3-dioxygenase (TDO2) and indoleamine 2,3-dioxygenase 1 (IDO1) using virtual screening. Molecules 24:4346

Sarris J, Murphy J, Mischoulon D, Papakostas GI, Fava M, Berk M, Ng CH (2016) Adjunctive nutraceuticals for depression: a systematic review and meta-analyses. Am J Psychiatry 173:575–587

Schulman G (2011) A nexus of progression of chronic kidney disease: tryptophan, profibrotic cytokines, and charcoal. J Ren Nutr 22:107–113

Shao J, Hou L, Liu J, Liu Y, Ning J, Zhao Q, Zhang Y (2021) Indoleamine 2,3-dioxygenase 1 inhibitor-loaded nanosheets enhance CAR-T cell function in esophageal squamous cell carcinoma. Front Immunol 12:661357

Sonner JK, Keil M, Falk-Paulsen M, Mishra N, Rehman A, Kramer M, Deumelandt K, Röwe J, Sanghvi K, Wolf L, von Landenberg A, Wolff H, Bharti R, Oezen I, Lanz TV, Wanke F, Tang Y, Brandao I, Mohapatra SR, Epping L, Grill A, Röth R, Niesler B, Meuth SG, Opitz CA, Okun JG, Reinhardt C, Kurschus FC, Wick W, Bode HB, Rosenstiel P, Platten M (2019) Dietary tryptophan links encephalogenicity of autoreactive T cells with gut microbial ecology. Nat Commun 10:4877

Strasser B, Geiger D, Schauer M, Gostner JM, Gatterer H, Burtscher M, Fuchs D (2016) Probiotic supplements beneficially affect tryptophan–kynurenine metabolism and reduce the incidence of upper respiratory tract infections in trained athletes: a randomized, double-blinded, placebo-controlled trial. Nutrients 8:E752

Sulo G, Vollset SE, Nygård O, Midttun Ø, Ueland PM, Eussen SJPM, Pedersen ER, Tell GS (2013) Neopterin and kynurenine–tryptophan ratio as predictors of coronary events in older adults, the Hordaland health study. Int J Cardiol 168:1435–1440

Tang K, Wang B, Yu B, Liu H-M (2022) Indoleamine 2,3-dioxygenase 1 (IDO1) inhibitors and PROTAC-based degraders for cancer therapy. Eur J Med Chem 227:113967

Tarun B, Alaniz RC, Wood TK, Arul J (2010) The bacterial signal indole increases epithelial-cell tight-junction resistance and attenuates indicators of inflammation. Proc Natl Acad Sci USA 107:228–233

Volavka J, Crowner M, Brizer D, Convit A, Van Praag H, Suckow RF (1990) Tryptophan treatment of aggressive psychiatric inpatients. Biol Psychiatry 28:728–732

Wang Y, Liu H, Gavin M, Witting PK, Johannes-Peter S, Michael H, Dechaboon C, Wu BJ, Ball HJ, Thomas SR, Vimal K, Celermajer DS, Mellor AL, Keaney JF Jr, Hunt NH, Roland S (2010) Kynurenine is an endothelium-derived relaxing factor produced during inflammation. Nat Med 16:279–285

Wirleitner B, Rudzite V, Neurauter G, Murr C, Kalnins U, Erglis A, Trusinskis K, Fuchs D (2003) Immune activation and degradation of tryptophan in coronary heart disease. Eur J Clin Investig 33:550–554

Wlodarska M, Luo C, Kolde R, d'Hennezel E, Annand JW, Heim CE, Krastel P, Schmitt EK, Omar AS, Creasey EA, Garner AL, Mohammadi S, O'Connell DJ, Abubucker S, Arthur TD, Franzosa EA, Huttenhower C, Murphy LO, Haiser HJ, Vlamakis H, Porter JA, Xavier RJ (2017) Indoleacrylic acid produced by commensal Peptostreptococcus species suppresses inflammation. Cell Host Microbe 22:25–37

Wurtman JR, Wurtman JJ (1995) Brain serotonin, carbohydrate-craving, obesity and depression. Obes Res 3(Suppl 4):477–480

Yu LC-H (2018) Microbiota dysbiosis and barrier dysfunction in inflammatory bowel disease and colorectal cancers: exploring a common ground hypothesis. J Biomed Sci 25:79

Yuki W, Shohei K, Atsushi Y, Yunosuke M, Kensaku N, Kazuo K, Yujiro A (2018) Indoleamine 2,3-dioxygenase 1 in coronary atherosclerotic plaque enhances tissue factor expression in activated macrophages. Res Prac Thrombosis Haemostasis 2:726–735

Yun TJ, Lee JS, Machmach K, Shim D, Choi J, Wi YJ, Jang HS, Jung I-H, Kim K, Yoon WK, Miah MA, Li B, Chang J, Bego MG, Pham TNQ, Loschko J, Fritz JH, Krug AB, Lee S-P, Keler T, Guimond JV, Haddad E, Cohen EA, Sirois MG, El-Hamamsy I, Colonna M, Oh GT, Choi J-H, Cheong C (2016) Indoleamine 2,3-dioxygenase-expressing aortic plasmacytoid dendritic cells protect against atherosclerosis by induction of regulatory T cells. Cell Metab 24:852–866

Yvonne R, Christina H, Gyöngyi B, Barbara E, Martin T, D'Haese JG, Bischoff SC, Bernd S (2015) Altered intestinal neuroendocrine gene expression in humans with obesity. Obesity 23:2278–2285

Zelante T, Iannitti RG, Cunha C, Luca AD, Giovannini G, Pieraccini G, Zecchi R, D'Angelo C, Massi-Benedetti C, Fallarino F, Carvalho A, Puccetti P, Romani L (2013) Tryptophan catabolites from microbiota engage aryl hydrocarbon receptor and balance mucosal reactivity via interleukin-22. Immunity 39:372–385

Zhang C, Yin A, Li H, Wang R, Wu G, Shen J, Zhang M, Wang L, Hou Y, Ouyang H, Zhang Y, Zheng Y, Wang J, Lv X, Wang Y, Zhang F, Zeng B, Li W, Yan F, Zhao Y, Pang X, Zhang X, Fu H, Chen F, Zhao N, Hamaker BR, Bridgewater LC, Weinkove D, Clement K, Dore J, Holmes E, Xiao H, Zhao G, Yang S, Bork P, Nicholson JK, Wei H, Tang H (2015) Dietary modulation of gut microbiota contributes to alleviation of both genetic and simple obesity in children. EBioMedicine 2:968–984

Zhang Y, Ruan Y, Zhang P, Wang L (2017) Increased indoleamine 2,3-dioxygenase activity in type 2 diabetic nephropathy. J Diabetes Complicat 31:223–227

Zhang Y, Shi H, Yang G, Yang Y, Li W, Song M, Shao M, Su X, Lv L (2021) Associations between expression of indoleamine 2, 3-dioxygenase enzyme and inflammatory cytokines in patients with first-episode drug-naive schizophrenia. Transl Psychiatry 11:595

Chapter 6
Industrial Production of Tryptophan

Chang Il Seo, So Young Kim, and Sung Woo Kim

Abstract The use of supplemental amino acids to reduce the use of traditional protein supplements in animal feeds has been increasingly recognized as a feasible way to improve the sustainability of modern animal agriculture. It is of great importance to improve the industrial-scale production efficiency of feed-grade supplemental amino acids in animal agriculture. L-Tryptophan (L-Trp) is one of the major essential and limiting amino acids for animals, and its nutritional and physiological roles in animal agriculture have been well documented, whereas the high cost of using L-Trp in feeds brings challenges in animal agriculture. This chapter summarizes the current knowledge and method of industrial production of tryptophan by introducing the biosynthesis of L-Trp and improvement of L-Trp productivity, as well as the optimization of bacteria-based fermentation strategy to reduce the cost of L-Trp production for animal agriculture.

Abbreviations

L-Trp	L-Tryptophan
GRAS	Gram-positive and generally recognized as safe
PP	Pentose phosphate
PRPP	Phosphoribosyl pyrophosphate

C. I. Seo
CJ Blossom Park, Suwon, Gyeonggi-do, South Korea
e-mail: changil.seo@cj.net

S. Y. Kim
CJ Blossom Park, Suwon, Gyeonggi-do, South Korea
e-mail: soyoung.kim@cj.net

S. W. Kim (✉)
North Carolina State University, Raleigh, NC, USA
e-mail: sungwoo_kim@ncsu.edu

Y. Yin et al. (eds.), *Tryptophan in Animal Nutrition and Human Health*,
https://doi.org/10.1007/978-981-97-4719-1_6

E4P	Erythrose 4-phosphate
PEP	Phosphoenolpyruvate
DHAP	3-deoxy-D-arabino-heptulosonate-7-phosphate
InGP	Indole glycerol phosphate
CdRP	1-(O-carboxyphenyl amino) 1-deoxyribulose 5-phosphate
PTS	Phosphotransferase system
PPP	Pentose phosphate pathway
GRAS	Generally recognized as safe

6.1 Feed-Grade Supplemental Amino Acids in Animal Agriculture

Animal agriculture largely relies on the use of feed-grade supplemental amino acids for production efficiency and sustainability. Feed-grade supplemental amino acids, including L-lysine, L-threonine, and L-tryptophan, are available from microbial fermentation, with the exception of DL-methionine. Microbial fermentation is a preferred method to produce feed-grade supplemental amino acids because of its efficiency for mass production and also because microbial fermentation produces only L-amino acids that are available to animals. In recent years, advances in fermentation technology have made it feasible to produce other feed-grade amino acids such as L-arginine (2007), L-valine (2009), L-isoleucine (2010), and L-histidine (2020). Since 2015, even methionine, which was previously available from chemical synthesis, can be produced by microbial fermentation making L-methionine available in animal agriculture. Currently, eight feed-grade supplemental amino acids are available in L-forms: L-lysine, L-threonine, L-tryptophan, L-methionine, L-valine, L-arginine, L-isoleucine, and L-histidine.

Advances in biotechnology enable the cost-down of amino acid production. Considering the benefits of the use of supplemental amino acids in animal agriculture not just for the economic advantages but also for the environmental benefits, an increased number of feed-grade supplemental amino acids will be available in the near future. This will allow nutritionists to formulate animal diets with reduced amounts of crude protein with minimal contents of extraneous amino acids exceeding animals' needs(Kim et al. 2019). It is rational to expect that animal feeds will contain amino acids precisely meeting the needs of animals according to their age, health status, production environment, and physiological status.

This chapter is about the industrial aspect of the production of L-tryptophan. Tryptophan is one of the major essential amino acids that is often limited in animal diets unless supplemented to meet the requirement. Tryptophan is needed by animals not only because it meets their nutritional needs but also because of its functional roles that are critical to animals (Koopmans et al. 2006; Shen et al. 2012a, b, 2015, 2021).

6.2 Strain Development

Corynebacterium glutamicum and *Escherichia coli* are the primary microorganisms used to produce amino acids. *Corynebacterium glutamicum* was the microorganism discovered to produce glutamic acid, which was the first amino acid produced through fermentation, and has been widely used to produce various L-amino acids. Since the 1950s, the use of *C. glutamicum* has been successful in commercial production of L-glutamate and L-lysine. *Corynebacterium glutamicum* is gram-positive and generally recognized as safe (GRAS), and thus, it has extensively been used in the production of raw materials for food and feed production. The use of *E. coli* has been studied to produce L-threonine since the 1950s.

6.2.1 Production of L-Tryptophan Using Microbes

With its rapid growth and advanced genetic manipulation technology, *E. coli* has been known as a useful microorganism. Although the *E. coli K12* strain is not GRAS, it is a safe microorganism widely used in the production of medicinal substances and amino acids alongside *C. glutamicum*. Subsequent to commercial production of L-threonine using *E. coli*, metabolic engineering approaches have been employed in *E. coli* to produce L-tryptophan.

To improve the amino acid production in microorganisms, major strategies include the relief of the feedback inhibition/regulation on the key enzymes, the strengthening of the precursor flux, the attenuation of competing pathways, and the increase of the efflux system (Ikeda 2006).

The tryptophan biosynthetic pathway can be divided into the following four modules: (1) the pentose phosphate (PP) pathway to provide phosphoribosyl pyrophosphate (PRPP) and erythrose 4-phosphate (E4P), (2) the aromatic/shikimate pathway, (3) the serine biosynthesis pathway, and (4) the terminal L-tryptophan biosynthetic pathway (Fig. 6.1).

The biosynthesis of L-tryptophan begins with the condensation of phosphoenolpyruvate (PEP), the glycolytic intermediates, and E4P from the PP pathway to produce 3-deoxy-D-arabino-heptulosonate-7-phosphate (DHAP). DHAP is the first intermediate of aromatic amino acid synthesis, and this condensation reaction takes place by the DHAP synthase isoenzymes encoded by the *aroF*, *aroG*, and *aroH* genes. Subsequently, DHAP is converted to chorismate by six enzymes: AroB, AroD, AroE, AroK/AroL, AroA, and AroC (Fig. 6.2).

The TrpEDCBA operon is responsible for catalyzing the terminal biosynthetic pathway of L-tryptophan. Anthranilate synthase, encoded by the *trpED* gene, synthesized the first intermediary metabolite, anthranilate from chorismate and L-glutamine. In addition, anthranilate phosphoribosyl transferase, encoded by the *trpD* gene, can independently convert anthranilate to anthranilate 5-phosphoribosyl pyrophosphate (PRA). Next, two stepwise reactions by PRA isomerase and indole

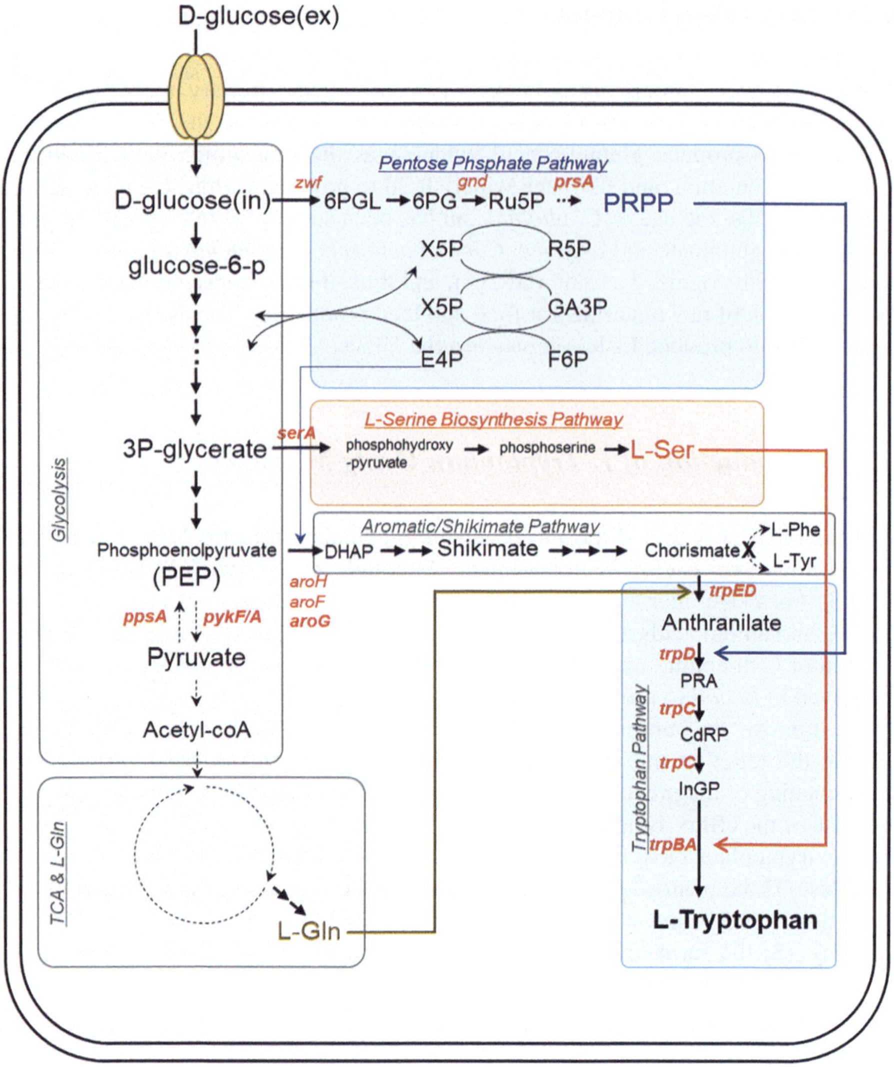

PRA : N-(5′-phosphoribosyl)-anthranilate
CdRP : 1-(ocarboxyphenylamino)−1′-deoxyribulose-5′-phosphate
IGP : indole glycerol phosphate

Fig. 6.1 Overview of L-tryptophan biosynthetic pathway in *E. coli*

glycerol phosphate (InGP) synthase, respectively, performed the synthesis of 1-(O-carboxyphenyl amino) 1-deoxyribulose 5-phosphate (CdRP) and InGP, respectively. Noteworthy, PRA isomerase and indole glycerol phosphate (InGP) synthase are encoded by the bifunctional *trpC* gene. Lastly, InGP and L-serine are condensed into L-tryptophan and D-glyceraldehyde 3-phosphate at the final step catalyzed by L-tryptophan synthase, encoded by the *trpBA* genes (Fig. 6.1).

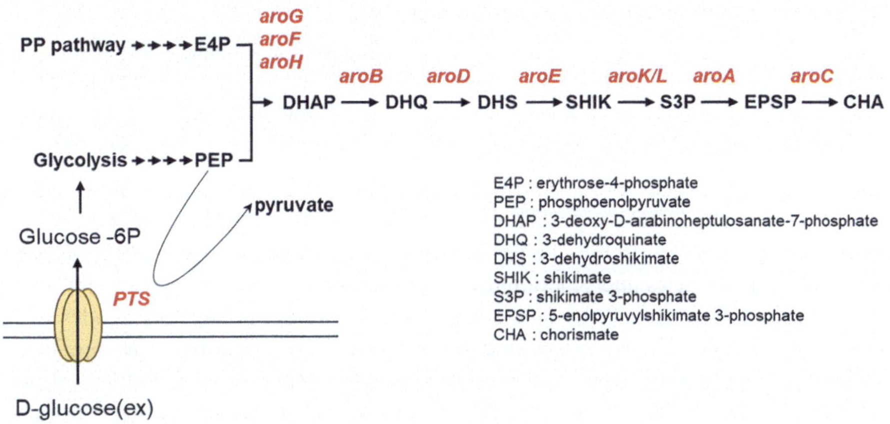

Fig. 6.2 The common aromatic amino acid pathway

6.2.2 Metabolic Pathway and Precursors of Aromatic Amino Acids for L-Tryptophan Biosynthesis

The major biosynthetic pathway of L-tryptophan starts from the stereo-specific condensation of PEP and E4P to generate DHAP that is catalyzed by DHAP synthase (Li et al. 2020). In this pathway, it is crucial to consider two important factors: the relief of DHAP synthase's feedback inhibition and the sufficient supply of PEP and E4P while maintaining a balance between them.

DHAP synthases, classified as A*ro*G, A*ro*F, and A*roH* are tightly regulated through feedback inhibition by aromatic amino acids. For this reason, previous studies have been conducted on the feedback-resistant mutation of DHAP synthase. The mutations to remove the feedback inhibition of AroG, such as AroG (L76V) and AroG (S180F), may be helpful to increase the carbon flux in the pathway, as AroG contributes about 80% of the overall DHAP activity (Ger et al. 1994; Kikuchi et al. 1997).

It is also important to increase the PEP and E4P pool, which are precursors, as well as this feedback-resistant DHAP synthase. Metabolic engineering strategies to improve the PEP pool can be divided into two categories. The first category involves controlling the local PEP-pyruvate metabolic node, which can be achieved by overexpressing PpsA (Liu et al. 2016), deleting PykAF (Liu et al. 2017), or deleting Ppc. The second category involves controlling the phosphoenolpyruvate: carbohydrate phosphotransferase system (PTS), which naturally utilizes one molecule of PEP during glucose uptake. As about 50% of the intracellular PEP is consumed by the PTS system, attenuating the PTS can dramatically increase PEP levels, such as by knocking out PtsG and PtsHI and replacing them with a PTS-independent transport system (non-PTS). Since the weakening of PTS may reduce the glucose uptake rate, various non-PTS transport systems that can import glucose without using PEP

Table 6.1 Transcription repression of aromatic amino acids

	Regulator of aromatic amino acid	
	TyrR	TrpR
The regulated genes of common aromatic pathway	aroG, aroF, aroK, aroL	aroH, aroL
The regulated genes of Tyr, Phe, Trp pathway	tyrA, tyrB	trpEDCBA

should be enhanced. This approach is exemplified by the introduction and overexpression of glucose facilitators and glucokinases from *Zymomonas mobilis*, as well as galactose permeases and glucokinases from *E. coli.*

To increase the intracellular concentration of E4P, other precursors besides PEP, transketolase (encoded by *tktA*), and transaldolase (encoded by *talB*) are often selected as engineering targets to strengthen the gene expression. In addition, it was reported that inactivation of phosphoglucose isomerase (encoded by *pgi*) was also advantageous for the supply of E4P by blocking glycolysis (Mascarenhas et al. 1991). Therefore, increasing the PEP and E4P pool as previously described would lead to an increase in the biosynthesis flux of aromatic amino acids.

In the downstream of DHAP synthase, genetic engineering approaches, such as increasing the copy number and replacing native promoters with strong promoters, can be used to strengthen the expression of aromatic pathway genes (aroB, aroD, aroE, aroK/L, aroA, and aroC), thereby improving the carbon flux of the aromatic amino acid pathway up to chorismate. TyrR and TrpR are transcriptional repressors that bind to the operator regions of aromatic pathways genes and lead to their downregulation. Hence, by knocking out the tryR and trpR genes, the repression is removed, allowing for increased expression of the aromatic genes and, subsequently, increased production of aromatic amino acids (Table 6.1).

6.2.3 Metabolic Pathway and Precursors of L-Tryptophan

The terminal L-tryptophan biosynthesis pathway requires the *Trp* operon and several major precursors including two amino acids, L-serine and L-glutamine. The Trp operon is composed of the gene cluster *trpEDCBA* which is regulated by the TrpR repression, and the attenuation relies on the structure of the leader sequence. Due to its central role in regulating tryptophan synthesis, the attenuation mechanism of the tryptophan operon has become the primary engineering target for optimizing tryptophan-producing strains.

The first intermediary metabolite is anthranilate, which is synthesized from chorismate catalyzed by TrpED, anthranilate synthase. This reaction requires L-glutamine as an amine donor. The substrate L-glutamine can be synthesized from L-glutamate by glutamine synthase, GlnA. Despite efforts to strengthen anthranilate synthase, increasing the copy number of the Trp operon in *E. coli* strains may have a negative effect without improving the expected productivity and yield of Trp (Lee et al. 2014). This is because the intracellular accumulation of the intermediate anthranilate inhibits the TrpC enzyme, indole glycerol phosphate (InGP) synthase by feed-forward regulation (Chen et al. 2018). Engineering the next enzyme reaction in the pathway, specifically enhancing the expression of TrpD, anthranilate phosphoribosyl transferase, and increasing the concentration of its substrate, phosphoribosyl pyrophosphate (PRPP), can alleviate the negative effects caused by the accumulation of anthranilate. This is because the increased expression of TrpD can promote the conversion of anthranilate to tryptophan, thereby reducing the buildup of anthranilate in the cell. Moreover, increasing the concentration of PRPP, which is derived from the pentose phosphate pathway (PPP) via overexpression of the *prs* gene(Li et al. 2020), can help the balance of the substrate availability, further improving the efficiency of the pathway toward tryptophan biosynthesis (Fig. 6.3).

Besides the accumulation of anthranilate, the anthranilate synthase reaction generates the pyruvate as a byproduct which can induce acetate accumulation and bacterial cell growth retardation. These factors highlight the primary challenge in the metabolic engineering of tryptophan, which involves addressing the anthranilate synthase reaction as the initial step. These are why metabolic engineering for tryptophan production was challenging previously, which addresses the anthranilate synthase reaction as the initial step. Overall, balancing each enzyme activity and substrate amount through the pathway reaction is the key to tryptophan production.

Lastly, L-tryptophan synthase (encoded by *trpBA*) is doing the condensation reaction of indole glycerol phosphate (InGP) and L-serine into L-tryptophan and D-glyceraldehyde 3-phosphate. The concentration of L-serine in glucose-fed, exponentially growing *E. coli* was very low about 0.068 mM (Bennett et al. 2008). To increase L-serine pools in *E. coli*, researchers have utilized L-serine feedback-resistant SerA gene variants, including SerA H344A/N364A (Al-Rabiee et al. 1996), as well as a knockout of the *sdaA* gene responsible for L-serine degradation.

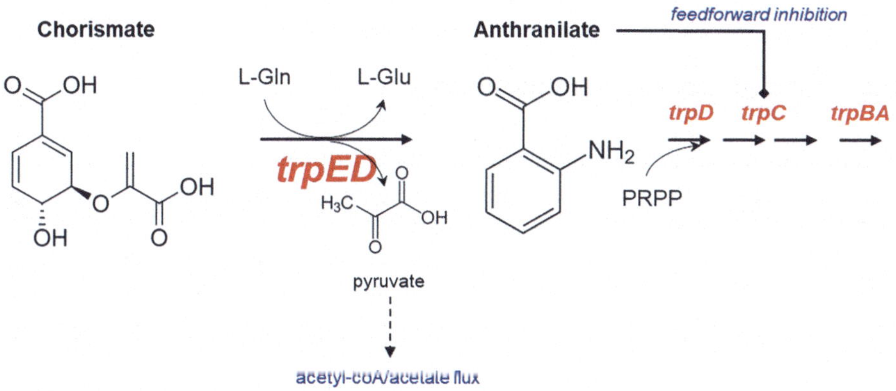

Fig. 6.3 The challenging anthranilate synthase reaction

6.2.4 Improvement of Tryptophan Productivity by Engineering Tryptophan Transporter

It is well demonstrated that the reduction of feedback control in the amino acid biosynthesis pathway increases the production of amino acids (Balderas-Hernández et al. 2009). To reduce the feedback control, keeping a low concentration of target amino acids inside microbial cells by rapid export is crucial to increase the production of target amino acids. This process will also prevent the influx of external amino acids into microbial cells. It is plausible that the regulation of the amino acid transport system through genetic modifications could lead to a decrease in the intracellular amino acid levels.

The main route for indole import is through the tryptophan permease Mtr (Pinero-Fernandez et al. 2011). Deletion of Mtr results in the loss of tryptophan-specific uptake. Mtr-deficient mutants accumulate low levels of intracellular L-tryptophan and exhibit high L-tryptophan production. Doroshenko et al. (2007) demonstrated that overexpression of YddG, an inner membrane protein involved in aromatic amino acid secretion, can enhance the excretion activity of L-tryptophan and L-phenylalanine. Overexpression of YddG was used to relieve feedback inhibition in L-phenylalanine biosynthesis and enable the excretion of this desired amino acid into the medium. Although numerous amino acid exporters have been identified in prokaryotic microorganisms, no specific efflux pump for L-tryptophan has been found to date.

Collectively, key requirements to improve the performance of tryptophan production strain are (1) strengthening the precursor flux to increase the PRPP, PEP, and serine, (2) relieving the feedback inhibition/regulation including TyrR and TrpR mutation for aromatic pathway and the tryptophan pathways, (3) balancing each intermediate amount by tuning the expression of each enzyme of the pathways, and (4) decreasing the intracellular tryptophan amount by enhancing the exporters and deleting the importers.

6.3 Production of L-Tryptophan

Tryptophan is currently produced through three different methods: (1) chemical synthesis, (2) protein hydrolysis, and (3) microbial processes (enzyme conversion and microbial fermentation). However, in general, extraction from protein hydrolysates is highly dependent on the availability of natural protein-rich resources, which can make it difficult to satisfy the increasing demand for amino acids (Ikeda 2006). In contrast, microbial fermentation has become the preferred method to produce feed-grade L-tryptophan due to its advantages of using cost-effective raw materials, producing high-purity products, and having a facile extraction process. Strain producers such as *E. coli* and *C. glutamicum* can be used to produce tryptophan in fermenters using cost-effective raw materials such as starch and raw sugars.

6.3.1 Fermenter

A fermenter, also known as a bioreactor, refers to any manufacturing device or system that supports a biologically active environment. For industrial amino acid fermentation, cylindrical aerated stirred tank fermenters in the size range of 100–500 kl are typically used (Fig. 6.4). Since tryptophan-producing strains are aerobic strains, air input and agitation speed are important factors for equipment in tryptophan production. In addition, there are various internal facilities such as a sparger, baffle, and impeller for rapid diffusion of oxygen within the fermenter. During the fermentation process, it is important to monitor parameters such as pH, temperature, and oxygen intensity that must be kept constant. Tryptophan-producing strains are generally cultured at 30–37 °C and near neutral pH. The temperature is controlled using cold water through a jacket or coils, and pH is adjusted using acids and alkalis.

Fig. 6.4 A photo showing the inside of an industrial-scale fermenter used in the large-scale production of amino acids. (A photo taken by the authors)

6.3.2 Tryptophan Fermentation Media

The elemental composition of a typical bacterial cell is carbon (50% of dry weight), oxygen (20% of dry weight), nitrogen (14% of dry weight), hydrogen (8% of dry weight), phosphorus (3% of dry weight), and other metal ions such as K, Mg, Ca, and Fe. Therefore, the important components in the tryptophan fermentation medium are carbon, nitrogen sources, phosphorus, and supplied oxygen.

Among media components, the most commonly used carbon source comes from raw sugar from sugar cane and glucose derived from corn starch, tapioca, or molasses. However, the use of crude raw materials in large quantities may cause problems in the tryptophan purification process due to increased impurities.

Ammonia is commonly used as a nitrogen source and to maintain pH during the fermentation process. Amino acid production requires the supply of ammonia in a molecular equivalent amount to the amino acids produced. If a tryptophan-producing strain has a requirement for a specific amino acid, the required amino acid input can act as an important factor for microbial cell growth and tryptophan production. In the tryptophan production strains, deficiencies in competing pathways such as the biosynthesis for phenylalanine and tyrosine are often necessary to increase tryptophan production. In that case, feeding control of phenylalanine and tyrosine is critical to optimize the balance of microbial cell growth and tryptophan production.

In the production of tryptophan, it is important to identify the lacking nutrients and optimize the input amount. Statistical techniques such as the Plackett-Burman design (Plackett and Burman 1946) and response surface methodology (Adinarayana and Ellaiah 2002; Park et al. 2002) are often used for effective medium optimization.

6.3.3 Tryptophan Fermentation Process

To scale up the tryptophan fermentation process, one of the first things to consider is how large a batch size is needed. From the agar plate to the large-scale main tank, it is important to effectively multiply tryptophan-producing strains. In general, between the agar plate and the main fermentation step, a seed culture step is performed for the effective growth of tryptophan-producing strains. Since the main purpose of the seed stages is the growth of microbial strains, there is a difference in medium and culture conditions from the main fermentation, where tryptophan production is the main purpose. There are four modes of fermentation: (1) batch fermentation, (2) fed-batch fermentation, (3) continuous fermentation, and (4) open fermentation. Industrial tryptophan fermentation is usually performed by means of batch or fed-batch processes, where batch fermentation is a closed culture system that contains limited amounts of nutrients and fed-batch is a semi-open system where nutrients are added aseptically, and the volume of liquid culture in the bioreactor increases as the culture is systematically added. Although the batch process is relatively simple as it does not require additional input of nutrients, in the industrial tryptophan production process, the fed-batch is mainly adopted to increase the productivity of tryptophan.

6.3.4 Monitoring and Control

In tryptophan fermentation, it is important to monitor specific substances as well as physical conditions such as temperature, pH, and oxygen. Concentrations of sugars, nutrients, and metal ions in the culture media affect tryptophan production, and the media feeding rate during the fermentation should be controlled also depending on the culture status. In order to increase tryptophan production, a method of monitoring the specific substances in real time using Fourier-transform near-infrared (FT-NIR) and automatically controlling them can be used.

6.4 Process of L-Tryptophan

6.4.1 Types of L-Tryptophan Products

Tryptophan products in the market are generally classified into high-purity powder products and low-content granular products depending on the customer applications. The characteristics of tryptophan products that are high-purity powder and low-content granules are summarized in Table 6.2.

Table 6.2 Comparison of the characteristics of tryptophan high-purity powder and low-content granular products

Characteristic	High-purity powder	Low-content granules
Description	White to yellowish crystalline powder	Yellow to brownish granules
Typical assay	> 98%	20–80%
Size distribution	Tens to hundreds of micrometers	Hundreds to thousands of micrometers
Flowability	Moderate good and fair	Excellent
Bulk density	Low	High
Dust generation	High	Low
Microscopic image		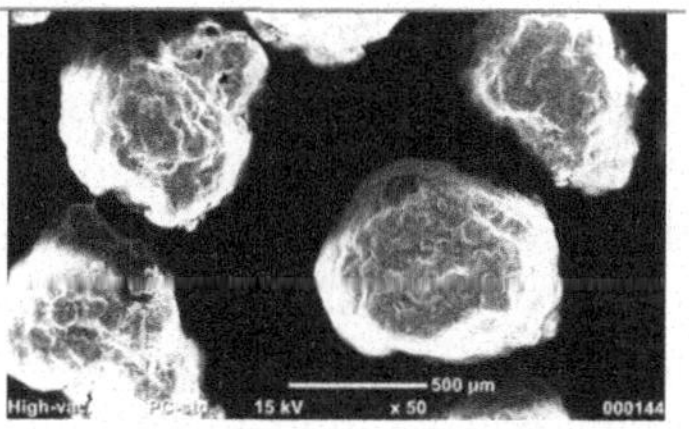

High-purity powder products are prepared in the order of (1) tryptophan fermentation, (2) cell separation, (3) membrane filtration, (4) active carbon treatment, (5) evaporative crystallization, (6) crystal-liquid separation, (7) chromatography, and (8) drying (GEA 2023; Novasep 2023). High-purity powder products have the advantage of high content but have the disadvantage of high cost due to low purification yield. In addition, since the particle size is relatively small, the flowability of the particles and increased dust generation are not desired by animal agriculture in feed manufacturing, feed delivery, feed storage, and actual feeding.

Low-content granular products are produced in the order of (1) tryptophan fermentation, (2) optional cell separation, (3) evaporation, and (4) granulation (Kwon et al. 2021). Low-content granular products typically have a particle size distribution of hundreds to thousands of micrometers without dust generation and enhanced followability. Low-content granular products have the disadvantage of low tryptophan content but have the advantage of a relatively simple process and reduced production cost. Since granular products contain various types of microbial proteins and metabolites derived from the fermentation broth, they should be confirmed for safety authorization such as generally recognized as safe (GRAS) to be sold commercially (FAD 2023).

The inclusion of microbial cells in low-content granular amino acid products can be beneficial to animal health. Recent research demonstrated the postbiotic effects of lysed *C. glutamicum* obtained from amino acid production and also the safety of low-content granular amino acids fed to nursery pigs (Cheng et al. 2021; Cheng et al., 2020).

6.4.2 Unit Processes

6.4.2.1 Cell Separation

Tryptophan broth contains microorganisms in solid form. To remove these solid impurities, a microfiltration or centrifugation process or a combination of both processes can be used (Novasep 2023; Fig. 6.5). Typically, *E. coli* cells used for tryptophan production should be removed from the fermentation broth.

6.4.2.2 Evaporation and Evaporative Crystallization

After the removal of microbial cells, the most impurities in tryptophan broth are water, and an evaporation process is essentially involved to remove it (Fig. 6.6). After the fermentation, tryptophan concentration is usually 10–60 g/L which is just above the solubility of tryptophan in water. During the evaporation process, tryptophan crystal is accumulated to make slurry concentrate. By controlling the evaporation speed, vacuum condition, and temperature, the crystal size of tryptophan can be optimized. The evaporation and crystallization process is used to produce both high-purity powder (GEA 2023; Novasep 2023) and low-content granules (Kwon et al. 2021).

Fig. 6.5 Various types of membranes used in amino acid purification. (Photos taken by the authors)

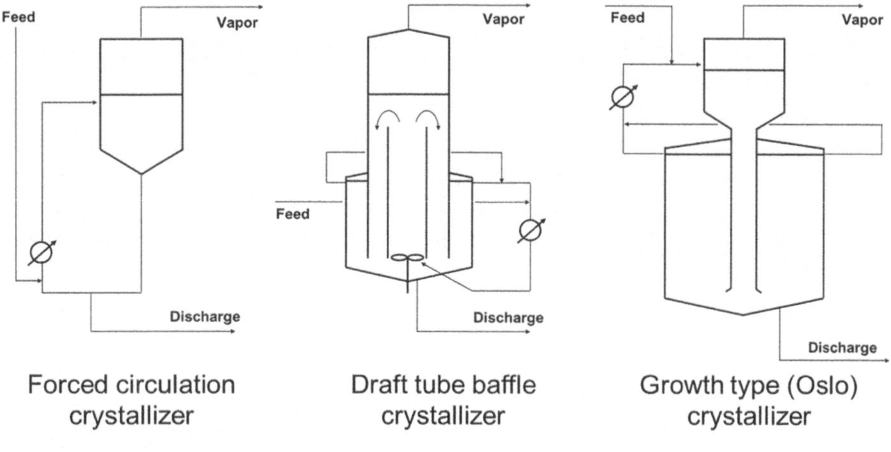

Fig. 6.6 Various types of evaporators and evaporative crystallizers used in amino acid production. (A drawing from the authors)

6.4.2.3 Crystal-Liquid Separation

High purity tryptophan crystals produced after the evaporation and crystallization process are recovered by various types of crystal-liquid separators. Typically, centrifugation or filtration separators can be used for crystal separation of amino acids. Mother liquid after the separation process is generally recycled to the evaporation process to increase the purification yield (GEA 2023; Novasep 2023).

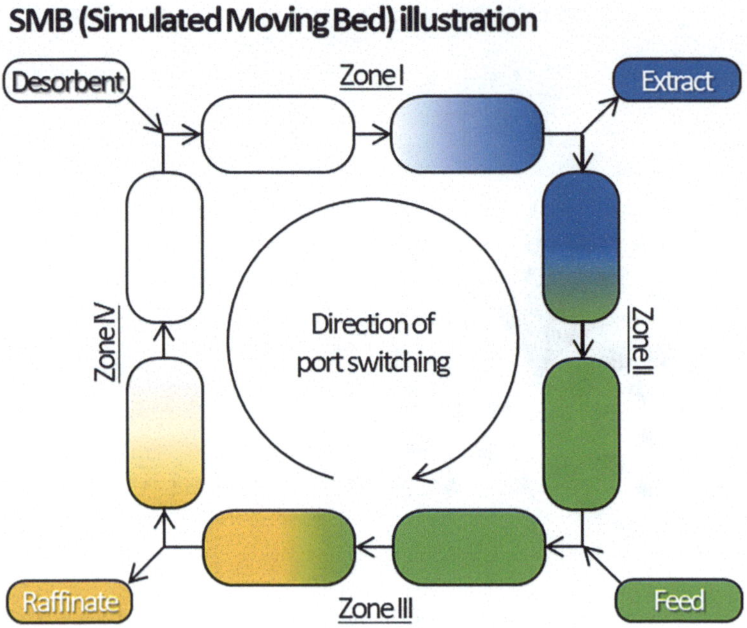

Fig. 6.7 Concept of industrial continuous chromatography in amino acid production. (A drawing from the authors)

6.4.2.4 Chromatography

Chromatography is a method of separating substances by the difference in movement speed of substances in a stationary phase (Fig. 6.7). Chromatography is a very powerful purification method, but it generally has the disadvantage of high cost. It also be used to purify the mother liquid before recycling it into the main purification process (Novasep 2023).

6.4.2.5 Drying and Granulation

The high-purity tryptophan crystals recovered in the crystal-liquid separator are dried using various dryers. Drying temperature and vacuum condition should be optimized to get good quality crystal powders. The fluidized bed dryer is one of the methods to produce tryptophan powders (Kwon et al. 2021; Fig. 6.8). In addition, a combination of granulation and drying processes can be used to make low-content granular type of tryptophan. In this case, mixing speed and ratio of solids and liquid is the key to getting the optimal size of tryptophan granules.

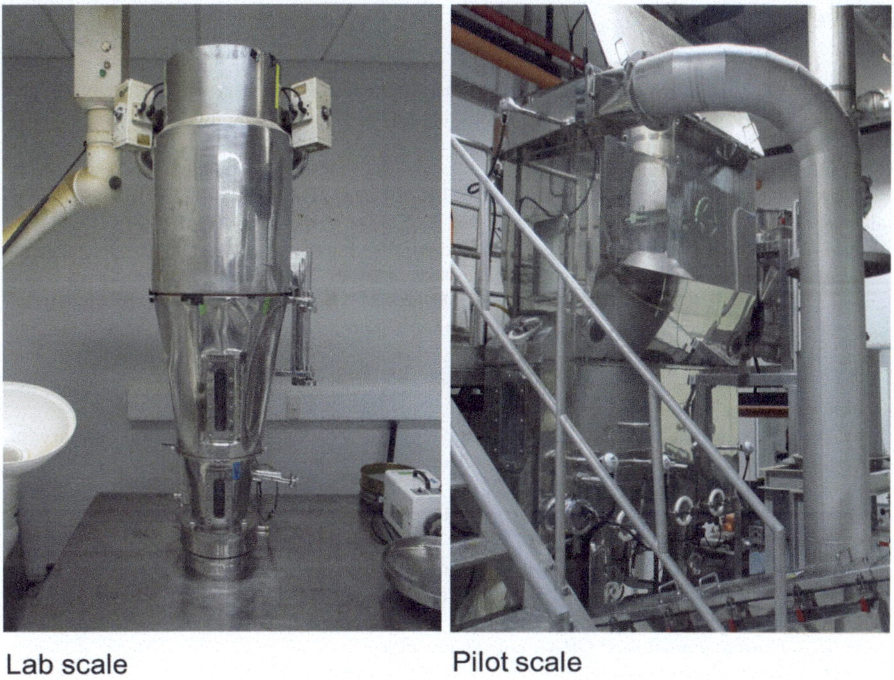

Fig. 6.8 Fluidized bed granulator used in amino acid production. (Photos taken by the authors)

6.5 Conclusion

Industrial-scale production of feed-grade supplemental amino acid has essential critical impacts on the success and sustainability of modern animal agriculture. Increased concerns with environmental pollution from animal agriculture are critical issues that animal nutritionists need to handle. The use of supplemental amino acids reducing the use of traditional protein supplements in animal feeds is an essential practice in animal nutrition. Considering both the nutritional and environmental benefits of the use of supplemental amino acids in animal agriculture, increasing the production efficiency of feed-grade supplemental amino acids is a great contribution to animal agriculture.

L-Tryptophan is one of the major essential and limiting amino acids for animals. Nutritional and functional benefits of L-tryptophan are well understood by animal nutritionists, whereas the cost of using L-tryptophan has been a limiting factor in the application. Reducing the production cost of L-tryptophan can be achieved by modification of the traditional processes. Granulated L-tryptophan is an example of this effort with increased benefits to animal growth and health supported by recent animal research.

References

Adinarayana K, Ellaiah P (2002) Response surface optimisation of the critical medium components for this production of alkaline protease by a newly isolated Bacillus sp. J Pharm Pharm Sci 5(3):272–278

Al-Rabiee R, Zhang Y, Grant GA (1996) The mechanism of velocity modulated allosteric regulation inD-3-phosphoglycerate dehydrogenase. J Biol Chem 271(38):23235–23238

Balderas-Hernández VE, Sabido-Ramos A, Silva P, Cabrera-Valladares N, Hernández-Chávez G, Báez-Viveros JL et al (2009) Metabolic engineering for improving anthranilate synthesis from glucose in Escherichia coli. Microb Cell Factories 8(1):19

Bennett BD, Yuan J, Kimball EH, Rabinowitz JD (2008) Absolute quantitation of intracellular metabolite concentrations by an isotope ratio-based approach. Nat Protoc 3(8):1299–1311

Chen L, Chen M, Ma C, Zeng A-P (2018) Discovery of feed-forward regulation in L-tryptophan biosynthesis and its use in metabolic engineering of E. Coli for efficient tryptophan bioproduction. Metab Eng 47:434–444

Cheng YC, Duarte ME, Kim SW (2020) 179 efficacy and safety of amino acids with biomass for growth and health of newly-weaned pigs. J Anim Sci 98(Suppl. 3):77

Cheng Y-C, Duarte ME, Kim SW (2021) Nutritional and functional values of lysed Corynebacterium glutamicum cell mass for intestinal health and growth of nursery pigs. J Anim Sci 99(12):skab331

Doroshenko V, Airich L, Vitushkina M, Kolokolova A, Livshits V, Mashko S (2007) YddG from Escherichia coli promotes export of aromatic amino acids. FEMS Microbiol Lett 275(2):312–318

FAD (2023) Current animal food GRAS notices inventory

GEA (2023) Crystallization of amino acids, GEA whitepaper

Ger Y-M, Chen S-L, Chiang H-J, Shiuan D (1994) A single Ser-180 mutation desensitizes feedback inhibition of the Phenylalanine-Sensitive 3-Deoxy-D-Arabino-Heptulosonate 7-Phosphate (DAHP) Synthetase in Escherichia coli1. J Biochem 116(5):986–990

Ikeda M (2006) Towards bacterial strains overproducing l-tryptophan and other aromatics by metabolic engineering. Appl Microbiol Biotechnol 69(6):615–626

Kikuchi Y, Tsujimoto K, Kurahashi O (1997) Mutational analysis of the feedback sites of phenylalanine-sensitive 3-deoxy-D-arabino-heptulosonate-7-phosphate synthase of Escherichia coli. Appl Environ Microbiol 63(2):761–762

Kim SW, Less JF, Wang L, Yan T, Kiron V, Kaushik SJ, Lei XG (2019) Meeting global feed protein demand: challenge, opportunity, and strategy. Ann Rev Animal Biosci 7(1):221–243

Koopmans SJ, Guzik AC, Van Der Meulen J, Dekker R, Kogut J, Kerr BJ, Southern LL (2006) Effects of supplemental L-tryptophan on serotonin, cortisol, intestinal integrity, and behavior in weanling piglets1,2. J Anim Sci 84(4):963–971

Kwon MK, Lee IS, Gwak WS, Yu JH, Hong JT, Kang J-H (2021) Method for preparing amino acid granules from fermented liquid. Patent (WO2021261952A1). https://patents.google.com/patent/WO2021261952A1/en

Lee KH, Park HM, Lee HH, Hwang YB, Lee SM (2014) Microorganisms of Escherichia coli having enhanced L-tryptophan production and method for producing L-tryptophan by using same. Patent (EP2803720). https://patents.google.com/patent/EP2803720A2/en

Li Z, Ding D, Wang H, Liu L, Fang H, Chen T, Zhang D (2020) Engineering Escherichia coli to improve tryptophan production via genetic manipulation of precursor and cofactor pathways. Synthe Syst Biotechnol 5(3):200–205

Liu X, Lin J, Hu H, Zhou B, Zhu B (2016) Site-specific integration and constitutive expression of key genes into Escherichia coli chromosome increases shikimic acid yields. Enzym Microb Technol 82:96–104

Liu L, Chen S, Wu J (2017) Phosphoenolpyruvate:glucose phosphotransferase system modification increases the conversion rate during l-tryptophan production in Escherichia coli. J Ind Microbiol Biotechnol 44(10):1385–1395

Mascarenhas D, Ashworth DJ, Chen CS (1991) Deletion of pgi alters tryptophan biosynthesis in a genetically engineered strain of Escherichia coli. Appl Environ Microbiol 57(10):2995–2999

Novasep (2023) Purification processes for amino-acids, Novasep whitepaper, https://www.novasep.com/processsolutions/fermentation-and-bio-based/industrial-processes/industrial-purification-processes-for-amino-acids.html. (accessed on May, 2023)

Park Y, Kang S, Lee J, Hong S, Kim S (2002) Xylanase production in solid state fermentation by aspergillus Niger mutant using statistical experimental designs. Appl Microbiol Biotechnol 58(6):761–766

Pinero-Fernandez S, Chimerel C, Keyser UF, Summers DK (2011) Indole transport across Escherichia coli membranes. J Bacteriol 193(8):1793–1798

Plackett RL, Burman JP (1946) The Design of Optimum Multifactorial Experiments. Biometrika, 33:305-325. https://doi.org/10.1093/biomet/33.4.305

Shen YB, Voilqué G, Kim JD, Odle J, Kim SW (2012a) Effects of increasing tryptophan intake on growth and physiological changes in nursery pigs1. J Anim Sci 90(7):2264–2275

Shen YB, Voilqué G, Odle J, Kim SW (2012b) Dietary L-tryptophan supplementation with reduced large neutral amino acids enhances feed efficiency and decreases stress hormone secretion in nursery pigs under social-mixing stress. J Nutr 142(8):1540–1546

Shen YB, Coffey MT, Kim SW (2015) Effects of short term supplementation of l-tryptophan and reducing large neutral amino acid along with l-tryptophan supplementation on growth and stress response in pigs. Anim Feed Sci Technol 207:245–252

Shen Y-B, Weaver AC, Kim SW (2021) Physiological effects of deoxynivalenol from naturally contaminated corn on cerebral tryptophan metabolism, Behavioral response, gastrointestinal immune status and health in pigs following a pair-feeding model. Toxins 13(6):393

Chapter 7
Methodology to Study Tryptophan Metabolism

Wenjing Wang, Chen Zhang, Yadong Zhao, Libin Wang, Mengxuan Tang, Yulong Yin, and Xia Xiong

Abstract This chapter introduces the current approaches to extract, separate, and detect tryptophan and its metabolites. Furthermore, high-throughput omics have also been included in this chapter to describe the large-scale study of tryptophan (Trp) metabolism. Among them, the detection of tryptophan and its metabolites can be achieved through high-performance liquid chromatography, capillary electrophoresis, near-infrared spectroscopy, amino acid analyzer, direct or indirect colo-

W. Wang · M. Tang
Laboratory of Animal Nutritional Physiology and Metabolic Process, Institute of Subtropical Agriculture, Chinese Academy of Science, Changsha, Hunan, China
e-mail: wangwenjing1020@stu.hunau.edu.cn; tmx20180012@stu.hunau.edu.cn

C. Zhang
Animal Nutritional Genome and Germplasm Innovation Research Center, College of Animal Science and Technology, Hunan Agricultural University, Changsha, Hunan, China
e-mail: zhch@hunau.edu.cn

Y. Zhao
School of Food and Pharmacy, Zhejiang Ocean University, Zhoushan, Zhejiang Province, China
e-mail: yadong@kth.se

L. Wang
College of Horticulture, Nanjing Agricultural University, Nanjing, Jiangsu, China
e-mail: wanglibin@njau.edu.cn

Y. Yin (✉)
Tianjin Institute of Industrial Biotechnology, Chinese Academy of Sciences, National Center of Technology Innovation for Synthetic Biology, Tianjin, China

Institute of Subtropical Agriculture, Chinese Academy of Science, Changsha, China
e-mail: yinyulong@isa.ac.cn

X. Xiong (✉)
CAS Key Laboratory of Agro-ecological Processes in Subtropical Region, Institute of Subtropical Agriculture, Hunan Province Key Laboratory of Animal Nutritional Physiology and Metabolic Process, National Engineering Laboratory for Pollution Control and Waste Utilization in Livestock and Poultry Production, Institute of Subtropical Agriculture, Chinese Academy of Sciences, Changsha, China
e-mail: xx@isa.ac.cn

Y. Yin et al. (eds.), *Tryptophan in Animal Nutrition and Human Health*,
https://doi.org/10.1007/978-981-97-4719-1_7

rimetry, biosensors, or mass spectrometry. In addition, high-throughput metabolomics analysis is commonly used for rapid detection of Trp and its metabolites.

7.1 Extraction and Separation Methods of Tryptophan

Acid hydrolysis is a pre-treatment method for amino acid detection, but it will destroy Trp in the sample. Currently, alkaline hydrolysis is generally used to pre-treat Trp samples separately, including microwave digestion and conventional hydrolysis methods. The specific operation steps are as follows:

7.1.1 Microwave Digestion

The sample of 0.05 ~ 0.1 g (accurate to 0.0001 g) was accurately weighed and then placed in a polytetrafluoroethylene digestion tube, where 5 mL sodium hydroxide solution (5 mol/L) was added, and nitrogen was filled for 1 min. Then, the sealing cover was spun tightly, and the tube was placed in a microwave digestion instrument for microwave hydrolysis. The hydrolysis conditions were as follows: the power was at 1200 W, the heating time was 10 min, the hydrolysis temperature was 170 °C, and the hydrolysis time was 30 min. The digestion tube was then taken out and cooled to room temperature. The hydrolysate was transferred to a 25-mL volumetric flask, neutralized with 6 mol/L hydrochloric acid to bring the pH to neutral, and diluted with water up to the scale. It was then centrifuged at 10,000 r/min for 10 min, filtered by a 0.45 μm filter, and determined by an amino acid analyzer.

7.1.2 Conventional Hydrolysis

The samples of 0.05 ~ 0.1 g (accurate to 0.0001 g) were accurately weighed and then placed in a polytetrafluoroethylene hydrolysis tube, where 1.5 mL of lithium hydroxide solution (4 mol/L) was added, and nitrogen was filled for 1 min. Then, the sealing cover was spun tightly, and the tube was placed in a 110 °C ± 2 °C constant temperature drying box for hydrolysis for 20 h. The hydrolysis tube was then taken out and cooled to room temperature. The hydrolysate was diluted with sodium citrate buffer solution (0.12 mol/L, pH = 3.45) or sodium acetate (0.0085 mol/L, pH = 4) in a 25-mL volumetric flask, centrifuged at 10,000 r/min for 10 min, filtered by 0.45 μm needle filter, and determined by an amino acid analyzer.

Samples with a fat content higher than 4% were processed as described above after defatting.

7.2 Tryptophan Detection Method

Trp detection methods include the traditional detection methods represented by chromatography techniques and the emerging biosensor analysis technology. Due to the complex composition of animal body fluids, there are many kinds of interfering substances, and some of them have large concentrations. Meanwhile, the concentration of Trp in general animal body fluids ranges from 20 to 95 μM (Duan et al. 2020), which is relatively low, making its quantitative detection a challenging job. This also makes traditional methods still have their place. In general, the traditional methods for the detection of Trp include chromatography, spectroscopy, electrophoresis, and amino acid analyzer (Chen et al. 1997; Hervé et al. 1996; Ilisz et al. 2009). These methods can provide precise concentrations of Trp with extremely high selectivity and accuracy over an extremely wide linear range (Steckl and Ray 2018).

Among them, high-performance liquid chromatography (HPLC) is mostly performed using pre-column derivatization reversed-phase HPLC, where amino acids are first converted into substances suitable for detection by UV or fluorescence detectors using derivatizing agents. HPLC has the advantages of high sensitivity, high separation efficiency, and automated operation; however, its analysis cost is high, and the analysis time is long (He et al. 2021; Zhu et al. 2021). On the other hand, with gas chromatography (GC), amino acids are first derived into substances that are easy to vaporize and then are quantitatively analyzed based on the different distribution coefficients of samples in the mobile phase and stationary phase. The advantages of GC are that it cannot only detect the content of amino acids, but also find new amino acids; the disadvantages are that the operation is more complicated, and there are more interfering factors. At present, the ultra-high-performance liquid chromatography-tandem mass spectrometry method can determine free Trp in an efficient and quick manner. Through the optimization of mass spectrometry, chromatographic conditions, and amino acid extraction conditions, 5 mM ammonium acetate solution with a volume fraction of 0.2% formic acid and methanol were used as mobile phases for gradient elution and detected in the positive ion scanning mode of the electrospray ion source. Compared with HPLC, this method is simpler in operation and shorter in analysis time and does not require derivatization of the sample; however, its analysis and equipment maintenance costs are high.

Capillary electrophoresis is a new type of analysis and separation technique developed in recent years. It combines the common characteristics of HPLC and gel electrophoresis, including high efficiency, fast speed, high resolution, and good repeatability, and can be used for the detection of Trp. According to different separation principles, capillary electrophoresis can be divided into capillary zone electrophoresis, capillary isoelectric focusing electrophoresis, capillary isotachophoresis, capillary gel electrophoresis, and micellar electrokinetic capillary electrophoresis. Among them, capillary zone electrophoresis and micellar electrokinetic capillary electrophoresis can be used for the analysis and detection of Trp.

The principle of Trp detection by near-infrared spectroscopy is to use the spectral changes generated by the transition of the hydrogen-containing group of Trp in a specific wavelength region, combined with statistical methods to indirectly detect the content of Trp. The advantage of this method is that it does not require the use of reagents or drugs for the pre-treatment of the analyte, and the analysis and maintenance costs are low. However, the near-infrared spectroscopy technology cannot analyze small batches of samples, cannot cover twenty standard amino acids, and requires specialized instruments.

An amino acid analyzer is an analytical instrument widely used in the fields of food science, agricultural environment, biochemistry, and medical and chemical industry. The basic principle of an amino acid analyzer is as follows: after the sample is pretreated to remove protein, the sample solution is separated by ion exchange column chromatography. Through post-column derivatization, the separated amino acids are reacted with ninhydrin solution, and then, the color reaction is observed. The detection is performed using a UV detector, and quantification is done using an external standard method. An amino acid analyzer can be divided into a general amino acid analyzer and an automatic amino acid analyzer, with the latter being more convenient. The amino acid analyzer method is accurate, reliable, and reproducible and can determine most types of free amino acids. However, its determination time is relatively long, the resolution is reduced due to the large post-column diffusion, the instrument structure is complex, and it is relatively expensive.

In conclusion, although traditional Trp detection methods have high sensitivity and strong specificity and can detect multiple amino acids simultaneously, the methods' equipment costs are high, and the detection processes often require well-trained operators and specialized laboratories. In addition to the high cost of traditional instruments, the time required for complete analysis is also very long, and the analysis can only be carried out in the laboratory, which cannot achieve rapid clinical detection.

In recent years, biosensors have been receiving increasing attention due to their low cost, good selectivity, and sensitivity, along with the ability to achieve rapid detections. Meanwhile, wearable devices and miniaturized implantable devices based on biosensor arrays have already been available for real-time dynamic monitoring of Trp in body fluids and for the prevention and treatment of diseases related to Trp metabolism. The following part summarizes the recent applications of advanced biosensors for the detection of Trp in animal body fluids and compares the characteristics of traditional methods with biosensor methods.

Idili et al. (2019) constructed an electrochemical sensor based on nucleic acid aptamers for the detection of Trp in urine. The aptamers that can recognize Trp and rhodium-based complexes, as introduced in previous work, were immobilized on the surface of the gold electrode, and then, the traditional blocking agent, mercaptohexanol, was used to block the unbound sites. When a complex existed in the solution, it was attracted by the aptamer and therefore adsorbed onto the electrode surface to increase the electrode surface's resistance, thereby detecting Trp. This sensor can detect Trp in the range of 0.1–10 mM in clinical applications (Idili et al. 2019).

Duan et al. (2020) constructed a fluorescent sensor based on a double-stranded DNA adapter for the detection of Trp in blood and urine. Firstly, fluorescein was modified at one end of a complementary DNA strand and quencher at the other.

Then, in the normal state, the fluorescence was quenched due to the proximity of the complementary strand. Next, continuous DNA strands and notched DNA strands were designed respectively. When the continuous DNA strand was added, the modified DNA strand was straightened due to the complementarity with the intermediate segment of the DNA strand modified with fluorescein, therefore resulting in fluorescence restoration. However, when the notched DNA strand was added, fluorescence restoration only occurred with Trp inhibitor and Trp added simultaneously. The detection limit of the sensor was 0.7 μM, and its detection range was 0 ~ 10 μM. The recovery rates of the sensor in actual serum and urine were 102.7% and 105.1%, respectively (Duan et al. 2020).

Xia et al. (2019) constructed an electrochemical sensor based on Trp oxidase for real-time and selective detection of Trp concentration in human sweat and blood. The newly synthesized amino acid (2-amino-3-(4-mercaptophenyl) propanoic acid) was merged with glycine enzyme and Trp enzyme, following which boron-dipyrromethene (Bodipy373) was used as a binding agent to immobilize the composition onto the electrode surface through a thiol-chlorine nucleophilic substitution reaction, thus achieving the detection of Trp. The sensor possessed a good linear range in the range of 0.02–0.8 mM. Meanwhile, the sensor achieved good practical applications in the actual detection of human sweat and blood (Xia et al. 2019).

Fan et al. (2020) constructed a colorimetric and surface-enhanced Raman scattering biosensor based on a diazotization reaction for the detection of Trp in human serum. First, Trp, nitric acid, and sodium nitrite were mixed into the test tube and stored at 0 ~ 5 °C for 15 min to initiate the diazotization reaction. Then, the ammonium sulfamate solution was rapidly added and vigorously shaken. Finally, N-(1-naphthyl) ethylenediamine dihydrochloride (NEDA) was added, and the solution turned purple after 20 min at room temperature. Due to the different concentrations of Trp added, the solution exhibited different color depths, resulting in different absorbance and peaks in the Raman spectrum, so it can be used for the detection of Trp. The detection range and detection limit of the sensor were 0.05 ~ 1000 mM and 20 nM, respectively. In the actual serum detection of healthy individuals and patients with colorectal cancer, the results were basically consistent with those of HPLC (Fan et al. 2020) (Fig. 7.1).

Li et al. (2020) constructed a fluorescent biosensor for the detection of Trp in human serum and living cells. Using pyridine (Py) derivative 3-aminoisonicotinic acid as carbon and nitrogen sources, pyridine-functionalized CDs (Py-CDs) were prepared by the one-pot hydrothermal method. When Trp existed in the solution, it bound to the active site—the pyridine ring—on Py-CDs, resulting in the quenching of the quantum dots' fluorescence. The detection range and detection limit of the sensor were 0.02–20 μM and 5.7 nM. The recovery rate in actual human serum and the relative standard deviation of the detection were 96.1–105.6% and 1.3–3.2%, respectively (Li et al. 2020).

Bhattacharjee et al. (2021) constructed an electrochemical sensor based on nano-mixed materials for the detection of Trp in bio-therapeutics and human serum. First, the silica nanospheres (SiO_2 NSs) were amino-functionalized. Then, with the presence of NaBH4, the silver nanomaterials were grafted onto the SiO_2 NSs by

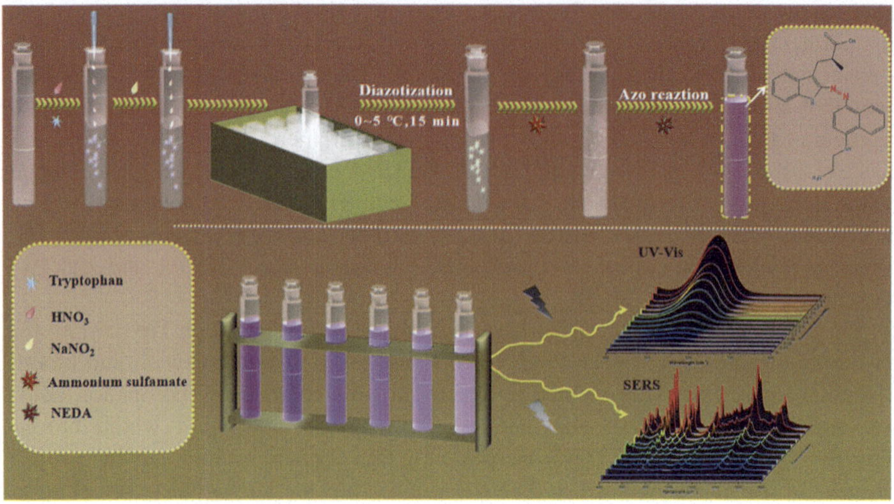

Fig. 7.1 Colorimetric and surface-enhanced Raman scattering biosensor based on diazotization reaction (Fan et al. 2020)

ultrasonic-assisted reduction to obtain silver-grafted silica nanohybrids (SiO_2 @ Ag-NHs). The nano-hybrid material was attached to the surface of the glassy carbon electrode to detect Trp. The sensor exhibited a good linear range within the range of 0.2 ~ 150 μM, with a detection limit of 0.127 μM. The biosensor was used for detections of Trp in Alamin-M Forte Capsules and human serum, and the recovery rates were 96.46 ~ 97.83% and 93.89 ~ 100.32%, respectively (Bhattacharjee et al. 2021).

Prabakaran et al. (2021) constructed a quartz crystal microbalance biosensor based on molecularly imprinted polymers for the detection of Trp in food and human serum. The molecularly imprinted polymer layer was obtained by polymerizing the monomer of methacrylic acid; meanwhile, ethylene glycol dimethacrylate was cross-linked into the molecularly imprinted polymer for the detection of Trp. The sensor had a detection range of 1 μg/mL–5 ng/mL and a detection limit of 0.72 ng/mL. The recovery rates for Trp detection in milk and urine were 97.6 ~ 100.4% and 98.0 ~ 104.0%, respectively (Prabakaran et al. 2021).

Gunavathana et al. (2022) constructed an electrochemical sensor based on polythiophene-loaded zinc oxide nanorods (PT@ZnO) for the detection of Trp in peanut samples. Polythiophene and zinc oxide nanorods were drop-coated onto the surface of the glassy carbon electrode for the detection of Trp. The detection range and detection limit of the sensor were 0.4 ~ 200 μM and 8.5 nM, respectively. The recovery rate for Trp detection in peanut samples was 96.59 ~ 103.90% (Dheepthi Gunavathana et al. 2022).

Prabakaran et al. (2022) constructed a surface acoustic wave sensor based on molecularly imprinted polymers for the detection of Trp. Using hydrophilic polyvinylidene fluoride (PVDF) as the material, the molecularly imprinted template of the Trp sensor was prepared by in situ polymerization of 2-hydroxyethyl methacrylate

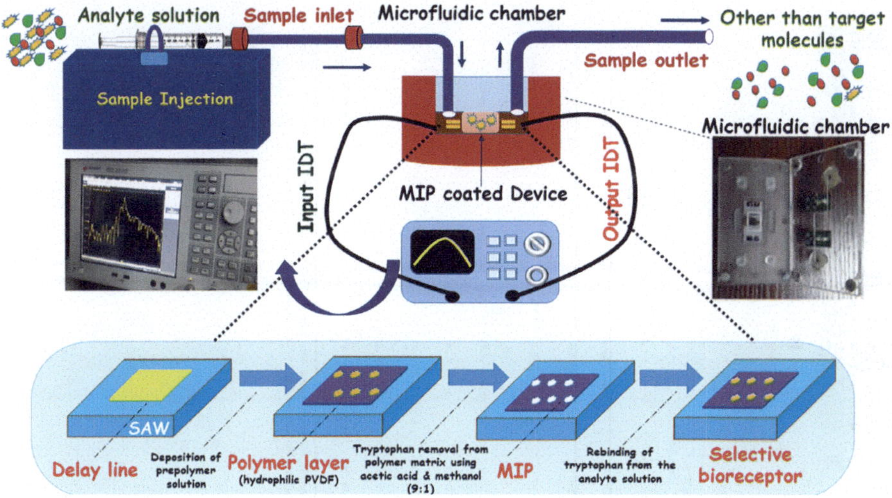

Fig. 7.2 Surface acoustic wave sensor based on molecularly imprinted polymers (Prabakaran et al. 2022)

(HEMA). The detection range and detection limit of the sensor were 0.5 ~ 150 ng/mL and 0.6 ng/mL, respectively (Prabakaran et al. 2022) (Fig. 7.2).

Additionally, Trp can also be detected through direct and indirect colorimetry.

1. Acidic-Ninhydrin for Trp Detection

   Gaitonde and Dovey found that Trp was converted into a yellow product (λmax = 390 nm; ε = 7120) when protein-bound Trp reacted with ninhydrin in a mixture of formic acid and hydrochloric acid at 100 °C for 10 min. This method can be applied to determine protein-bound Trp.

2. Rapid Colorimetry for Free-Trp Detection

   Wu et al. developed a method based on the Trp enzyme for the conversion of Trp into indole, which then followed a reaction with hydroxylamine to produce a pink product (λmax = 530 nm). Afterward, free Trp, protein-bound Trp, and Trp in biological samples were quantified by the 96-well flux assay method. Peptide aptamers, as an emerging recognition probe, are gaining popularity in the construction of biosensors due to their good properties such as acid and alkali resistance, precise recognition of small molecules, good biocompatibility, easy modification and inexpensive synthesis (Zhao et al. 2021; Song et al. 2021; Wang et al. 2022; Wang et al. 2023). Wang et al. (Wang et al. 2024) developed a colorimetric biosensor for the rapid detection of L-Trp in solutions by coupling a novel, screened, and validated peptide aptamer with Au nanoparticles via the Au-S bond. The biosensor is easy to operate, and the reaction time was as short as 10 s. In addition, it showed a wide linear detection range of 1-1000 μmol (Figs. 7.3 and 7.4).

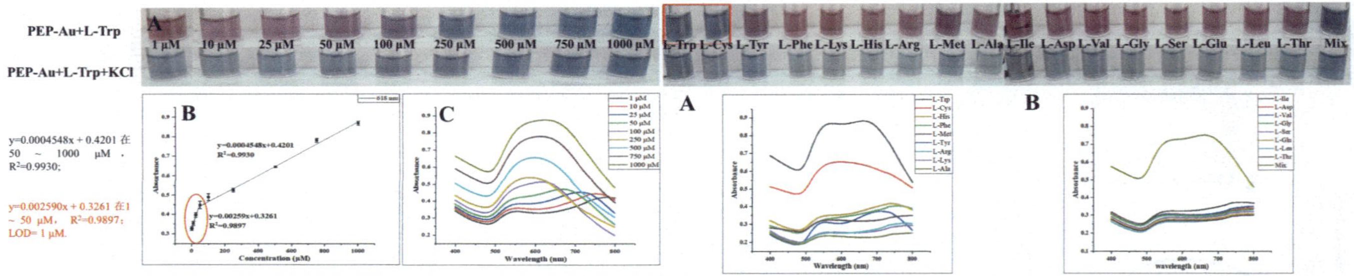

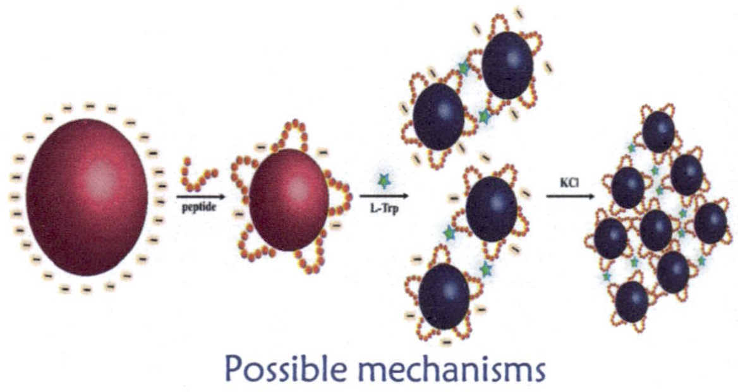

Signal Type	Recognition Element	Detection Limit	Linear Range	Detection time (s)	Reference
surface acoustic wave	HEMA-PVDF	9.80×10^{-10} M	2.45×10^{-9} M–7.35×10^{-8} M	3600	(Prabakaran et al., 2022)
Amperometric	PT-ZnO NRs	8.50×10^{-9} M	4.00×10^{-6} M–2.00×10^{-4} M	-	Dheepthi Gunavathana et al., 2022)
Amperometric	EGDMA-MIP	3.57×10^{-9} M	1.00×10^{-8} M–1.00×10^{-3} M	2400	(Prabakaran et al., 2021)
Fluorescence	Py-CDs	5.70×10^{-9} M	0–4.00×10^{-5} M	15	(Li et al., 2020)
Colorimetric	SERS-Au NRs	2.00×10^{-8} M	5.00×10^{-7} M–1.00×10^{-3} M	600	(Fan et al., 2020)
Fluorescence	TrpR	7.90×10^{-7} M	0–1.00×10^{-4} M	1200	(Duan et al., 2020)
Colorimetric	**PEP-Au**	**1.00×10^{-6} M**	**1.00×10^{-6} M–1.00×10^{-3} M**	**10**	**This work**

HEMA: 2-hydroxyethyl methacrylate; PVDF: poly (vinylidene fluoride); PT: polythiophene; NRs: nanorods; EGDMA: ethylene glycol dimethacrylate; MIP: molecularly imprinted polymer; CDs: carbon quantum dot; SERS: Surface-Enhanced Raman Scattering; Au NRs: gold nanorods; TrpR: L-Trp repressor

Fig. 7.3 Peptide aptamer-based colorimetric biosensor for L-Trpdetection (Wang et al. 2024)

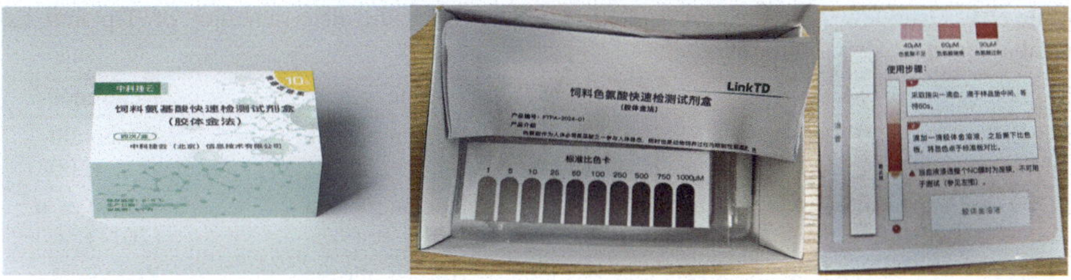

Fig. 7.4 Products for L-Trp rapid test (A photo taken by the authors)

7.3 Detection Methods for Trp Metabolites

7.3.1 Mass Spectrography

Recently, the development of highly complex technologies has greatly improved the analysis of Trp and its metabolites, which are typically present in the form of femtomolar concentrations. Table 7.1 represents the latest analytical methods developed in the past 10 years. It is worth noting that, compared to traditional UV and fluorescence detection, the updated methods cover a wider range of analytes due to the increased specificity of tandem mass spectrometry (MS/MS) and high-resolution mass spectrometry (HRMS).

7.3.2 Isotopic Labeling Based on Chromatography/Mass Spectrometry

Arnhard et al. validated an HPLC/MS method for simultaneous quantification of Trp as well as its metabolites including kynureninase, kynurenic acid, and quinolinic acid. The concentration of metabolites in 100 human plasma samples was determined by using stable isotope-labeled internal standards.

Another detection method for animal systems was developed by Sano et al., who designed a method for the detection of 15 N2-labeled Trp, l-kynurenine, serotonin, and quinolinic acid in rat plasma. These compounds were first converted into acylated derivatives by using pentafluoropropionic anhydride and pentafluoro-1-propanol. The derivatives were then analyzed by gas chromatography/mass spectrometry (GC/MS). This method is used to determine the isotope enrichment of plasma Trp in pregnant rats with continuous infusion, which can promote our understanding of the four Trp metabolic pathways in animals and humans.

Table 7.1 Overview of the latest analytical methods for the determination of Trp and its metabolites in biological fluids

Method	Metabolites	Sample	References
LC-ECL multiplex system	TRP, KYN	Human plasma	Fitzgerald et al. (2008)
LC-UV/FD	TRP, KYN, 5-HTP, 5-HT	Human plasma	Comai et al. (2011)
LC-UV	TRP, KYN, 5-HT	Human plasma	Zhen et al., (2011)
LC-MS/MS	TRP, KYN, KA, 3-HAA, AA, QA	Mouse plasma	Möller et al. (2012)
LC-UV/FD	TRP, 5-HT, 5-HIAA, KYN, KA	Rabbit brain tissue and amniotic fluid	Lesniak et al. (2013)
LC-MS/MS	KYN, KA, 3-HK, AA	Mouse plasma	Orsatti et al. (2015)
LC-MS/MS GC-MS/MS	KYN, KA, XA, 3-HK, 3-HAA	Human serum	Favennec et al. (2015)
LC-MS/MS	TRP, KYN, 5-HT, OA, XA, KA, IAA, indoxyl sulfate	Human serum	Karu et al. (2016)
LC-MS/MS	TRP, 5-HT, 5-HIAA, KYN, KA, 3-HK, AA, XA	Human urine, human plasma, and mouse prefrontal cortex	Marcos et al. (2016)
LC-MS/MS	TRP, KYN, 3-HK, 3-HAA, KA, XA, QA, PA, 5-HT	Mouse plasma, cerebrospinal fluid, brain tissue	Fuertig et al. (2016)
LC-MS/MS	KYN, AA, KA, 3-HK, XA3-HAA, 5-HTP, 5-HT	Human plasma	Boulet et al. (2017)
LC-MS/MS	TRP, KYN, KA, XA, 5-HT	Human plasma	Tong et al. (2018)
LC-HRMS	TRP, 5-HIAA, 5-HT, IAA, KA, MLT, PA, XA, tryptamine, 3-HAA, 3-HK, 5-HTP, IAM	Mouse urine, serum, intestinal contents, and liver	Lefèvre et al. (2019)

3-HAA 3-hydroxy-2-aminobenzoic acid, *3-HK* 3-hydroxykynureninase, *5-HIAA* 5-hydroxyindoleacetic acid, *5-HT* 5-hydroxytryptamine, *5-HTP* 5-hydroxyTrp, *AA* aminobenzoic acid, *IAA* indocyanine-3-acetic acid, *IAM* indocyanine-3-acetamide, *KA* kynurenic acid, *KYN* kynureninase, *MLT* melatonin, *QA* quinolinic acid, *PA* picolinic acid, *TRP* Trp, *XA* xanthurenic acid, *LC* liquid chromatography, *ECL* electrochemiluminescence multi-detection system, *FD* fluorescence detection, *UV* ultraviolet detection, *MS/MS* tandem mass spectrometry, *GC* gas chromatography, *HRMS* high-resolution mass spectrometry

7.4 High-Throughput Metabolomic Analysis Method for Tryptophan

7.4.1 Trp Metabolomics Analysis

7.4.1.1 Targeted Metabolomics Analysis

Trp metabolites are clear, and the metabolic pathways are also well defined. To

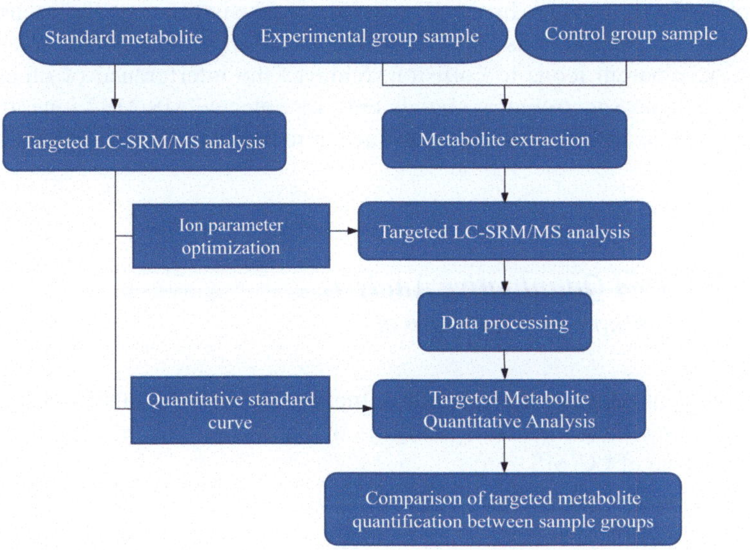

Fig. 7.5 The workflow of the targeted and quantitative metabolomics

study the regulation of Trp by body and intestinal flora, targeted and quantitative metabolomics technology can be used to detect Trp metabolites in the test samples and analyze metabolic pathways.

Targeted metabolomics analysis refers to the specific and targeted detection and analysis of a specific group of metabolites based on the standard sample (Roberts et al. 2012). The commonly used technical solutions include conventional liquid chromatograph mass spectrometer (LC-MS) analysis and selected reaction monitoring/multiple reaction monitoring (SRM/MRM). In the conventional analysis method, the LC-MS method was first utilized to perform large-scale metabolomic analysis on the samples to be analyzed, and then, the qualitative and quantitative analysis of the target metabolite group in the samples was realized by comparing the analysis results with those of standard samples under the same analysis conditions (Fig. 7.5). And SRM/MRM technology is a targeted and specific detection and analysis method for specific metabolite groups, with the reference of standard samples. It can obtain absolute quantitative results of target metabolites and has the characteristics of high specificity, sensitivity, and accuracy (Vidova and Spacil 2017).

7.4.1.2 SRM/MRM Principle

SRM/MRM technology selects data in a targeted manner for mass spectrometry signal acquisition on the basis of known or assumed reaction ion information, records signals for ion pairs that meet the criteria, and removes interference from ion signals that do not meet the criteria. The workflow is shown in Fig. 7.5

(Mermelekas et al. 2015). In the quantitative analysis process, the technology first screens the parent ions specific to the target metabolite, then selectively induces these parent icons to collision, removes the interference of other ions, and collects mass spectrometry signals only for selected MS/MS2 ions, thereby obtaining a more specific, sensitive, and accurate analysis of the target metabolic molecule.

7.4.2 *Targeted Quantitative Analysis of Tryptophan Metabolism*

High-throughput relative or absolute quantitative analysis of 25 major metabolites (Table 7.2) in the three metabolic pathways of Trp was performed using the SRM/MRM technique of LC-MS.

Table 7.2 Metabolites of tryptophan

Number	Metabolites	Metabolic pathway
1	Tryptophan	Kynurenine pathway
2	L-kynurenine	
3	Picolinic acid	
4	3-Hydroxyanthranilic acid	
5	Xanthurenate	
6	Quinolinic acid	
7	N-formyl-kynurenine	
8	Cinnavalininate	
9	3-hydroxyl-L-kynurenine	
10	Kynurenate	
11	NAD+	
12	Melatonin	Serotonin pathway
13	5-hydroxyl-L-tryptophan	
14	Serotonin	
15	Tryptamine	AhR pathway
16	Indoxyl-β-D-glucuronide	
17	Indole	
18	5-Hydroxyindole-3-acetic acid	
19	Indole-3-carboxaldehyde	
20	3-methyl-indole	
21	Indole-3-propionic acid	
22	Indole-3-lactic acid	
23	Indoleacetate (IAA)	
24	Indole-3-acetaldehyde	
25	3-Indoxyl sulfate potassium salt	

Table 7.3 Trp targeted and quantitative metabolomics testing sample requirements

Sample type	Regular sample quantity	Minimum sample quantity	Suggested biological replication
Blood (serum, plasma)	≥200 μL	≥100 μL	Clinical sample: ≥30 cases per group; animal source: ≥10 cases per group; plant source: ≥3 examples per group; cell samples: ≥6 cases per group
Urine	≥200 μL	≥100 μL	
Animal tissue	≥100 mg	≥60 mg	
Cell	10^7	10^7	
Cell supernatant	≥1 mL	≥500 μL	
Follicular fluid, cerebrospinal fluid, lymph fluid, etc.	≥200 μL	≥100 μL	
Feces	≥200 mg	≥100 mg	

7.4.3 Experimental Methods and Data Processing

7.4.3.1 Preparation of Samples

Animal blood, urine, tissues, extracellular fluid, and feces can all be used for studying Trp metabolism. See Table 7.3 for requirements for sample testing.

7.4.3.2 Metabolite Extraction

1. Chyme Samples

   The sample was thawed, and 100 ± 5 mg of the sample was taken into a 2-mL centrifuge tube, 10 μL of the internal standard solution was added, 0.5 mL of extract (water:acetonitrile:methanol = 1:2:2) was added, and the mixture was shaken for 1 min. Then, the sample was centrifuged at 14,000 g for 10 min. Four hundred microliters of the supernatant was taken and blown dry with nitrogen. Then, it was re-dissolved with 100 μL acetonitrile-water (1:1, v/v) and centrifuged again at 14,000 g for 10 min. The supernatant was taken for analysis.

2. Tissue Samples

   The sample was weighed at 100 ± 5 mg and placed in a 2-mL centrifuge tube after being thawed on ice. Ten microliters of internal standard solution and 0.5 mL water-acetonitrile-methanol (1:2:2, v/v/v) were added. Several ceramic beads with a diameter of 1.0 mm were added and then homogenated using BeadBeater. After being centrifuged under 14,000 g, 400 μL supernatant was collected and dried under nitrogen gas. The residue was re-dissolved in 100 μL acetonitrile-water (1:1, v/v) and then centrifuged under 14,000 g. The supernatant was then injected for HPLC-MS/MS analysis.

3. Liquid Samples

A 100 ± 5 μL liquid sample was placed in a 2-mL centrifuge tube after being thawed on ice. Ten microliters of internal standard solution and 0.4 mL acetonitrile-methanol (1:1, v/v) were added and then mixed by vortex. After being centrifuged under 14,000 g, 400 μL supernatant was collected and dried under nitrogen gas. The residue was re-dissolved in 100 μL acetonitrile-water (1:1, v/v) and then centrifuged under 14,000 g. The supernatant was then injected for HPLC-MS/MS analysis.

7.4.3.3 Chromatography/Mass Spectrometry Analysis

1. High-Performance Liquid Chromatography (HPLC) Conditions

The separation was performed on a UPLC system (Agilent 1290 Infinity UHPLC) on a C-18 column (Waters, CSH C18 1.7 μm, 2.1 mm × 100 mm column) by gradient elution. Eluent A was acetonitrile, and eluent B was water consisting of 20 mM ammonium formate buffer (pH 3.7). The gradient elution program was as follows: 0 min = 15% B, 2 min = 15% B, 9 min = 98% B, 11 min = 98% B, 11.5 min = 15% B, and 14 min = 15% B. Before injecting the next sample, the column was equilibrated with the initial mobile phase for 5 min. The flow rate was constant at 0.4 mL/min, and the column temperature was set at 50 °C.

Mass Spectrometry

A 5500 QTRAP (AB SCIEX) was performed in positive and negative switch modes. The ESI positive source conditions were as follows: source temperature: 550 °C; ion source gas1 (Gas1): 55; ion source gas2 (Gas2): 55; curtain gas (CUR): 40; and ion sapary voltage floating (ISVF): +4500 V. The ESI negative source conditions were as follows: source temperature: 550 °C; Gas1: 55; Gas2: 55; CUR: 40; and ISVF: − 4500 V. MRM method was used for mass spectrometry quantitative data acquisition.

2. Data Processing and Analysis

Quantitative analysis with MultQuant or Analyst and the analytic process is shown in Table 7.4.

Table 7.4 Types and content of data analysis

Analysis type	Analysis content	Remarks
Quality control	QC RSD result standard curve	RSD% of each analyte <30%, indicating stable and reliable experimental data; all test samples should have good linearity within the linear distance, with correlation coefficients greater than 0.99.
Differential analysis	Univariate statistical analysis *t*-test	
	Box plot	Box plot is a statistical graph used to display the dispersion of a set of data. A set of data ranges from large to small, and its upper edge (maximum value), upper quartile Q3, median, lower quartile Q1, and lower edge (minimum value) are calculated, respectively. The minimum value and the maximum value formation spacing can reflect the degree of data variability.
Functional analysis	Hierarchical cluster analysis	When the selected candidate metabolites are reasonable and accurate, samples from the same group can cluster together in the same cluster. Meanwhile, metabolites clustered in the same cluster have similar expression patterns, indicating that they may be involved in closely related metabolic pathways.
	KEGG pathway analysis	
Advanced bioinformatics analysis	Regression analysis	
	Trend clustering analysis	
	Machine learning ROC	

References

Bhattacharjee S, Subha P, Paul Das M, Ganesh M-R, Shim Y-B, Neppolian B, Das J (2021) Fabrication of silver-grafted silica nanohybrids via aminosilane-inspired surface functionalization for enhanced electrochemical performance towards gastric cancer biomarker. Appl Surf Sci 541:148517

Boulet L, Faure P, Flore P, Montérémal J, Ducros V (2017) Simultaneous determination of tryptophan and 8 metabolites in human plasma by liquid chromatography/tandem mass spectrometry. J Chromatogr B Analyt Technol Biomed Life Sci 1054:36–43

Chen GN, Lin RE, Zhao ZF, Duan JP, Zhang L (1997) Electrogenerated chemiluminescence for determination of indole and tryptophan. Anal Chim Acta 341(2):251–256

Comai S, Cavalletto L, Chemello L, Bernardinello E, Ragazzi E, Costa CV, Bertazzo A (2011) Effects of PEG-interferon alpha plus ribavirin on tryptophan metabolism in patients with chronic hepatitis C. Pharmacol Res 63(1):85–92

Dheepthi Gunavathana S, Girija S, Wilson J, Cyrac Peter A (2022) ZnO nanorods bonded polythiophene nanocomposite: an enhanced electrochemical voltammetric biosensing of L-tryptophan. Bull Mater Sci 45(2):57

Duan X, Chen Z, Tang S, Ge M, Wei H, Guan Y, Zhao G (2020) A strategy employing a TF-splinting duplex Nanoswitch to achieve single-step, enzyme-free, signal-on detection of l-Tryptophan. ACS Sens 5(3):837–844

Fan M, Lu D, You R, Chen C, Lu Y, Wu Y et al (2020) Highly sensitive detection of tryptophan (Trp) in serum based on diazo-reaction coupling with surface-enhanced Raman scattering and colorimetric assay. Anal Chim Acta 1119:52–59

Favennec M, Hennart B, Caiazzo R, Leloire A, Yengo L, Verbanck M et al (2015) The kynurenine pathway is activated in human obesity and shifted toward kynurenine monooxygenase activation. Obesity (Silver Spring) 23(10):2066–2074

Fitzgerald P, Cassidy Eugene M, Clarke G, Scully P, Barry S, Quigley Eamonn MM et al (2008) Tryptophan catabolism in females with irritable bowel syndrome: relationship to interferon-gamma, severity of symptoms and psychiatric co-morbidity. Neurogastroenterol Motil 20(12):1291–1297

Fuertig R, Ceci A, Camus SM, Bezard E, Luippold AH, Hengerer B (2016) LC-MS/MS-based quantification of kynurenine metabolites, tryptophan, monoamines and neopterin in plasma, cerebrospinal fluid and brain. Bioanalysis 8(18):1903–1917

He Y, Zhou L, Deng L, Feng Z, Cao Z, Yin Y (2021) An electrochemical impedimetric sensing platform based on a peptide aptamer identified by high-throughput molecular docking for sensitive l-arginine detection. Bioelectrochemistry 137:107634

Hervé C, Beyne P, Jamault H, Delacoux E (1996) Determination of tryptophan and its kynurenine pathway metabolites in human serum by high-performance liquid chromatography with simultaneous ultraviolet and fluorimetric detection. J Chromatogr B Biomed Appl 675(1):157–161

Idili A, Gerson J, Parolo C, Kippin T, Plaxco KW (2019) An electrochemical aptamer-based sensor for the rapid and convenient measurement of L-tryptophan. Anal Bioanal Chem 411(19):4629–4635

Ilisz I, Fodor G, Berkecz R, Iványi R, Szente L, Péter A (2009) Enantioseparation of beta-substituted tryptophan analogues with modified cyclodextrins by capillary zone electrophoresis. J Chromatogr A 1216(15):3360–3365

Karu N, McKercher C, Nichols DS, Davies N, Shellie RA, Hilder EF, Jose MD (2016) Tryptophan metabolism, its relation to inflammation and stress markers and association with psychological and cognitive functioning: Tasmanian chronic kidney disease pilot study. BMC Nephrol 17(1):171

Lefèvre A, Mavel S, Nadal-Desbarats L, Galineau L, Attucci S, Dufour D et al (2019) Validation of a global quantitative analysis methodology of tryptophan metabolites in mice using LC-MS. Talanta 195:593–598

Lesniak WG, Jyoti A, Mishra MK, Louissaint N, Romero R, Chugani DC et al (2013) Concurrent quantification of tryptophan and its major metabolites. Anal Biochem 443(2):222–231

Li W, Huang S, Wen H, Luo Y, Cheng J, Jia Z et al (2020) Pyridine functionalized carbon dots for specific detection of tryptophan in human serum samples and living cells. Microchem J 154:104579

Marcos J, Renau N, Valverde O, Aznar-Laín G, Gracia-Rubio I, Gonzalez-Sepulveda M et al (2016) Targeting tryptophan and tyrosine metabolism by liquid chromatography tandem mass spectrometry. J Chromatogr A 1434:91–101

Mermelekas G, Vlahou A, Zoidakis J (2015) SRM/MRM targeted proteomics as a tool for biomarker validation and absolute quantification in human urine. Expert Rev Mol Diagn 15(11):1441–1454

Möller M, Du Preez JL, Harvey BH (2012) Development and validation of a single analytical method for the determination of tryptophan, and its kynurenine metabolites in rat plasma. J Chromatogr B Analyt Technol Biomed Life Sci 898:121–129

Orsatti L, Speziale R, Orsale MV, Caretti F, Veneziano M, Zini M et al (2015) A single-run liquid chromatography mass spectrometry method to quantify neuroactive kynurenine pathway metabolites in rat plasma. J Pharm Biomed Anal 107:426–431

Prabakaran K, Jandas PJ, Luo J, Fu C, Wei Q (2021) Molecularly imprinted poly(methacrylic acid) based QCM biosensor for selective determination of L-tryptophan. Colloids Surf A Physicochem Eng Asp 611:125859

Prabakaran K, Jandas PJ, Luo J, Fu C (2022) A highly sensitive surface acoustic wave sensor modified with molecularly imprinted hydrophilic PVDF for the selective amino acid detection. Sensors Actuators A Phys 341:113525

Roberts LD, Souza AL, Gerszten RE, Clish CB (2012) Targeted metabolomics. Curr Protoc Mol Biol Chapter 30 Unit 30(32):31–24. https://doi.org/10.1002/0471142727.mb3002s98

Song Z, Ma Y, Chen M, Ambrosi A, Ding C, Luo X (2021) Electrochemical biosensor with enhanced antifouling capability for COVID-19 nucleic acid detection in complex biological media. Anal Chem 93(14):5963–5971. https://doi.org/10.1021/acs.analchem.1c00724

Steckl AJ, Ray P (2018) Stress biomarkers in biological fluids and their point-of-use detection. ACS Sens 3(10):2025–2044

Tong Q, Song J, Yang G, Fan L, Xiong W, Fang J (2018) Simultaneous determination of tryptophan, kynurenine, kynurenic acid, xanthurenic acid and 5-hydroxytryptamine in human plasma by LC-MS/MS and its application to acute myocardial infarction monitoring. Biomed Chromatogr 32(4), e4156

Vidova V, Spacil Z (2017) A review on mass spectrometry-based quantitative proteomics: targeted and data independent acquisition. Anal Chim Acta 964:7–23

Wang W, He Y, Gao Y, Gao H, Deng L, Gui Q, Cao Z, Yin Y, Feng Z (2022) A peptide aptamer based electrochemical amperometric sensor for sensitive L-glutamate detection. Bioelectrochemistry 146:108165. https://doi.org/10.1016/j.bioelechem.2022.108165

Wang W, He Y, Deng L, Wang H, Liu X, Gui Q-w, Cao Z, Feng Z, Xiong B, Yin Y (2023) Peptide aptamer-based polyaniline-modified amperometric biosensor for L-lysine detection in real serum samples. Measurement 221:113468. https://doi.org/10.1016/j.measurement.2023.113468

Wang W, He Y, He S, Liu X, Gui Q-w, Deng L, Wang H, Cao Z, Feng Z, Xiong B, Yin Y (2024) Peptide aptamer-based colorimetric sensor for the detection of L-tryptophan in porcine serum. Microchem J 197:109896. https://doi.org/10.1016/j.microc.2024.109896

Xia L, Han MJ, Zhou L, Huang A, Yang Z, Wang T et al (2019) S-click reaction for isotropic orientation of oxidases on electrodes to promote electron transfer at low potentials. Angew Chem Int Ed Engl 58(46):16480–16484

Zhu KJ, Zhou L, Wu L, Feng SF, Hu HY, He JL et al (2021) An enzyme-free Amperometric sensor based on self-assembling Ferrocene-conjugated oligopeptide for specific determination of L-arginine. Chin J Chem 39(10):2755–2762

Zhao S, Liu N, Wang W, Xu Z, Wu Y, Lu X (2021) An electrochemical biosensor for alpha-fetoprotein detection in human serum based on peptides containing isomer D-Amino acids with enhanced stability and antifouling property. Biosens Bioelectron 190:113466. https://doi.org/10.1016/j.bios.2021.113466

Chapter 8
Conclusions and Perspectives

Xuan Zhao, Yulong Yin, and Yaoyao Xia

In the exploration of tryptophan (Trp), we will see its intricate story in the fields of animal nutrition and human health unfolds with interconnected insights. From its origins as a dietary amino acid to its profound roles in animals and humans metabolism, tryptophan is not only a nutrient, but also a dynamic force that shapes host health.

The source of Trp, primarily derived from the alimentary protein such as cereals, nuts, milk, and legumes, lays the foundation for its essential role in both animal and human physiology. A number of physiological functions of Trp and its metabolites have been elaborated. Trp is primarily metabolized through three pathways: the Kyn, 5-HT, and indole pathways. Gut microbiota can directly or indirectly participate in and regulate Trp metabolism through various bioactive metabolites in these pathways. It should be noted that the major pathways are differentially affected in different physiology of inter-organs but remain tightly interconnected. For example, the 5-HT plays an important role in the development of intestinal structure and function by mediating the motility, secretion, and function of intestinal epithelial cells. Additionally, 5-HT is metabolized into melatonin in the brain through N-acetyltransferase and hydroxyindole oxygen methyltransferase, regulating sleep patterns and rhythms. Trp supplementation affects intestinal immune homeostasis dependently of gut microbiota. Moreover, metabolites from the Kyn and indole pathways are capable of regulating disease progression in the central nervous system. For example, Trp is metabolized by gut microbes to produce various ligands (e.g., indoles) that bind to the aryl hydrocarbon receptor or pregnane X

X. Zhao
College of Animal Science and Technology, Southwest University, Chongqing, China
e-mail: zhaoxuann@swu.edu.cn

Y. Yin (✉)
Tianjin Institute of Industrial Biotechnology, Chinese Academy of Sciences, National Center of Technology Innovation for Synthetic Biology, Tianjin, China

Institute of Subtropical Agriculture, Chinese Academy of Science, Changsha, China
e-mail: yinyulong@isa.ac.cn

Y. Xia (✉)
College of Animal Science and Technology, Southwest University, Chongqing, China
e-mail: yaoyaoxia20230750@swu.edu.cn

Y. Yin et al. (eds.), *Tryptophan in Animal Nutrition and Human Health*,
https://doi.org/10.1007/978-981-97-4719-1_8

receptor to modulate a series of physiological processes maintaining intestinal homeostasis. Thus, the expanding knowledge of the key functions of intestinal Trp metabolism has revealed promising therapeutic targets for a variety of diseases by either directly targeting the receptors of Trp metabolites or indirectly manipulating the structure of the gut microbiota.

Trp acts as a precursor for protein synthesis and plays a crucial role in animal nutrition. With the development of microbial engineering technology, Trp has been industrially produced in large quantities through microbial fermentation, thus providing a strong guarantee for the demand for dietary amino acids and promoting the development of amino acid–based nutrition. In addition to being a foundational nutrient for farmed animals to maximize their growth potential, Trp plays an important role in regulating the intestinal barrier function and amino acid transportation, thereby influencing the absorption and utilization of proteins and amino acids. Furthermore, Trp could act on the feeding center of the hypothalamus to affect the secretion of hunger and satiety related peptides, thereby regulating animal feeding behavior and promoting growth and development. Trp availability is vital for supporting growth, reproduction, humoral immunity, and the overall health of various animal species. The implications of Trp in animal nutrition extend far beyond individual health and resonate in the broader context of sustainable agriculture. As we concluded here, Trp may be a key component that affects farmed animals and a powerful guarantee for the sustainability of agricultural practices. For humans, the significance of Trp metabolism extends beyond conventional dietary considerations. Besides to serving substrates for protein synthesis, Trp produces biological compounds through metabolic pathways contributing to the maintenance of human health and the prevention or treatments of various diseases. Trp also exerts its function in the regulation of mood, sleep, appetite, memory, and sexual behavior. The activation of the Kyn pathway, along with the subsequent imbalance in the formation of downstream metabolites, is implicated in a variety of central and peripheral diseases including depression, suicide, Alzheimer's disease, schizophrenia, cancer, and immunologic deficiency syndromes. In addition to these neurological diseases, Trp metabolic disorders are also related to digestive, respiratory, and circulatory systems. Several Trp metabolites can be utilized as potential biomarkers to support clinical applications. Targeting Trp metabolism represents a novel and promising therapeutic strategy for treating several diseases. In concluding the exploration of tryptophan's role in human health, the realization dawns on its impact on the complexities of mental and physical well-being. Trp serves as a nexus where dietary food intake intersects with physiological destiny, dictating the intricate balance between availability and utilization of nutrition in multiple physiological states.

Industrial-scale production of Trp has essential impacts on the sustainability of modern animal agriculture. In animal agriculture, Trp serves as a catalyst in the formulation of balanced feeds, optimizing the growth and productivity of farmed animals. Microbial fermentation has become the preferred method to produce feed-grade L-Trp due to its advantages of using cost-effective raw materials and producing high-purity products. Strain producers such as *E. coli* can be used to produce L-Trp in fermenters. Simultaneously, Trp and its metabolites are used as feed additives in animal husbandry or substrates in the synthesis of medications for disease treatment or prevention, highlighting its economic and therapeutic significance. The methods used to reveal the structure and characteristics of Trp, from high-performance liquid chromatography to molecular biology tools and techniques, represent the cutting edge of scientific inquiry. These approaches allow us to gain a deeper understanding of the function and regulatory mechanisms of Trp metabolism. As our understanding evolves, technological advancements enable more precise quantification, detailed analyses, and a deeper appreciation of Trp metabolism–based nutrition within living organisms. Future efforts may be supported by the implementation of state-of-the-art analytical tools to assess Trp and its metabolites in a tissue-specific manner in animal agriculture.

The application of Trp in animal nutrition and human health takes on new dimensions. Advances in precision nutrition should be explored to tailor Trp supplementation based on the specific requirements of different organisms, production stages, and environmental conditions. This approach can optimize the production efficiency and reduce environmental impacts. The personalized approach to nutrition, tailored to metabolic profiles, emerges as a frontier yet to be fully explored. The application of nutrigenomics can shed light on the variations influencing Trp metabolism. This knowledge can pave the way for personalized nutrition, ensuring that individuals, whether animals or humans, receive tailored dietary interventions for optimal health outcomes. Trp has the potential to be used as a personalized intervention strategy that combines nutritional science with precision medicine to improve health outcomes. However, the compartmentalized characteristics of Trp metabolism should be considered when developing therapeutic strategies and identifying biomarkers. Additionally, the presence of intermediate metabolites and the tight enzymatic regulation in Trp metabolism increases the complexity of drug targeting and monitoring. Further research into the intersection of animal nutrition, biology, and medicine by integrating the functions of Trp and its metabolites could provide insights into holistic approaches to improve both animal and human health. Combining different methods, particularly cross-fusion of multi-omics and systems biology seems a promising strategy to identify core microbes and Trp transporters involved in the modulation of Trp metabolism. Future research and collaborative efforts on Trp nutrition would contribute to the development of innovative nutritional and therapeutic strategies that positively impact the health of both farmed animals and humans.

The manufacturer's authorised representative in the EU is Springer Nature Customer Service Centre GmbH, Europaplatz 3, 69115 Heidelberg, Germany. If you have any concerns regarding our products, please contact ProductSafety@springernature.com

Printed and bound by CPI Group (UK) Ltd, Croydon, CR0 4YY
01/10/2026
02225923-0003